MILITARY AIRCRAFT MARKINGS 1993

PETER R. MARCH

A-494 a R Netherlands AF Alouette III. *PRM*

Contents

Photographs by Peter R. March (PRM) unless otherwise credited

This fourteenth edition published 1993

ISBN 0 7110 2130 9

Published by Ian Allan Ltd
and phototypeset and printed by Ian Allan Printing Ltd at its works
at Coombelands in Runnymede, England

Front cover:
Royal Navy Harrier T4N ZB604 operated by 889 Naval Air Squadron from RNAS Yeovilton. *Peter R. March (PRM)*

Back cover top:
This A&AEE Westland Sea King HC4 ZG829 crashed at Boscombe Down on 20 October 1992. *PRM*

Back cover bottom:
The 53rd Fighter Squadron operates this F-15C Eagle with the 36th Fighter Wing, USAFE from Bitburg. *PRM*

Introduction

This fourteenth edition of *abc Military Aircraft Markings*, a companion to *abc Civil Aircraft Markings*, again sets out to list in alphabetical and numerical order all the aircraft which carry a United Kingdom military serial, and which are based, or might be seen, in the UK. The term ***aircraft*** used here covers powered, manned aeroplanes, helicopters and gliders. Included are all the current Royal Air Force, Royal Navy, Army Air Corps, Ministry of Defence (Procurement Executive), manufacturers' test aircraft and civilian-owned aircraft with military markings.

Aircraft withdrawn from operational use but which are retained in the UK for ground training purposes or otherwise preserved by the Services and in museums and collections are listed. The serials of some incomplete aircraft have been included, such as the cockpit sections of machines displayed by the RAF Exhibition Flight, aircraft used by airfield fire sections and for service battle damage repair training (BDRT), together with significant parts of aircraft held by preservation groups and societies. Many of these aircraft are allocated, and sometimes wear, a secondary identity, such as an RAF Support Command 'M' maintenance number. These numbers are listed against those aircraft to which they have been allocated.

A serial 'missing' is either because it was never issued as it formed part of a 'black-out block', or because the aircraft is written off, scrapped, sold, abroad or allocated an alternative marking. Aircraft used as targets on MoD ranges to which access is restricted, and un-manned target drones, are omitted, as are UK military aircraft that have been permanently grounded overseas and unlikely to return to Britain.

In the main, the serials listed are those markings presently displayed on the aircraft. Aircraft which bear a false serial are quoted in *italic type*. The manufacturer and aircraft type are given, together with recent alternative, previous, secondary or civil identity shown in round brackets. Complete records of multiple previous identities are only included where space permits. The operating unit and its based location, along with any known unit and base code markings in square brackets, are given as accurately as possible. The unit markings are normally carried boldly on the sides of the fuselage or on the aircraft's fin. In the case of RAF and AAC machines currently in service, they are usually one or two letters or numbers, while the RN continues to use a well-established system of three-figure codes between 000 and 999 together with a fin letter code denoting the aircraft's operational base. RN squadrons, units and bases are allocated blocks of numbers from which individual aircraft codes are issued. To help identification of RN bases and landing platforms on ships, a list of tail-letter codes with their appropriate name, helicopter code number, ship pennant number and type of vessel is included; as is a helicopter code number/ships' tail-letter code grid cross-reference.

Codes change, for example when aircraft move between units, and therefore the markings currently painted on a particular aircraft might not be those shown in this edition because of subsequent events. The effect of the on-going 'Options for Change' RAF re-organisation accounts for the large number of changes in this edition to Tornado GR1s/F3s, Jaguars, Phantoms, Pumas and Chinooks. They will continue to be noticeable for several years. Those airframes which may not appear in the next edition because of sale, accident, etc, have their fates, where known, given in italic type in the *locations* column.

The Irish Army Air Corps fleet is listed, together with the serials of other overseas air arms whose aircraft might be seen visiting the UK from time to time. The serial numbers are as usually presented on the individual machine or as they are normally identified. Where possible, the aircraft's base and operating unit have been shown.

USAF, US Army and US Navy aircraft based in the UK and in Western Europe, and of types which regularly visit the UK from the USA, are each listed in separate sections by aircraft type. The serial number actually displayed on the aircraft is shown in full, with additional Fiscal Year (FY) or full serial information also provided. Where appropriate, details of the operating wing, squadron allocation and base are added. The USAF is, like the RAF, continuing a major reorganisation which is producing new unit titles, many squadron changes and the closure of bases worldwide. Only details that concern changes effected by December 1992 are shown.

Veteran and Vintage aircraft which carry overseas military markings but which are based in the UK have been separately listed showing their principal means of identification.

Information shown is believed to be correct at 31 January 1993, and significant changes can be monitored through the monthly 'Military Markings' column in *Aircraft Illustrated.*

Acknowledgements

The compiler again wishes to thank the many people who have taken trouble to send comments, criticism and other useful information following the publication of the previous editions of *abc Military Aircraft Markings*. In particular the following correspondents: A.Allen, S. J. Bond, D. Braithwaite, P. F. Burton, J. G. Chree, D. P. Curtis, T. Hall, A.Helden, I.Logan, P.- J. Martin, A.P. March, P.Moonen, I. Polson, M.-Powney, R. Robinson, B. V. Taylor, M. K. Thompson, P. Wiggins.

This compilation has relied heavily on the publications of the following aviation groups and societies: *Air North, British Aviation Review* (British Aviation Research Group), *Clwyd Air Letter* (Clwyd Aviation Group), *Norfolk Air Review* (Norfolk Aviation Group), *North-West Air News* (Air Britain, Merseyside Branch), *Osprey* (Solent Aviation Society), *Prestwick Airport Letter* (Prestwick Airport Aviation Group), *Scottish Air News* (Central Scotland Aviation Group), *Stansted Aviation Newsletter* (The Stansted Aviation Society), *Strobe* (The East of England Aviation Group) and *Ulster Air Mail* (Ulster Aviation Society).

This new edition of *abc Military Aircraft Markings* would not have been possible without considerable research and checking by Howard Curtis, Wal Gandy and Brian Strickland to whom I am indebted.

PRM **January 1993**

V9281 Westland Lysander III (G-BCWL) operated by Wessex Aviation & Transport from Henstridge. *Peter R. March (PRM)*

DE992/G-AXXV is a DH Tiger Moth based at Membury. *PRM*

Abbreviations

AAC	Army Air Corps
A&AEE	Aeroplane & Armament Experimental Establishment
AAS	Aeromedical Airlift Squadron
ABS	Air Base Squadron
ABW	Air Base Wing
ACCGS	Air Cadets Central Gliding School
ACCS / ACCW	Airborne Command and Control Squadron/Wing
ACR	Armoured Cavalry Regiment
AEF	Air Experience Flight
AES	Air Engineering School
AEW	Airborne Early Warning
AFRES	Air Force Reserve
AFSC	Air Force System Command
AG	Airlift Group
AHB	Attack Helicopter Battalion
AIU	Accident Investigation Unit
AKG	Aufklärürngsgeschwader (Reconnaissance Wing)
ALW	Airlift Wing
AMD-BA	Avions Marcel Dassault-Breguet Aviation
AMG	Aircraft Maintenance Group
AMS	Air Movements School
ANG	Air National Guard
APS	Aircraft Preservation Society
ARRS / ARRW	Aerospace Rescue and Recovery Squadron/Wing
ARS	Air Refuelling Squadron
ARW	Air Refuelling Wing
ARWS	Advanced Rotary Wing Squadron
AS	Aggressor Squadron
AS	Airlift Squadron
ASF	Aircraft Servicing Flight
AS&RU	Aircraft Salvage and Repair Unit
ATC	Air Training Corps
ATCC	Air Traffic Control Centre
ATS	Aircrewman Training Squadron
AvCo	Aviation Company
AW	Armstrong Whitworth Aircraft
AWG	Amt für Wehrgeophysik
AW&CS / AW&CW	Airborne Warning & Control Squadron/Wing
BAC	British Aircraft Corporation
BAe	British Aerospace PLC
BAOR	British Army of the Rhine
BAPC	British Aviation Preservation Council
BATUS	British Army Training Unit Support
BBMF	Battle of Britain Memorial Flight
BDRF	Battle Damage Repair Flight
BDRT	Battle Damage Repair Training
Be	Beech
Bf	Bayerische Flugzeugwerke
BFME	British Forces Middle East
BFWF	Basic Fixed Wing Flight
BGA	British Gliding & Soaring Association
BHC	British Hovercraft Corporation
BNFL	British Nuclear Fuels Ltd
BP	Boulton & Paul
B-V	Boeing-Vertol
BW	Bomber Wing
CAC	Commonwealth Aircraft Corporation
CARG	Cotswold Aircraft Restoration Group
CASA	Construcciones Aeronautics SA
CATCS	Central Air Traffic Control School
CBAS	Commando Brigade Air Squadron
CC	County Council
CCF	Combined Cadet Force
CDE	Chemical Defence Establishment
CEAM	Centre d'Expérimentation Aériennes Militaires
CEV	Centre d'Essais en Vol
CFS	Central Flying School
CGMF	Central Glider Maintenance Flight
CIFAS	Centre d'Instruction des Forces Aériennes Strategiques
CinC	Commander in Chief
Co	Company
CSDE	Central Servicing Development Establishment
CT	College of Technology
CTE	Central Training Establishment
CTTS	Civilian Technical Training School
CV	Chance-Vought
D-BD	Dassault-Breguet Dornier
Det	Detachment
DH	de Havilland
DHC	de Havilland Canada
DRA	Defence Research Agency
DTI	Department of Trade and Industry
EABDR	Engineering and Battle Damage Repair
EAP	European Aircraft Project
ECS/ ECW	Electronic Countermeasures Squadron/Wing
EDA	Escadre de Detection Aéroportée
EE	English Electric
EFTS	Elementary Flying Training Squadron
EHI	European Helicopter Industries
EMA	East Midlands Airport
EoN	Elliot's of Newbury
ERV	Escadre de Ravitaillement en Vol
ETPS	Empire Test Pilots' School
ETS	Engineering Training School
EWAU	Electronics Warfare Avionics Unit
FAA	Fleet Air Arm/Federal Aviation Administration
FACF	Forward Air Control Flight
FBS	Flugbereitschaftstaffel
FBW	Fly by wire
FCS	Facility Checking Squadron
FE	Further Education
FEWSG	Fleet Electronic Warfare Support Group
FG	Fighter Group
FGF	Flying Grading Flight
FH	Fairchild-Hiller
FI	Falkland Islands
Flt	Flight
FMA	Fabrica Militar de Aviones
FOL	Forward Operating Location
FONA	Flag Officer Naval Aviation
FRADU	Fleet Requirements and Air Direction Unit
FRL	Flight Refuelling Ltd
FS	Fighter Squadron
FSCTE	Fire School Central Training Establishment
FTS	Flying Training School
FW	Foster Wikner
FY	Fiscal Year
F3 OCU	Tornado F3 Operational Conversion Unit
GAL	General Aircraft Ltd
GD	General Dynamics
GI	Ground Instruction
HFR	Heeresfliegerregiment (Corps transport regiment)
HFWS	Heeresflieger Waffenschule
HMS	Her Majesty's Ship

HOCU	Harrier OCU
HP	Handley-Page
HQ	Headquarters
HS	Hawker Siddeley
HSF	Harrier Servicing Flight
IAF	Israeli Air Force
IAM	Institute of Aviation Medicine
IHM	International Helicopter Museum
IWM	Imperial War Museum
JATE	Joint Air Transport Establishment
JbG	Jagd Bomber Geschwader (Fighter Bomber Wing)
JG	Jagd Geschwader (Fighter Wing)
JMU	Jaguar Maintenance Unit
JTU	Joint Trials Unit
LOFTU	Lynx Operational Flying Trials Unit
LTG	Luft Transport Geschwader (Air Transport Wing)
LTV	Ling-Temco-Vought
LVG	Luftwaffen Versorgungs Geschwader (Air Force Maintenance Wing)
McD	McDonnell Douglas
MFG	Marine Flieger Geschwader (Naval Air Wing)
MGSP	Mobile Glider Servicing Party
MH	Max Holste
MIB	Military Intelligence Battalion
MiG	Mikoyan — Gurevich
MoD(PE)	Ministry of Defence (Procurement Executive)
Mod	Modified
MR	Maritime Reconnaissance
MRF	Meteorological Research Flight
MS	Morane-Saulnier
MU	Maintenance Unit
NA	North American
NACDS	Naval Air Command Driving School
NAEWF	NATO Airborne Early Warning Force
NAF	Naval Air Facility
NASU	Naval Air Support Unit
NE	North-East
NHTU	Naval Hovercraft Trials Unit
NI	Northern Ireland
NMSU	Nimrod Major Servicing Unit
OCU	Operational Conversion Unit
OEU	Operation Evaluation Unit
OPITB	Offshore Petroleum Industry Training Board
OTD	Overseas Training Division
OTS	Operational Training Squadron
PAX	Passenger procedural trainer
PRU	Photographic Reconnaissance Unit
RAeS	Royal Aeronautical Society
RAF	Royal Aircraft Factory/Royal Air Force
RAFC	Royal Air Force College
RAFGSA	Royal Air Force Gliding and Soaring Association
RAOC	Royal Army Ordnance Corps
RCAF	Royal Canadian Air Force
RE	Royal Engineers
Regt	Regiment
REME	Royal Electrical & Mechanical Engineers
RGE	Ridge Gliding Establishment
RM	Royal Marines
RMC of S	Royal Military College of Science
RN	Royal Navy
RNAS	Royal Naval Air Station
RNAY	Royal Naval Aircraft Yard
RNEC	Royal Naval Engineering College
RNEFTS	Royal Naval Elementary Flying Training School
RNGSA	Royal Navy Gliding and Soaring Association
ROF	Royal Ordnance Factory
R-R	Rolls-Royce
RS	Reid & Sigrist/Reconnaissance Squadron
RS&RE	Royal Signals and Radar Establishment
RSV	Reparto Sperimentale Volo
RW	Reconnaissance Wing
SA	Scottish Aviation
Saab	Svenska Aeroplan Aktieboleg
SAC	Strategic Air Command
SAE	School of Aircraft Engineering
SAH	School of Air Handling
SAL	Scottish Aviation Limited
SAOEU	Strike/Attack Operational Evaluation Unit
SAR	Search and Rescue
Saro	Saunders-Roe
SARTU	Search and Rescue Training Unit
SCF	Scout Conversion Flight
SEPECAT	Société Européenne de Production de l'avion Ecole de Combat et d'Appui Tactique
SHAPE	Supreme Headquarters Allied Forces Europe
SIF	Servicing Instruction Flight
SKTU	Sea King Training Unit
SNCAN	Société Nationale de Constructions Aéronautiues du Nord
SOS	Special Operations Squadron
SoTT	School of Technical Training
SPAD	Société Pour les Appareils Deperdussin
Sqn	Squadron
SSF	Station Servicing Flight
SSTF	Small Ships Trials Flight
SW	Strategic Wing
TASS	Tactical Air Support Squadron
TCW	Tactical Control Wing
TDCS	Tactical Deployment Control Squadron
T&EE	Test & Evaluation Establishment
TMTS	Trade Management Training School
TSLw	Technische Schule der Luftwaffe
TTTE	Tri-national Tornado Training Establishment
TW	Test Wing
TWU	Tactical Weapons Unit
UAS	University Air Squadron
UK	United Kingdom
UKAEA	United Kingdom Atomic Energy Authority
UNFICYP	United Nations' Forces in Cyprus
US	United States
USAF	United States Air Force
USAFE	United States Air Forces in Europe
USAREUR	US Army Europe
USEUCOM	United States European Command
USN	United States Navy
VGS	Volunteer Gliding School
VQ	Air Reconnaissance Squadron
VR	Logistic Support Squadron
VS	Vickers-Supermarine
WG	Wing
WLT	Weapons Loading Training
WRG	Weather Reconnaissance Group
WRS	Weather Reconnaissance Squadron
WS	Westland
WW2	World War II

A Guide to the Location of Operational Bases in the UK

This section is to assist the reader to locate the places in the United Kingdom where operational military aircraft are based. The term 'aircraft' also includes helicopters and gliders.

The alphabetical order listing gives each location in relation to its county and to its nearest classified road(s) ('by' means adjoining; 'of' means proximate to), together with its approximate direction and mileage from the centre of a nearby major town or city.

Some civil airports are included where active military units are also based, but **excluded** are MoD sites with non-operational aircraft (eg 'gate guardians'), the bases of privately-owned civil aircraft which wear military markings, and museums.

User	*Base name*	*County/Region*	*Location*	*Distance/direction from (town)*
DRA	Aberporth	Dyfed	N of A487	6m ENE of Cardigan
USAF	Alconbury	Cambridgeshire	E by A1/A14	4m NW of Huntingdon
RAF	Aldergrove/Belfast	Co Antrim	W by A26	13m W of Belfast Airport
RAF	Benson	Oxfordshire	E by A423	1m NE of Wallingford
A&AEE	Boscombe Down	Wiltshire	S by A303, W of A338	6m N of Salisbury
RAF	Boulmer	Northumberland	E of B1339	4m E of Alnwick
RAF	Brawdy	Dyfed	N of A487	9m NW of Haverfordwest
RAF	Brize Norton	Oxfordshire	W of A4095	5m SW of Witney
RAF	Cambridge Airport/ Teversham	Cambridgeshire	S by A1303	2m E of Cambridge
RAF	Catterick	Yorkshire North	E by A1	7m WNW of Northallerton
RAF	Chivenor	Devon	S of A361	4m WNW of Barnstaple
RAF	Church Fenton	Yorkshire North	S of B1223	7m WNW of Selby
RAF	Coltishall	Norfolk	W of B1150	9m NNE of Norwich
RAF	Coningsby	Lincolnshire	S of A153, W by B1192	10m NW of Boston
RAF	Cosford	Shropshire	W of A41, N of A464	9m WNW of Wolverhampton
RAF	Cottesmore	Leicestershire	W of A1, N of B668	9m NW of Stamford
RAF	Cranwell	Lincolnshire	N by A17, S by B1429	5m WNW of Sleaford
RNAS	Culdrose	Cornwall	E by A3083	1m SE of Helston
RAF	Dishforth	Yorkshire North	E by A1	4m E of Ripon
BAe	Dunsfold	Surrey	W of A281, S of B2130	9m S of Guildford
RAF	Exeter Airport	Devon	S by A30	4m ENE of Exeter
DRA	Farnborough	Hampshire	W of A325, N of A323	2m W of Farnborough
BAe/R-R	Filton	Avon	E by M5 jn 17, W by A38	4m N of Bristol
RAF	Finningley	Yorkshire South	W of A614, S of B1396	6m ESE of Doncaster
RNAY	Fleetlands	Hampshire	E by A32	2m SE of Fareham
RAF	Glasgow Airport	Strathclyde	N by M8 jn 28	7m W of city
RAF	Halton	Buckinghamshire	N of A4011, S of B4544	4m ESE of Aylesbury
RAF	Henlow	Bedfordshire	E of A600, W of A6001	1m SW of Henlow
RAF	Honington	Suffolk	E of A134, W of A1088	6m S of Thetford
RAF	Hullavington	Wiltshire	W of A429	1m N of M4 jn 17
RAF	Kenley	Surrey	W of A22	1m W of Warlingham
RAF	Kinloss	Grampian	E of B9011, N of B9089	3m NE of Forres
RAF	Kirknewton	Lothian	E by B7031, N by A70	8m SW of Edinburgh
USAF	Lakenheath	Suffolk	W by A1065	8m W of Thetford
RAF	Leconfield	Humberside	E of A164	N of Beverley
RAF	Leeming	Yorkshire North	E by A1	5m SW of Northallerton
RNAS	Lee-on-Solent	Hampshire	S by B3334	3m S of Fareham
RAF	Leuchars	Fife	E of A919	7m SE of Dundee
RAF	Linton-on-Ouse	Yorkshire North	E of B6265	10m NW of York
T&EE	Llanbedr	Gwynedd	W of A496	7m NNW of Barmouth
RAF	Lossiemouth	Grampian	W of B9135, S of B9040	4m N of Elgin
RAF	Lyneham	Wiltshire	W of A3102, S of A420	10m WSW of Swindon
RAE	Machrihanish	Strathclyde	W of A83	3m W of Campbeltown
RAF	Manston	Kent	N by A253	3m W of Ramsgate
RAF	Marham	Norfolk	N by A1122	6m W of Swaffham
AAC	Middle Wallop	Hampshire	S by A343	6m SW of Andover
USAF/ USN	Mildenhall	Suffolk	S by A1101	9m NNE of Newmarket
AAC	Netheravon	Wiltshire	E of A345	5m N of Amesbury
RAF	Newton	Nottinghamshire	N of A52, W of A46	7m E of Nottingham
RAF	Northolt	Greater London	N by A40	3m E of M40 jn 1

UK Operational Bases — continued

User	*Base name*	*County/Region*	*Location*	*Distance/direction from (town)*
RAF	Odiham	Hampshire	E of A32	2m S of M3 jn 5
RAF	Plymouth Airport/ Roborough	Devon	E by A386	4m NNE of Plymouth
RNAS	Portland	Dorset	E by A354	3m S of Weymouth
RN	Predannack	Cornwall	W by A3083	7m S of Helston
RN	Prestwick Airport	Strathclyde	E by A79	3m N of Ayr
RAF	St Athan	South Glamorgan	N of B4265	13m WSW of Cardiff
RAF	St Mawgan/Newquay Airport	Cornwall	N of A3059	4m ENE of Newquay
RAF	Samlesbury	Lancashire	N by A677, S by A59	5m E of Preston
RAF	Scampton	Lincolnshire	W by A15	6m N of Lincoln
RAF	Sealand	Clwyd	W by A550	6m WNW of Chester
RAF	Shawbury	Shropshire	W of B5063	7m NNE of Shrewsbury
RAF	South Cerney	Gloucestershire	W by A419	3m SE of Cirencester
RAF	Swansea Airport/ Fairwood Common	West Glamorgan	W by A4118	6m W of Swansea
RAF	Swanton Morley	Norfolk	W of B1147	4m NNE of Dereham
RAF	Swinderby	Lincolnshire	SE by A46	9m SW of Lincoln
RAF/ Shorts	Sydenham/ Belfast City Airport	Co Down	W by A2	2m E of city
RAF	Syerston	Nottinghamshire	W by A46	5m SW of Newark
RAF	Ternhill	Shropshire	SW by A41	3m SW of Market Drayton
DRA	Thurleigh/Bedford	Bedfordshire	E of A6, W of B660	7m N of Bedford
AAC	Topcliffe	Yorkshire North	E of A167, W of A168	3m SW of Thirsk
RAF	Turnhouse/Edinburgh	Lothian	N of A8	6m W of Edinburgh Airport
RAF	Upavon	Wiltshire	S by A342	14m WNW of Andover
USAF	Upper Heyford	Oxfordshire	N of B430, W of M40	12m N of Oxford jn 10
RAF	Valley	Gwynedd	S of A5 on Anglesey	5m SE of Holyhead
RAF	Waddington	Lincolnshire	E by A607, W by A15	5m S of Lincoln
BAe	Warton	Lancashire	S by A584	8m SE of Blackpool
RAF	Wattisham	Suffolk	N of B1078	5m SSW of Stowmarket
DRA	West Freugh	Dumfries & Galloway	S by A757, W by A715	5m SE of Stranraer
RAF	Weston-on-the-Green	Oxfordshire	E by A43	9m N of Oxford
RAF	Wittering	Northamptonshire	W by A1, N of A47	3m S of Stamford
RAF	Woodvale	Merseyside	W of A565	5m SSW of Southport
RAF	Wyton	Cambridgeshire	E of A141, N of B1090	3m NE of Huntingdon
WS	Yeovil	Somerset	N of A30, S of A3088	1m W of Yeovil
RNAS	Yeovilton	Somerset	S by B3151, S of A303	5m N of Yeovil

JV928 PBY-5A Catalina (G-BLSC) is based at Duxford. *PRM*

British Military Aircraft Serials

The Committee of Imperial Defence through its Air Committee introduced a standardised system of numbering aircraft in November 1912. The Air Department of the Admiralty was allocated the first batch 1-200 and used these to cover aircraft already in use and those on order. The Army was issued with the next block from 201-800, which included the number 304 which was given to the Cody Biplane now preserved in the Science Museum. By the outbreak of World War 1 the Royal Navy was on its second batch of serials 801-1600 and this system continued with alternating allocations between the Army and Navy until 1916 when number 10000, a Royal Flying Corps BE2C, was reached.

It was decided not to continue with five digit numbers but instead to start again from 1, prefixing RFC aircraft with the letter A and RNAS aircraft with the prefix N. The RFC allocations commenced with A1 an FE2D and before the end of the year had reached A9999 an Armstrong Whitworth FK8. The next group commenced with B1 and continued in logical sequence through the C, D, E and F prefixes. G was used on a limited basis to identify captured German aircraft, while H was the last block of wartime-ordered aircraft. To avoid confusion I was not used, so the new postwar machines were allocated serials in the J range. A further minor change was made in the serial numbering system in August 1929 when it was decided to maintain four numerals after the prefix letter, thus omitting numbers 1 to 999. The new K series therefore commenced at K1000, which was allocated to an AW Atlas.

The Naval N prefix was not used in such a logical way. Blocks of numbers were allocated for specific types of aircraft such as seaplanes or flying-boats. By the late 1920s the sequence had largely been used up and a new series using the prefix S was commenced. In 1930 separate naval allocations were stopped and subsequent serials were issued in the 'military' range which had by this time reached the K series. A further change in the pattern of allocations came in the L range. Commencing with L7272 numbers were issued in blocks with smaller blocks of serials between not used. These were known as blackout blocks. As M had already been used as a suffix for Maintenance Command instructional airframes it was not used as a prefix. Although N had previously been used for naval aircraft it was used again for serials allocated from 1937.

With the build-up to World War 2 the rate of allocations quickly accelerated and the prefix R was being used when war was declared. The letters O and Q were not allotted, and nor was S which had been used up to S1865 for naval aircraft before integration into the RAF series. By 1940 the serial Z9999 had been reached, as part of a blackout block, with the letters U and Y not used to avoid confusion. The option to recommence serial allocation at A1000 was not taken up; instead it was decided to use an alphabetical two-letter prefix with three numerals running from 100 to 999. Thus AA100 was allocated to a Blenheim IV.

This two-letter, three-numeral serial system which started in 1940 continues today with the current issue being in the ZH range. The letters C, I, O, Q, U and Y were, with the exception of NC, not used. For various reasons the following letter combinations were not issued: DA, DB, DH, EA, GA to GZ, HA, HT, JE, JH, JJ, KR to KT, MR, NW, NZ, SA to SK, SV, TN, TR and VE. The first postwar serials issued were in the VP range while the end of the WZs had been reached by the Korean War. At the current rate of issue the Z range will last well into the next century.

Note: Whilst every effort has been made to ensure the accuracy of this publication, no part of the contents has been obtained from official sources. The compiler will be pleased to continue to receive comments, corrections and further information for inclusion in subsequent editions of *Military Aircraft Markings*. A monthly up-date of additions and amendments is published in *Aircraft Illustrated.*

WB588 DH Chipmunk T10 (G-AOTD) is flown by the Shuttleworth Collection. *PRM*

SM969/G-BRAF VS Spitfire XVIIIe is one of the late Doug Arnold's fleet of Warbirds. *PRM*

British Military Aircraft Markings

A serial in *italics* denotes that it is not the genuine marking for that airframe.

Serial	*Type*	*Owner or Operator*	*Notes*
164	Bleriot Type XI (BAPC 106)	RAF Museum, Hendon	
168	Sopwith Tabloid Scout Replica (G-BFDE)	RAFM Restoration Centre, Cardington	
304	Cody Biplane (BAPC 62)	Science Museum, South Kensington	
433	Bleriot Type XXVII (BAPC 107)	RAF Museum, Hendon	
687	RAF BE2b Replica (BAPC 181)	RAF Museum, Hendon	
2345	Vickers FB5 Gunbus Replica (G-ATVP)	RAF Museum, Hendon	
2699	RAF BE2C	Imperial War Museum, Lambeth	
3066	Caudron GIII (G-AETA)	RAF Museum, Hendon	
5492	Sopwith LC-1T Triplane Replica (G-PENY)	Privately owned, Sywell	
5844	Avro 504K Replica (BAPC 42)	RAF Museum, stored St Athan	
5894	DH2 Replica (G-BFVH) [FB2]	Privately owned, Chalmington	
5964	DH2 Replica (BAPC 112)	Museum of Army Flying, Middle Wallop	
6232	RAF BE2C Replica (BAPC 41)	RAF, stored St Athan	
8359	Short 184	FAA Museum, RNAS Yeovilton	
A1325	RAF BE2e	Mosquito Aircraft Museum, London Colney	
A1742	Bristol Scout D Replica (BAPC 38)	RAF, stored Wroughton	
A4850	RAF SE5A Replica (BAPC 176)	South Yorkshire APS, Firbeck	
A7317	Sopwith Pup Replica (BAPC 179)	Midland Air Museum, Coventry	
A8226	Sopwith 1_ Strutter Replica (G-BIDW)	RAF Museum, Hendon	
A9507	Airco DH5 Replica [E] (N950JS)	*Returned to USA Jan 1992*	
B1807	Sopwith Pup (G-EAVX) [A7]	Privately owned, Keynsham, Avon	
B2458	Sopwith Camel Replica [R] (G-BPOB)	Privately owned, Booker	
B4863	Eberhardt SE5E (G-BLXT) [G]	Museum of Army Flying, Middle Wallop	
B6291	Sopwith Camel F1 (G-ASOP)	Shuttleworth Collection, Old Warden	
B6401	Sopwith Camel F1 Replica (G-AWYY/C1701)	FAA Museum, RNAS Yeovilton	
B7270	Sopwith Camel F1 Replica (G-BFCZ)	Brooklands Aviation Museum	
C1904	RAF SE5A Replica (G-PFAP) [Z]	Privately owned, Finmere	
C3011	Phoenix Currie Super Wot [S] (G-SWOT)	Privately owned, Dunkeswell	
C4451	Avro 504J Replica (BAPC210)	Southampton Hall of Aviation	
C4912	Bristol M1C Replica (BAPC 135)	Northern Aeroplane Workshops	
C4994	Bristol M1C Replica (G-BLWM)	RAF Museum, Hendon	
D2700	RAF SE5A Replica [A] (BAPC208)	Prince's Mead Shopping Precinct, Farnborough	
D3419	Sopwith Camel F1 Replica (BAPC 59)	RAF, stored St Athan	
D5329	Sopwith Dolphin	RAF Museum Store, Cardington	
D7560	Avro 504K	Museum of Army Flying, Middle Wallop	
D7889	Bristol F2b Fighter (G-AANM/BAPC 166)	Privately owned, St Leonards-on-Sea	
D8084	Bristol F2B Fighter (G-ACAA)	The Fighter Collection, Duxford	
D8096	Bristol F2B Fighter (G-AEPH) [D]	Shuttleworth Collection, Old Warden	
E373	Avro 504K Replica (BAPC 178)	Privately owned, Eccleston, Lancs	
E449	Avro 504K (G-EBJE)	RAF Museum, Hendon	
E2466	Bristol F2b (BAPC 165)	RAF Museum, Hendon	
E2581	Bristol F2b Fighter	Imperial War Museum, Duxford	
F141	RAF SE5A Replica (G-SEVA) [G]	Privately owned, Boscombe Down	
F235	RAF SE5A Replica (G-BMDB) [B]	*Written off 2 August 1992*	
F344	Avro 504K Replica	RAF Museum Store, Henlow	
F760	SE5A Microlight Replica [A]	Privately owned, Redhill	
F904	RAF SE5A (G-EBIA)	Shuttleworth Collection, Old Warden	
F938	RAF SE5A (G-EBIC)	RAF Museum, Hendon	
F943	RAF SE5A Replica (G-BIHF) [S]	Privately owned, Booker	
F943	RAF SE5A Replica (G-BKDT)	Yorkshire Air Museum, Elvington	
F1010	DH9A [19]	RAF Museum, Hendon	
F3556	RAF RE8	Imperial War Museum, Duxford	

Notes	*Serial*	*Type*	*Owner or Operator*
	F4013	Sopwith Camel Replica	Privately owned, Coventry
	F4516	Bristol F2B Fighter (G-ACAA)	*Repainted as D8084*
	F5447	RAF SE5A Replica (G-BKER) [N]	Privately owned, Cumbernauld
	F5459	RAF SE5A Replica (BAPC 142) [11-Y]	Flambards Village Theme Park, Helston
	F5459	RAF SE5A Replica (G-INNY) [Y]	Privately owned, Old Sarum
	F6314	Sopwith Camel F1 [B]	RAF Museum, Hendon
	F8010	RAF SE5A Replica (G-BDWJ) [Z]	Privately owned, Booker
	F8614	Vickers Vimy Replica (G-AWAU)	RAF Museum, Hendon
	G1381	Avro 504K Replica [G] (BAPC 177)	*Repainted as G-ACCA*
	H1968	Avro 504K Replica (BAPC 42)	RAF, stored St Athan
	H2311	Avro 504K (G-ABAA)	*Repainted as G-ABAA*
	H3426	Hawker Hurricane Replica (BAPC 68)	Midland Air Museum, Coventry
	H5199	Avro 504N (BK892, 3118M, G-ACNB, G-ADEV)	Shuttleworth Collection, Old Warden
	J7326	DH Humming Bird (G-EBQP)	Privately owned, Hemel Hempstead
	J8067	Pterodactyl 1a	Science Museum, South Kensington
	J9941	Hawker Hart 2 (G-ABMR)	RAF Museum, Hendon
	K1786	Hawker Tomtit (G-AFTA)	Shuttleworth Collection, Old Warden
	K2050	Hawker Fury Replica (G-ASCM)	Privately owned, Andrewsfield
	K2059	Isaacs Fury (G-PFAR)	Privately owned, Dunkeswell
	K2567	DH Tiger Moth (G-MOTH)	Russavia Collection, Bishops Stortford
	K2572	DH Tiger Moth (G-AOZH) (really NM129)	Privately owned, Shoreham
	K2572	DH Tiger Moth Replica	The Aeroplane Collection, Warmingham
	K3215	Avro Tutor (G-AHSA)	Shuttleworth Collection, Old Warden
	K3584	DH 82B Queen Bee (BAPC 186)	Mosquito Aircraft Museum, London Colney
	K3731	Isaacs Fury Replica (G-RODI)	Privately owned, Dunkeswell
	K4232	Avro Rota I (SE-AZB)	RAF Museum, Hendon
	K4235	Avro Rota I (G-AHMJ)	Shuttleworth Collection, Old Warden
	K4972	Hawker Hart Trainer IIA (1764M)	RAF Cosford Aerospace Museum
	K5054	Supermarine Spitfire Replica (BAPC 190)	Southampton Hall of Aviation
	K5054	Supermarine Spitfire Replica (G-BRDV)	Privately owned, Hullavington
	K5054	Supermarine Spitfire Replica (BAPC 214)	Privately owned, Andover
	K5414	Hawker Hind (G-AENP/BAPC 78) [XV]	Shuttleworth Collection, Old Warden
	K6038	Westland Wallace II (2365M)	RAF Museum Store, Cardington
	K7271	Hawker Fury II Replica (BAPC 148)	RAF Cosford Aerospace Museum, stored
	K8042	Gloster Gladiator II (8372M)	RAF Museum, Hendon
	K8203	Hawker Demon (G-BTVE/2292M)	Privately owned, Hatch
	K9926	VS Spitfire IX Replica (BAPC 217)	RAF Bentley Priory, on display
	K9942	VS Spitfire IA (8383M) [SD-V]	RAF Museum, Hendon
	L1070	VS Spitfire IX Replica [XT-A] (BAPC 227)	RAF Turnhouse, on display
	L1096	VS Spitfire IX Replica [PR-O] (BAPC 229)	RAF Digby, on display
	L1592	Hawker Hurricane I [KW-Z]	Science Museum, South Kensington
	L1592	Hawker Hurricane I Replica (BAPC 63) [KW-Z]	Kent Battle of Britain Museum, Hawkinge
	L1710	Hawker Hurricane IIc Replica [AL-D] (BAPC 219)	RAF Biggin Hill, on display
	L2301	VS Walrus I (G-AIZG)	FAA Museum, RNAS Yeovilton
	L2940	Blackburn Skua I (remains)	FAA Museum, RNAS Yeovilton
	L5343	Fairey Battle I [VO-S]	RAF Museum, Hendon
	L6906	Miles Magister I (G-AKKY/T9841) (BAPC 44)	Museum of Berkshire Aviation, Woodley
	L7775	Vickers Wellington IA (fuselage)	Privately owned, Moreton-in-Marsh
	L8756	Bristol Bolingbroke IVT (RCAF 10001) [XD-E]	RAF Museum, Hendon
	N248	Supermarine S6A	Southampton Hall of Aviation
	N546	Wright Quadruplane Type 1 Replica (BAPC 164)	Southampton Hall of Aviation
	N1671	Boulton Paul Defiant I (8370M) [EW-D]	RAF Museum, Hendon
	N1854	Fairey Fulmar II (G-AIBE)	FAA Museum, RNAS Yeovilton
	N2078	Sopwith Baby	FAA Museum, RNAS Yeovilton

Serial	Type	Owner or Operator	Notes
N2308	Gloster Gladiator I (G-AMRK) (really L8032) [HP-B]	Shuttleworth Collection, Old Warden	
N2980	Vickers Wellington IA [R]	Brooklands Museum, Weybridge	
N3194	VS Spitfire IX Replica [GR-Z] (BAPC 220)	RAF Biggin Hill, on display	
N3289	VS Spitfire Replica [OW-K] (BAPC65)	Kent Battle of Britain Museum, Hawkinge	
N4389	Fairey Albacore I [4M] (really N4172)	FAA Museum, RNAS Yeovilton	
N4877	Avro Anson I (G-AMDA) [VX-F]	Skyfame Collection, Duxford	
N5180	Sopwith Pup (G-EBKY)	Shuttleworth Collection, Old Warden	
N5182	Sopwith Pup Replica (G-APUP)	RAF Museum, Hendon	
N5195	Sopwith Pup (G-ABOX)	Museum of Army Flying, Middle Wallop	
N5226	Gloster Sea Gladiator II [H] (really N5903)	Shuttleworth Collection, FAA Museum, RNAS Yeovilton	
N5419	Bristol Scout Replica (N5419)	RAF Museum Restoration Centre, Cardington	
N5492	Sopwith Triplane Replica (BAPC 111)	FAA Museum, RNAS Yeovilton	
N5628	Gloster Gladiator II	RAF Museum, Hendon	
N5912	Sopwith Triplane (8385M)	RAF Museum, Hendon	
N6004	Short Stirling I	RAeS Medway Branch, Rochester	
N6290	Sopwith Triplane Replica (G-BOCK)	Shuttleworth Collection, Old Warden	
N6452	Sopwith Pup Replica (G-BIAU)	FAA Museum, RNAS Yeovilton	
N6466	DH Tiger Moth (G-ANKZ)	Privately owned, Barton	
N6720	DH Tiger Moth (7014M) [RUO-B]	No 1940 Sqn ATC, Levenshulme	
N6797	DH Tiger Moth (G-ANEH)	Privately owned, Chilbolton	
N6812	Sopwith Camel	Imperial War Museum, Lambeth	
N6847	DH Tiger Moth (G-APAL)	Privately owned, Little Gransden	
N6848	DH Tiger Moth (G-BALX)	Privately owned, Sedlescombe	
N6965	DH Tiger Moth (G-AJTW) [FL-J]	Privately owned, Tibenham	
N6985	DH Tiger Moth (G-AHMN)	Museum of Army Flying, Middle Wallop	
N9191	DH Tiger Moth (G-ALND)	Privately owned, Shobdon	
N9389	DH Tiger Moth (G-ANJA)	Privately owned, Shipmeadow, Suffolk	
N9899	Supermarine Southampton I	RAF Museum Restoration Centre, Cardington	
P1344	HP Hampden [PL-K]	RAF Museum Restoration Centre, Cardington	
P2617	Hawker Hurricane I (8373M) [AF-F]	RAF Museum, Hendon	
P3059	Hawker Hurricane Replica [SD-N] (BAPC64)	Kent Battle of Britain Museum, Hawkinge	
P3175	Hawker Hurricane I (Wreckage)	RAF Museum, Hendon	
P3386	Hawker Hurricane IIc Replica (BAPC 218)	RAF Bentley Priory, on display	
P3395	Hawker Hurricane IV [KX-B] (really KX829)	Birmingham Museum of Science and Technology	
P3554	Hawker Hurricane I (Composite)	Wiltshire Historic Aviation Grp, Salisbury	
P4139	Fairey Swordfish II [5H] (really HS618/A2001)	FAA Museum, RNAS Yeovilton	
P5865	CCF Harvard 4 (G-BKCK) [LE-W]	Privately owned, North Weald	
P6382	Miles M.14A Hawk Trainer 3 (G-AJRS)	Shuttleworth Collection, Old Warden	
P7350	VS Spitfire IIA (G-AWIJ) [YT-F]	RAF Battle of Britain Memorial Flight, Coningsby	
P7540	VS Spitfire IIA [DU-W]	Dumfries & Galloway Aviation Museum, Tinwald Downs	
P8140	VS Spitfire I Replica (BAPC 71)	Norfolk & Suffolk Aviation Museum, Flixton	
P8448	VS Spitfire IX Replica [UM-D] (BAPC 225)	RAF Swanton Morley, on display	
P9390	VS Spitfire I Replica (BAPC 71)	*Repainted as P8140*	
P9444	VS Spitfire IA [RN-D]	Science Museum, South Kensington	
R1914	Miles Magister (G-AHUJ)	Privately owned, Strathallan	
R3971	HP Halifax II (cockpit)	Cotswold Aviation Research Grp, Innsworth	
R4897	DH Tiger Moth (G-ERTY)	Privately owned, Meppershall	
R4907	DH Tiger Moth (G-ANCS)	Privately owned, Felthorpe	
R5250	DH Tiger Moth (G-AODT)	Privately owned, Swanton Morley	
R5868	Avro Lancaster I (7325M) [PO-S]	RAF Museum, Hendon	
R6915	VS Spitfire I	Imperial War Museum, Lambeth	
R9125	Westland Lysander III (8377M) [LX-L]	RAF Museum, Hendon	

Notes	*Serial*	*Type*	*Owner or Operator*
	S1287	Fairey Flycatcher Replica (G-BEYB) [5]	Privately owned, Stockbridge
	S1579	Hawker Nimrod Replica [571] (G-BBVO)	Privately owned, Dunkeswell
	S1595	Supermarine S6B	Science Museum, South Kensington
	T5424	DH Tiger Moth (G-AJOA)	Privately owned, Chiseldon
	T5493	DH Tiger Moth (G-ANEF)	Privately owned, Cranwell North
	T5672	DH Tiger Moth (G-ALRI)	Privately owned, Chalmington
	T5854	DH Tiger Moth (G-ANKK)	Privately owned, Halfpenny Green
	T5879	DH Tiger Moth (G-AXBW)	Privately owned, Tongham
	T5968	DH Tiger Moth (G-ANNN)	Privately owned, Kilkerran
	T6099	DH Tiger Moth (G-AOGR/XL714)	*Repainted as G-AOGR*
	T6256	DH Tiger Moth	Privately owned, Cranfield
	T6269	DH Tiger Moth (G-AMOU) [FOR-T]	Privately owned, Coventry
	T6296	DH Tiger Moth (8387M)	RAF Museum, Hendon
	T6313	DH Tiger Moth (G-AHVU)	Privately owned, Liphook
	T6818	DH Tiger Moth (G-ANKT) [44]	Shuttleworth Collection, Old Warden
	T6991	DH Tiger Moth (G-ANOR/DE694)	Privately owned, Paddock Wood
	T7230	DH Tiger Moth (G-AFVE)	Privately owned, Denham
	T7281	DH Tiger Moth (G-ARTL)	Privately owned, Egton, nr Whitby
	T7404	DH Tiger Moth (G-ANMV)	Privately owned, Booker
	T7909	DH Tiger Moth (G-ANON)	Privately owned, Sherburn-in-Elmet
	T7997	DH Tiger Moth (G-AOBH)	Privately owned, Benington
	T8191	DH Tiger Moth	RN Historic Flight, stored Lee-on-Solent
	T9707	Miles Magister (G-AKKR/8378M/T9708)	Greater Manchester Museum of Science and Industry
	T9738	Miles Hawk Trainer III (G-AKAT)	Privately owned, Breighton
	V1075	Miles M14A Magister (G-AKPF)	Privately owned, Shoreham
	V3388	Airspeed Oxford (G-AHTW)	Imperial War Museum, Duxford
	V6028	Bristol Bolingbroke IVT (G-MKIV)	British Aerial Museum, Duxford
	V7350	Hawker Hurricane I (fuselage)	Brenzett Aviation Museum
	V7467	Hawker Hurricane IIc Replica [LE-D] (BAPC 223)	RAF Coltishall, on display
	V7767	Hawker Hurricane Replica (BAPC 72)	Midland Air Museum, Coventry
	V9281	WS Lysander III (G-BCWL) [RU-M]	Privately owned, Henstridge
	V9300	WS Lysander III (G-LIZY) [MA-J]	British Aerial Museum, Duxford
	V9441	WS Lysander IIIA (RCAF2355/ G-AZWT) [AR-A]	Privately owned, stored Strathallan
	W1048	HP Halifax II (8465M) [TL-S]	RAF Museum, Hendon
	W2718	VS Walrus I (G-RNLI)	Privately owned, Micheldever
	W4041	Gloster E28/39 [G]	Science Museum, South Kensington
	W4050	DH Mosquito I	Mosquito Aircraft Museum, London Colney
	W5856	Fairey Swordfish II (G-BMGC) [A2A]	RN Historic Flight, BAe Brough
	X4277	VS Spitfire XVIe [XT-M] (really TB382) (7244M)	RAF Exhibition Flight, Abingdon
	X4474	VS Spitfire XVIe [QV-I] (really TE311) (7241M)	RAF Exhibition Flight, Abingdon
	X4590	VS Spitfire I (8384M) [PR-F]	RAF Museum, Hendon
	X7688	Bristol Beaufighter 1F (3858M) (G-DINT)	Privately owned, Hatch
	Z7015	Hawker Sea Hurricane IB (G-BKTH)	Shuttleworth Collection, Duxford
	Z7197	Percival Proctor III (G-AKZN/ 8380M)	RAF Museum, Hendon
	Z7381	Hawker Hurricane XII [XR-T] (G-HURI)	The Fighter Collection, Duxford
	AA908	VS Spitfire Replica Vb (BAPC 230) [UM-W]	Privately owned, Malton
	AB130	VS Spitfire IV (parts)	Privately owned, Ludham
	AB910	VS Spitfire VB [MD-E]	RAF BBMF, Coningsby
	AE436	HP Hampden (parts)	RAFM Restoration Centre, Cardington
	AL246	Grumman Martlet I	FAA Museum, RNAS Yeovilton
	AM561	Lockheed Hudson V (remains)	Cornwall Aero Park, Helston
	AP506	Cierva C30A (G-ACWM)	International Helicopter Museum, Weston-super-Mare
	AP507	Cierva C30A (G-ACWP) [KX-P]	Science Museum, South Kensington
	AR213	VS Spitfire IA (G-AIST) [PR-O]	Privately owned, Booker
	AR501	VS Spitfire LFVC (G-AWII) [NN-A]	Shuttleworth Collection, Old Warden
	AR614	VS Spitfire Vc	Old Flying Machine Company, Duxford

Serial	Type	Owner or Operator	Notes
BB807	DH Tiger Moth (G-ADWO)	Southampton Hall of Aviation	
BB814	DH Tiger Moth (G-AFWI)	*Repainted as G-AFWI*	
BE417	Hawker Hurricane XII (G-HURR)[AE-K]	Privately owned, Brooklands	
BE421	Hawker Hurricane IIC Replica (BAPC 205)[XP-G]	RAF Museum, Hendon	
BL370	VS Spitfire V	Privately owned, Oxford	
BL614	VS Spitfire VB (4354M) [ZD-F]	Greater Manchester Museum of Science and Industry	
BL628	VS Spitfire Vb (G-BTTN)	Privately owned, Thruxton	
BM597	VS Spitfire FVB (5718M) [PR-O] (G-MKVB)	Privately owned, Audley End	
BN230	Hawker Hurricane IIC (5466M) (really LF751) [FT-A]	RAF Manston, Memorial Pavilion	
BR600	VS Spitfire IX Replica (BAPC 222)[SH-V]	RAF Uxbridge, on display	
BR600	VS Spitfire IX Replica (BAPC 224)	Ambassador Hotel, Norwich	
BW853	Hawker Sea Hurricane XIIA	Privately owned, Sudbury	
BW881	Hawker Sea Hurricane XIIA	Privately owned, Milden	
DD931	Bristol Beaufort VIII	RAF Museum, Hendon	
DE208	DH Tiger Moth (G-AGYU)	Privately owned, Audley End	
DE363	DH Tiger Moth (G-ANFC)	Military Aircraft Preservation Group,Hadfield, Derbyshire	
DE373	DH Tiger Moth T2 (A680/A2127)	Privately owned	
DE623	DH Tiger Moth (G-ANFI)	Privately owned, St Athan	
DE673	DH Tiger Moth (G-ADNZ/6948M)	Privately owned, Hampton	
DE970	DH Tiger Moth (G-AOBJ)	Privately owned, Cardiff	
DE992	DH Tiger Moth (G-AXXV)	Privately owned, Membury	
DF128	DH Tiger Moth (G-AOJJ) [RCO-U]	Privately owned, Abingdon	
DF155	DH Tiger Moth (G-ANFV)	Privately owned, Lossiemouth	
DF198	DH Tiger Moth (G-BBRB)	Privately owned, Biggin Hill	
DG202	Gloster F9/40 Meteor (5758M) [G]	RAF Cosford Aerospace Museum	
DG590	Miles Hawk Major	RAF Museum Restoration Centre, Cardington	
DK431	Fairey Firefly I (Z2033/G-ASTL)	Imperial War Museum, Duxford	
DP872	Fairey Barracuda II (fuselage)	FAA Museum, Yeovilton	
DR613	FW Wicko GM1 (G-AFJB)	Privately owned, Berkswell	
DR628	Beech D.17s (N18V) [PB-1]	Privately owned, North Weald	
DV372	Avro Lancaster I (nose only)	Imperial War Museum, Lambeth	
EE416	Gloster Meteor III (nose only)	Science Museum, South Kensington	
EE425	Gloster Meteor F3 (nose only)	Rebel Air Museum, Earls Colne	
EE531	Gloster Meteor F4 (7090M)	Midland Air Museum, Coventry	
EE549	Gloster Meteor F4 (7008M)	Tangmere Military Aviation Museum	
EF545	VS Spitfire V	Privately owned, High Wycombe	
EJ693	Hawker Tempest V [SA-J]	Privately owned	
EM720	DH Tiger Moth (G-AXAN)	Privately owned, Staverton	
EM727	DH Tiger Moth (G-AOXN)	Privately owned, Yeovil	
EM903	DH Tiger Moth (G-APBI)	Privately owned, Audley End	
EN224	VS Spitfire F XII (G-FXII)	Privately owned, Newport Pagnell	
EN343	VS Spitfire PR21 Replica (BAPC 226)	RAF Benson, on display	
EN398	VS Spitfire IX Replica (BAPC184) [JE-J]	Aces High Ltd, North Weald	
EP120	VS Spitfire LFVB (5377M/8070M) [QV-H]	Privately owned, Duxford	
EX280	NA Harvard IIA (G-TEAC) [G]	*Repainted as 889696 (USN)*	
EX976	NA Harvard III	FAA Museum, RNAS Yeovilton	
EZ259	NA Harvard III (G-BMJW)	Privately owned, Bracknell	
EZ407	NA Harvard III	RN Historic Flight, stored Lee-on-Solent	
FB226	Bonsall Mustang Replica [MT-A] (G-BDWM)	Privately owned, Netherthorpe	
FE695	NA Harvard IIB (G-BTXI)	Privately owned, Duxford	
FE905	NA Harvard IIB (LN-BNM/12392)	Newark Air Museum, Winthorpe	
FE992	NA Harvard IIB (G-BDAM) [KT]	Privately owned, Duxford	
FH153	NA Harvard IIB (G-BBHK) [GW-A]	Privately owned, Exeter	
FR870	Curtiss Kittyhawk III (NL1009N) [GA-S]	The Fighter Collection, Duxford	
FS728	NA Harvard IIB (G-BAFM)	Privately owned, Denham	
FS890	NA Harvard IIB (7554M)	A&AEE, stored Boscombe Down	
FT239	NA Harvard IV (G-BIWX)	Privately owned, North Weald	

Notes	Serial	Type	Owner or Operator
	FT375	NA Harvard IIB	MoD(PE), A&AEE Boscombe Down
	FT391	NA Harvard IIB (G-AZBN)	Privately owned, Shoreham
	FX301	NA Harvard III (G-JUDI) (really EX915)	Privately owned, Bryngwyn Bach, Clwyd
	FX360	NA Harvard IIB (really KF435)	Booker Aircraft Museum
	FX442	NA Harvard IIB	Privately owned, Bournemouth
	FX760	Curtiss Kittyhawk IV [GA-?]	RAF Museum, Hendon
	HB275	Beech C-45 Expeditor II (N5063N)	Privately owned, North Weald
	HB751	Fairchild Argus III (G-BCBL)	Privately owned, Little Gransden
	HH379	GAL48 Hotspur II (rear fuselage only)	Museum of Army Flying, Middle Wallop
	HJ711	DH Mosquito NFII [VI-C]	Night Fighter Preservation Team, Elvington
	HM354	Percival Proctor III (G-ANPP)	Privately owned, Stansted
	HM580	Cierva C-30A (G-ACUU)	Imperial War Museum, Duxford
	HR792	HP Halifax GR II	Yorkshire Air Museum, Elvington
	HS503	Fairey Swordfish IV (BAPC 108)	RAF Museum, stored Henlow
	HX922	DH Mosquito TT35 (G-AWJV) [EG-F] (really TA634)	Mosquito Aircraft Museum, London Colney
	JV482	Grumman Wildcat V	Ulster Aviation Society, Langford Lodge
	JV928	PBY-5A Catalina (G-BLSC) [Y]	Plane Sailing, Duxford
	KB889	Avro Lancaster B10 (G-LANC) [NA-I]	Imperial War Museum, Duxford
	KB976	Avro Lancaster B10 (G-BCOH)	Warbirds of GB, Bournemouth
	KB994	Avro Lancaster B10 (fuselage)	Warbirds of GB, Bournemouth
	KD431	CV Corsair IV [E2-M]	FAA Museum, RNAS Yeovilton
	KE209	Grumman Hellcat II	FAA Museum, RNAS Yeovilton
	KE418	Hawker Tempest (rear fuselage)	RAF Museum Store, Cardington
	KF183	NA Harvard IIB [3]	MoD(PE) A&AEE Boscombe Down
	KF487	NA Harvard IIB	Privately owned. Duxford (for spares)
	KF488	NA Harvard IIB (nose only)	Privately owned, Bournemouth
	KF532	NA Harvard IIB (cockpit section)	Newark Air Museum, Winthorpe
	KG374	Douglas Dakota C-4 [YS] (really KN645/8355M)	RAF Cosford Aerospace Museum
	KJ351	HP Horsa II composite [23] (really TL659) (BAPC 80)	Museum of Army Flying, Middle Wallop
	KK995	Sikorsky Hoverfly I [E]	RAF Museum, Hendon
	KL161	NA B-25D Mitchell (N88972)[VO-B]	The Fighter Collection, Duxford
	KN448	Douglas Dakota C4 (nose only)	Science Museum, South Kensington
	KN751	Consolidated Liberator VI [F]	RAF Cosford Aerospace Museum
	KP208	Douglas Dakota C-4 [YS]	Airborne Forces Museum, Aldershot
	KZ191	Hawker Hurricane IV (frame only)	Privately owned, North Weald
	KZ321	Hawker Hurricane IV (G-HURY) [JV-N]	The Fighter Collection, Duxford
	LA198	VS Spitfire F21 (7118M) [RAI-G]	RAF, stored St Athan
	LA226	VS Spitfire F21 (7119M)	RAF, stored Shawbury
	LA255	VS Spitfire F21 (6490M) [JX-U]	RAF No 1 Sqn, Wittering
	LA546	VS Seafire F46 (nose only)	Privately owned, Colchester
	LA564	VS Seafire F46	Privately owned, Newport Pagnell
	LB294	Taylorcraft Plus D (G-AHWJ)	Museum of Army Flying, Middle Wallop
	LB312	Taylorcraft Plus D (G-AHXE)	Privately owned, Shoreham
	LB375	Taylorcraft Plus D (G-AHGW)	Privately owned, Coventry
	LF363	Hawker Hurricane IIC	RAF BBMF, Coningsby, under restoration
	LF738	Hawker Hurricane IIC (5405M)	RAF, RAeS Medway Branch, Rochester
	LF858	DH Queen Bee (G-BLUZ)	Privately owned, Hatch
	LH208	Airspeed Horsa I (parts only) (8596M)	Museum of Army Flying, Middle Wallop
	LS326	Fairey Swordfish II (G-AJVH)[L2]	RN Historic Flight, RNAS Yeovilton
	LZ551	DH Sea Vampire I [G]	FAA Museum, RNAS Yeovilton
	LZ766	Percival Proctor III (G-ALCK)	Skyfame Collection, Duxford
	LZ842	VS Spitfire IX	Privately owned, Battle, Sussex
	MD497	WS51 Widgeon (G-ANLW) [NE-X]	*Repainted as G-ANLW*
	MF628	Vickers Wellington T10	RAF Museum, Hendon
	MH434	VS Spitfire HF.IXB (G-ASJV)[AV-H]	Privately owned, Duxford
	MH486	VS Spitfire IX Replica (BAPC 206) [FF-A]	RAF Museum, Hendon
	MH603	VS Spitfire IX	Privately owned, Stretton, Cheshire
	MH777	VS Spitfire IX Replica (BAPC 221) [RF-N]	RAF Northolt, on display

Serial	*Type*	*Owner or Operator*	*Notes*
MJ627	VS Spitfire TIX (G-ASOZ/G-BMSB)	Privately owned, Baginton	
MJ730	VS Spitfire HFIXe (G-HFIX) [GZ-?]	Privately owned, East Midlands Airport	
MK356	VS Spitfire IX (5690M)	RAF BBMF, St Athan	
MK732	VS Spitfire LFIX (G-HVDM)	Privately owned, Guernsey	
MK805	VS Spitfire LFIX Replica [SH-B]	Privately owned, Lowestoft	
MK912	VS Spitfire LFIXe (SM-29) (G-BRRA) [MN-P]	Privately owned, Paddock Wood	
ML407	VS Spitfire TIX (G-LFIX) [OU-V]	Privately owned, Goodwood	
ML417	VS Spitfire LFIXE (G-BJSG)[21-T]	Privately owned, Duxford	
ML427	VS Spitfire IX (6457M) [I-ST]	Birmingham Museum of Science & Industry	
ML796	Short Sunderland V	Imperial War Museum, Duxford	
ML814	Short Sunderland V (G-BJHS)	Privately owned, Calshot	
ML824	Short Sunderland V [NS-Z]	RAF Museum, Hendon	
MN235	Hawker Typhoon IB	RAF Museum, Hendon	
MP425	Airspeed Oxford I (G-AITB)	Newark Air Museum, Winthorpe	
MT438	Auster III (G-AREI)	Museum of Army Flying, Middle Wallop	
MT719	VS Spitfire LF VIIIc (G-VIII)	Privately owned, Micheldever	
MT847	VS Spitfire FRXIVe (6960M)[AX-H]	RAF Cosford Aerospace Museum	
MV154	VS Spitfire HF VIII (G-BKMI/A58-671)	Privately owned, Filton	
MV262	VS Spitfire XIV (G-CCVV) [42-G]	Privately owned, Booker	
MV293	VS Spitfire XIV (G-SPIT/G-BGHB) (previously painted as *MV363*) [O1-C]	Privately owned, Duxford	
MV370	VS Spitfire XIV (G-FXIV) [EB-Q]	Privately owned, Duxford	
MW100	Avro York C1 (G-AGNV/TS798)	*Repainted as TS798*	
MW376	Hawker Tempest II (G-BSHW) (IAFHA564)	*Repainted as MW800*	
MW401	Hawker Tempest II (G-PEST) (IAF HA604)	Privately owned, Brooklands	
MW404	Hawker Tempest II (HA557)	Privately owned	
MW467	VS Spitfire Vb Replica (BAPC 202) [R-D]	Maes Astro Craft Village, Llanbedr	
MW758	Hawker Tempest II (HA580)	Privately owned, Chichester	
MW763	Hawker Tempest II (G-TEMT) (HA586)	Privately owned, Brooklands	
MW800	Hawker Tempest II (G-BSHW) [HF-V] (IAFHA564)	Privately owned, Spanhoe Lodge, Northumbria	
NF370	Fairey Swordfish II	Imperial War Museum, Duxford	
NF389	Fairey Swordfish III [5B]	RN Historic Flight, Brough	
NF875	DH Dragon Rapide 6 (G-AGTM) [603/CH]	Privately owned, Audley End	
NH238	VS Spitfire IX (N238V/G-MKIX)	Warbirds of GB, Bournemouth	
NH799	VS Spitfire FRXIV	Privately owned, Duxford	
NJ673	Auster 5D (G-AOCR)	Privately owned, Wyberton	
NJ695	Auster 4 (G-AJXV)	Privately owned, Tollerton	
NJ703	Auster 5 (G-AKPI)	Privately owned, Skegness	
NJ719	Auster 5 (G-ANFU) (really TW385)	North-East Aircraft Museum, Usworth	
NL750	DH Tiger Moth (G-AHUF/A2123)	Privately owned, Popham	
NL879	DH Tiger Moth (G-AVPJ)	Privately owned, Wellesbourne Mountford	
NL985	DH Tiger Moth (7015M)	Privately owned, Sywell	
NP181	Percival Proctor IV (G-AOAR)	Privately owned, Biggin Hill	
NP184	Percival Proctor IV (G-ANYP) [K]	Privately owned, Chatteris	
NP294	Percival Proctor IV [TB-M]	Lincolnshire Aviation Museum, East Kirkby	
NP303	Percival Proctor IV (G-ANZJ)	Privately owned, Byfleet, Surrey	
NV778	Hawker Tempest V (8386M)	RAFM Restoration Centre, Cardington	
NX611	Avro Lancaster VII (G-ASXX/8375M) [YF-C]	Lincolnshire Aviation Heritage Centre, East Kirkby	
PA474	Avro Lancaster I [PM-M2]	RAF BBMF, Coningsby	
PF179	HS Gnat T1 (XR541/8602M)	Privately owned, Worksop	
PK624	VS Spitfire F22 (8072M) [RAU-T]	RAF, stored St Athan	
PK664	VS Spitfire F22 (7759M) [V6-B]	RAF, stored St Athan	
PK683	VS Spitfire F24 (7150M)	Southampton Hall of Aviation	
PK724	VS Spitfire F24 (7288M)	RAF Museum, Hendon	
PL344	VS Spitfire IX (G-IXCC)	Privately owned, Booker	
PL965	VS Spitfire XI (G-MKXI)	Privately owned, Duxford	
PL983	VS Spitfire XI (G-PRXI)	Warbirds of GB, Bournemouth	

Notes	Serial	Type	Owner or Operator
	PM631	VS Spitfire PRXIX [N]	RAF BBMF, Coningsby
	PM651	VS Spitfire PRXIX (7758M) [X]	RAF, stored St Athan
	PN323	HP Halifax VII (nose only)	Imperial War Museum, Lambeth
	PP566	Fairey Firefly I (fuselage)	South Yorkshire Air Museum, Firbeck
	PP972	VS Seafire LFIIIc (G-BUAR)[6M-D]	Privately owned, East Midlands
	PR536	Hawker Tempest II (HA457) [OQ-H]	RAF Museum, Hendon
	S853	VS Spitfire PRXIX [C]	RAF BBMF, Coningsby
	PS915	VS Spitfire PR XIX (7548M/7711M)[P]	RAF BBMF, Coningsby
	PT462	VS Spitfire TIX (G-CTIX)	Privately owned, Duxford
	PV202	VS Spitfire TRIX (G-TRIX) [VZ-M]	Privately owned, Leavesden
	PZ865	Hawker Hurricane IIc (G-AMAU) [J]	RAF BBMF, Coningsby
	RA848	Slingsby Cadet TX1	Privately owned, Leeds
	RA854	Slingsby Cadet TX1	Privately owned, Harrogate
	RA897	Slingsby Cadet TX1	Newark Air Museum store, Hucknall
	RD253	Bristol Beaufighter TF10 (7931M)	RAF Museum, Hendon
	RF342	Avro Lincoln B2 (G-29-1/G-APRJ)	Warbirds of GB, Bournemouth
	RF398	Avro Lincoln B2 (8376M)	RAF Cosford Aerospace Museum
	RG333	Miles Messenger IIA (G-AIEK)	Privately owned, Felton, Bristol
	RG333	Miles Messenger IIA (G-AKEZ)	Privately owned, Chelmsford
	RH377	Miles Messenger 4A (G-ALAH)	Privately owned, Stretton, Cheshire
	RH746	Bristol Brigand TF1	North-East Aircraft Museum, Usworth
	RL962	DH Dragon Rapide (G-AHED)	RAF Museum Store, Cardington
	RM221	Percival Proctor IV (G-ANXR)	Privately owned, Biggin Hill
	RM689	VS Spitfire XIV (G-ALGT) [MN-E]	*Crashed Woodford, 27 June 1992*
	RN201	VS Spitfire XIV (SG-31/G-BSKP)	Privately owned, Paddock Wood
	RN218	Isaacs Spitfire Replica (G-BBJI) [N]	Privately owned, Langham
	RR232	VS Spitfire HFIXc(G-BRSF)	Privately owned, Lancing, Sussex
	RR299	DH Mosquito T3 (G-ASKH) [HT-E]	British Aerospace, Hawarden
	RT486	Auster AOP5 (G-AJGJ)	Privately owned, Popham
	RT520	Auster 5 (G-ALYB)	South Yorkshire Air Museum, Firbeck
	RT610	Auster 5A (G-AKWS)	Privately owned, Exeter
	RW382	VS Spitfire XVIe (7245M/8075M) [NG-C] (G-XVIA)	Privately owned, Duxford
	RW386	VS Spitfire XVIe (6944M/G-BXVI) [RAK-A]	Warbirds of GB, Biggin Hill
	RW388	VS Spitfire XVIe (6946M) [U4-U]	Stoke-on-Trent City Museum, Hanley
	RW393	VS Spitfire XVIe (7293M) [XT-A]	RAF, stored St Athan
	RX168	VS Seafire LFIIIc	Privately owned, High Wycombe
	SL542	VS Spitfire XVIe (8390M) [4M-N]	Privately owned, St Athan
	SL674	VS Spitfire XVIe [RAS-H] (8392M)	RAF, stored St Athan
	SM520	VS Spitfire IX	Privately owned, Duxford
	SM832	VS Spitfire XIV (G-WWII)	Privately owned, Duxford
	SM845	VS Spitfire XVIII (G-BUOS)	Privately owned, Witney, Oxon
	SM969	VS Spitfire XVIIIe (G-BRAF)[D-A]	Warbirds of GB, Lelystad
	SX137	VS Seafire XVII	FAA Museum, RNAS Yeovilton
	SX300	VS Seafire XVII (A646/A696/A2054)	Privately owned, Warwick
	SX336	VS Seafire XVII (A2055) (G-BRMG)	Privately owned, Twyford, Berks
	TA122	DH Mosquito FBVI [UP-G]	Mosquito Aircraft Museum, London Colney
	TA634	DH Mosquito B35 (G-AWJV)	Mosquito Aircraft Museum, London Colney
	TA639	DH Mosquito TT35 (7806M) [AZ-E]	RAF Cosford Aerospace Museum
	TA719	DH Mosquito TT35 (G-ASKC) [6T]	Skyfame Collection, Duxford
	TA805	VS Spitfire IX	Privately owned, Battle, East Sussex
	TB252	VS Spitfire XVIe (G-XVIE) [GW-H]	Privately owned, Audley End
	TB752	VS Spitfire XVIe (7256M/7279M/ 8086M) [KH-Z]	RAF Manston, Memorial Pavilion
	TB885	VS Spitfire LFXVIe	Shoreham Aircraft Preservation Society
	TD248	VS Spitfire XVIe (7246M) (G-OXVI) [D]	Privately owned, Audley End
	TE184	VS Spitfire LF XVIe (G-MXVI)	Privately owned, East Midlands
	TE392	VS Spitfire XVIe (7000M/8074M)	Warbirds of GB, Biggin Hill
	TE462	VS Spitfire XVIe (7243M)	Royal Scottish Museum of Flight, East Fortune
	TE476	VS Spitfire LFXVIe (G-XVIB)	Privately owned, Booker

Serial	Type	Owner or Operator	Notes
TE517	VS Spitfire LFIX (2046/G-BIXP/ G-CCIX)	Privately owned, Booker	
TE566	VS Spitfire LFXVIe (G-BLCK) [DU-A]	Privately owned, Audley End	
TG263	Saro SR.A1 (G-12-1) [P]	Imperial War Museum, Duxford	
TG511	HP Hastings T5 (8554M)	RAF Cosford Aerospace Museum	
TG517	HP Hastings T5 [517]	Newark Air Museum, Winthorpe	
TG528	HP Hastings C1A	Imperial War Museum, Duxford	
TJ118	DH Mosquito TT35 (nose only)	Mosquito Aircraft Museum store	
TJ138	DH Mosquito B35 (7607M) [VO-L]	RAF Museum, Hendon	
TJ343	Auster 5 (G-AJXC)	Privately owned, Hook	
TJ398	Auster AOP5 (BAPC 70)	Aircraft Preservation Society of Scotland, East Fortune	
TJ569	Auster 5 (G-AKOW)	Museum of Army Flying, Middle Wallop	
TJ672	Auster 5 (G-ANIJ)	Privately owned, RAF Swanton Morley	
TJ704	Auster 5 [JA] (G-ASCD)	Yorkshire Air Museum, Elvington	
TK718	GAL Hamilcar I	Museum of Army Transport, Beverley	
TK777	GAL Hamilcar I (fuselage only)	Museum of Army Flying, Middle Wallop	
TL615	Airspeed Horsa II	Robertsbridge Aviation Society	
TP280	VS Spitfire XVIIIe (HS654/ G-BTXE)	*To USA, August 1992*	
TP298	VS Spitfire XIV (fuselage)	Privately owned, Ludham	
TS291	Slingsby Cadet TX1 (BGA852)	Royal Scottish Museum of Flt, East Fortune	
TS423	Douglas Dakota 3 (G-DAKS)	Aces High Ltd, North Weald	
TV959	DH Mosquito T3 [AF-V]	Imperial War Museum, Duxford, stored	
TV959	DH Mosquito Replica	Privately owned, Heald Green, Cheshire	
TW117	DH Mosquito T3 (7805M)	*To Gardermoen, Norway, 3 Feb 1992*	
TW439	Auster 5 (G-ANRP)	Privately owned, Dorchester	
TW448	Auster 5 (G-ANLU)	Privately owned, Hedge End	
TW467	Auster 5 (G-ANIE) [AW]	Privately owned, Sywell	
TW511	Auster 5 (G-APAF)	Privately owned, Skegness	
TW536	Auster AOP6 (7704M/G-BNGE) [TS-V]	Privately owned, Middle Wallop	
TW591	Auster 6A (G-ARIH) [N]	Privately owned, Burnaston	
TW641	Auster AOP6 (G-ATDN)	Privately owned, Biggin Hill	
TX183	Avro Anson C19 (G-BSMF)	Privately owned, Arbroath	
TX213	Avro Anson C19 (G-AWRS)	North-East Aircraft Museum, Usworth	
TX214	Avro Anson C19 (7817M)	RAF Cosford Aerospace Museum	
TX226	Avro Anson C19 (7865M)	Imperial War Museum, Duxford	
TX228	Avro Anson C19	City of Norwich Aviation Museum	
TX235	Avro Anson C19	Caernarfon Mountain Aviation Museum	
VD165	Slingsby T7 Kite	Russavia Collection, Dunstable	
VF301	DH Vampire F1 (7060M) [RAL-B]	Midland Air Museum, Coventry	
VF516	Auster AOP6 (G-ASMZ) [T]	Privately owned, Crediton	
VF526	Auster T7 (G-ARXU) [T]	Privately owned, Keevil	
VF548	Beagle Terrier 1 (G-ASEG)	Privately owned, Dunkeswell	
VH127	Fairey Firefly TT4	FAA Museum, RNAS Yeovilton	
VL348	Avro Anson C19 (G-AVVO)	Newark Air Museum, Winthorpe	
VL349	Avro Anson C19 (G-AWSA)	Norfolk & Suffolk Aviation Museum, Flixton	
VM325	Avro Anson C19	Midland Air Museum, Coventry	
VM360	Avro Anson C19 (G-APHV)	Royal Scottish Museum of Flight, East Fortune	
VM791	Slingsby Cadet TX3 (really XA312) (8876M)	No 135 Redhill & Reigate Sqn ATC, RAF Kenley	
VN148	Grunau Baby II b (BAPC 33) (BGA 2400)	Russavia Collection, Duxford	
VN485	VS Spitfire F24 (7326M)	Imperial War Museum, Duxford	
VP519	Avro Anson C19 (nose)(G-AVVR)	Privately owned, Dukinfield	
VP952	DH Devon C2 (8820M)	RAF Cosford Aerospace Museum	
VP955	DH Devon C2 (G-DVON)	Privately owned, Bournemouth	
VP957	DH Devon C2 (nose) (8822M)	No 1137 Sqn ATC, Belfast	
VP959	DH Devon C2 [L]	MoD(PE), DRA West Freugh	
VP967	DH Devon C2 (G-KOOL)	East Surrey Technical College, Redhill	
VP968	DH Devon C2	A&AEE Boscombe Down, derelict	
VP971	DH Devon C2 (8824M)	FSCTE, RAF Manston	
VP975	DH Devon C2 [M]	Science Museum, Wroughton	
VP976	DH Devon C2 (8784M)	RAF Northolt Fire Section	
VP977	DH Devon C2 (G-ALTS)	DRA West Freugh Fire Section	
VP978	DH Devon C2 (8553M)	RAF Brize Norton, instructional use	
VP981	DH Devon C2	RAF BBMF, Coningsby	
VR137	Westland Wyvern TF1	FAA Museum, RNAS Yeovilton	

Notes	Serial	Type	Owner or Operator
	VR192	Pervical Prentice T1 (G-APIT)	Second World War Aircraft Preservation Society, Lasham
	VR249	Percival Prentice T1 (G-APIY) [FA-EL]	Newark Air Museum, Winthorpe
	VR259	Percival Prentice T1 (G-APJB)	Privately owned, Coventry
	VR930	Hawker Sea Fury FB11 (8382M)	RN Historic Flight, stored Lee-on-Solent
	VS356	Percival Prentice T1 (G-AOLU)	Privately owned, Stonehaven
	VS517	Avro Anson T20	RN, Lee-on-Solent, stored
	VS562	Avro Anson T21 (8012M)	Maes Arto Craft Village, Llanbedr
	VS610	Percival Prentice T1 (G-AOKL)[K-L]	Privately owned, Southend
	VS623	Percival Prentice T1 (G-AOKZ) [KQ-F]	Midland Air Museum, Coventry
	VT260	Gloster Meteor F4 (8813M) [67]	South Yorks Aircraft Preservation Society, Firbeck
	VT409	Fairey Firefly AS5(mostly WD889)	North-East Aircraft Museum, Usworth
	VT812	DH Vampire F3 (7200M) [N]	RAF Museum, Hendon
	VT935	Boulton Paul P111A (VT769)	Midland Air Museum, Coventry
	VT987	Auster AOP6 (G-BKXP)	Privately owned, Gransden, Cambs
	VV106	Supermarine 510 (7175M)	RN, Lee-on-Solent, stored
	VV119	Supermarine 535 (nose only) (7285M)	Lincolnshire Aviation Museum, East Kirkby
	VV217	DH Vampire FB5 (7323M)	North-East Aircraft Museum, Usworth
	VV901	Avro Anson T21	Pennine Aviation Museum, Bacup
	VW453	Gloster Meteor T7 (8703M)	RAF Innsworth, on display
	VW985	Auster AOP6 (G-ASEF)	Privately owned, Upper Arncott, Oxon
	VX118	Auster 6A (G-ASNB)	Privately owned, Keevil
	VX147	Alon Aircoupe (G-AVIL)	Privately owned, Headcorn
	VX185	EE Canberra B(I)8 (nose only) (7631M)	Science Museum, South Kensington
	VX250	DH Sea Hornet 21 [48] (rear fuselage)	Mosquito Aircraft Museum, London Colney
	VX272	Hawker P1052 (7174M)	RN, Lee-on-Solent, stored
	VX275	Slingsby Sedbergh TX1 (8884M) (BGA 572)	RAF Museum Restoration Centre, Cardington
	VX461	DH Vampire FB5 (7646M)	RAF Cosford Aerospace Museum, stored
	VX573	Vickers Valetta C2 (8389M)	RAF Cosford Aerospace Museum, stored
	VX577	Vickers Valetta C2	North-East Aircraft Museum, Usworth
	VX580	Vickers Valetta C2	Norfolk & Suffolk Aviation Museum Flixton
	VX595	WS51 Dragonfly HR1 [29]	Gosport Aviation Society, HMS *Sultan*
	VX653	Hawker Sea Fury FB11 (G-BUCM)	The Fighter Collection, Duxford
	VX926	Auster T7 (G-ASKJ)	Privately owned, Gransden
	VZ304	DH Vampire FB5 [A-T](7630M)	Vintage Aircraft Team, Bruntingthorpe#
	VZ304	DH Vampire FB5 (7630M)	Vintage Aircraft Team, Cranfield
	VZ345	Hawker Sea Fury T20S	RN Historic Flight, Yeovilton
	VZ462	Gloster Meteor F8	Second World War Aircraft Preservation Society, stored
	VZ467	Gloster Meteor F8 (G-METE)	Privately owned, RAF Cosford
	VZ477	Gloster Meteor F8 (nose)(7741M)	Midland Air Museum, Coventry
	VZ608	Gloster Meteor FR9	Newark Air Museum, Winthorpe
	VZ634	Gloster Meteor T7 (8657M)	Newark Air Museum, Winthorpe
	VZ638	Gloster Meteor T7 (G-JETM) [HF]	Privately owned, Charlwood, Surrey
	VZ728	RS4 Desford Trainer (G-AGOS)	Coalville Museum of Science & Technology
	VZ962	WS51 Dragonfly HR1 [904]	International Helicopter Museum, Weston-super-Mare
	VZ965	WS51 Dragonfly HR5	FAA Museum, at RNAS Culdrose
	WA473	VS Attacker F1 [102/J]	FAA Museum, RNAS Yeovilton
	WA576	Bristol Sycamore 3 (G-ALSS/7900M)	Dumfries & Galloway Aviation Museum, Tinwald Downs
	WA577	Bristol Sycamore 3 (G-ALST/7718M)	North-East Aircraft Museum, Usworth
	WA591	Gloster Meteor T7 (7917M) [W]	
	WA630	Gloster Meteor T7 [69](nose)	Robertsbridge Aviation Museum
	WA634	Gloster Meteor T7/8	RAF Cosford Aerospace Museum
	WA638	Gloster Meteor T7 (spares)	Martin Baker Aircraft, Chalgrove
	WA662	Gloster Meteor T7 (spares)	Martin Baker Aircraft, Chalgrove
	WA984	Gloster Meteor F8 [A]	Tangmere Military Aviation Museum
	WB188	Hawker Hunter F3 (7154M)	Tangmere Military Aviation Museum
	WB271	Fairey Firefly AS5 [204/R]	RN Historic Flight, RNAS Yeovilton
	WB440	Fairey Firefly AS6 (nose only)	South Yorks Air Museum, Firbeck

Serial	Type	Owner or Operator	Notes
WB491	Avro Ashton 2 (nose only) (TS897/G-AJJW)	Wales Aircraft Museum, Cardiff	
WB550	DH Chipmunk T10 [F]	RAF EFTS, Swinderby	
WB556	DH Chipmunk T10	RAFGSA, Bicester	
WB560	DH Chipmunk T10	RAF No 4 AEF, Exeter	
WB565	DH Chipmunk T10 [X]	AAC BFWF, Middle Wallop	
WB567	DH Chipmunk T10	RAF No 12 AEF, Turnhouse	
WB569	DH Chipmunk T10 [2]	RAF No 1 AEF, Manston	
WB575	DH Chipmunk T10 [907]	RN Flying Grading Flt, Plymouth	
WB584	DH Chipmunk T10 PAX (7706M)	No 327 Sqn ATC, Kilmarnock	
WB585	DH Chipmunk T10 (G-AOSY) [RCU-X]	Privately owned, Blackbushe	
WB586	DH Chipmunk T10 [X]	RAF No 3 AEF Colerne	
WB588	DH Chipmunk T10 (G-AOTD) [D]	Shuttleworth Collection, Old Warden	
WB615	DH Chipmunk T10 [E]	AAC BFWF, Middle Wallop	
WB624	DH Chipmunk T10 PAX	The Aeroplane Collection, Long Marston	
WB626	DH Chipmunk T10 PAX	Privately owned, Swanton Morley	
WB627	DH Chipmunk T10 [N]	RAF No 5 AEF, Cambridge	
WB647	DH Chipmunk T10 [R]	AAC BFWF, Middle Wallop	
WB652	DH Chipmunk T10 [V]	RAF No 5 AEF, Cambridge	
WB654	DH Chipmunk T10 [T]	RAF No 3 AEF, Colerne	
WB657	DH Chipmunk T10 [908]	RN Flying Grading Flt, Plymouth	
WB660	DH Chipmunk T10 (G-ARMB)	Privately owned, Shipdham	
WB670	DH Chipmunk T10 PAX (8361M)	No 1312 Sqn ATC, Southend	
WB671	DH Chipmunk T10 [910]	RN Flying Grading Flt, Plymouth	
WB685	DH Chipmunk T10 PAX	North-East Aircraft Museum, Usworth	
WB693	DH Chipmunk T10 [S]	AAC BFWF, Middle Wallop	
WB697	DH Chipmunk T10 [95]	RAF No 10 AEF, Woodvale	
WB702	DH Chipmunk T10 (G-AOFE)	Privately owned, Denham	
WB703	DH Chipmunk T10 (G-ARMC)	Privately owned, White Waltham	
WB732	DH Chipmunk T10 (G-AOJZ/*G-ASTD)*	Air Service Training Ltd, Perth	
WB739	DH Chipmunk T10 [8]	RAF No 8 AEF, Shawbury	
WB754	DH Chipmunk T10 [H]	AAC BFWF, Middle Wallop	
WB758	DH Chipmunk T10 (7729M) [P]	Privately owned, Torbay	
WB763	DH Chipmunk T10 (G-BBMR) [14]	Southall Technical College	
WB971	Slingsby T21B (BGA 3324)	Privately owned, Lleweni Parc, Clwyd	
WB981	Slingsby T21B (BGA3238)	Privately owned, Aston Down	
WB983	Slingsby T21B (BGA3175)	*Repainted as BGA3175*	
WB990	Slingsby T21B (BGA 3148)	Privately owned, Lleweni Parc, Clwyd	
WD286	DH Chipmunk T10 [J] (G-BBND)	Privately owned, Bourn	
WD288	DH Chipmunk T10 (G-AOSO) [38]	Privately owned, Charlton Park, Wilts	
WD289	DH Chipmunk T10 [N]	RAF EFTS, Swinderby	
WD292	DH Chipmunk T10 (G-BCRX)	Privately owned, Old Sarum	
WD293	DH Chipmunk T10 PAX (7645M)	No 1367 Sqn ATC, Caerleon	
WD305	DH Chipmunk T10 (G-ARGG)	Privately owned, Coventry	
WD310	DH Chipmunk T10 [H]	RAF EFTS, Swinderby	
WD318	DH Chipmunk T10 PAX (8207M)	RAF No 1 SoTT, Halton	
WD325	DH Chipmunk T10 [N]	AAC BFWF, Middle Wallop	
WD331	DH Chipmunk T10 [A] [6]	RAF EFTS, Swinderby	
WD355	DH Chipmunk T10 PAX	No 1955 Sqn ATC, Wells, Somerset	
WD356	DH Chipmunk T10 (7625M)	Privately owned, St Ives, Cambridgeshire	
WD363	DH Chipmunk T10 (G-BCIH) [5]	Privately owned, Andrewsfield	
WD370	DH Chipmunk T10 PAX	No 176 Sqn ATC, Hove	
WD373	DH Chipmunk T10 [12]	RAF No 2 AEF, Bournemouth	
WD374	DH Chipmunk T10 [903]	RN Flying Grading Flt, Plymouth	
WD379	DH Chipmunk T10 (really WB696/G-APLO) [K]	Privately owned, Jersey	
WD386	DH Chipmunk T10 PAX	No 1284 Sqn ATC, Tenby	
WD390	DH Chipmunk T10 [68]	RAF No 9 AEF, Finningley	
WD413	Avro Anson C 21 (G-BFIR/7881M)	Privately owned, Teesside	
WD646	Gloster Meteor TT20 (8189M) [R]	No 2030 Sqn ATC, Sheldon	
WD686	Gloster Meteor NF11	Muckleburgh Collection, Weybourne	
WD790	Gloster Meteor NF11 (8743M) (nose only)	North-East Aircraft Museum, Usworth	
WD931	EE Canberra B2 (nose only)	RAF Cosford Aerospace Museum	
WD935	EE Canberra B2 (8440M) [cockpit]	Privately owned, Egham	
WD954	EE Canberra B2 (nose only)	Privately owned, Rayleigh, Essex	
WD955	EE Canberra T17A [EM]	RAF No 360 Sqn, Wyton	
WE113	EE Canberra T4 [BJ]	*Scrapped at RAF Wyton*	

Notes	Serial	Type	Owner or Operator
	WE122	EE Canberra TT18 [845] (nose only)	Privately owned, Chelmsford
	WE139	EE Canberra PR3 (8369M)	RAF Museum, Hendon
	WE168	EE Canberra PR3 (8049M) (nose)	Privately owned, Colchester
	WE173	EE Canberra PR3 (8740M)	*Scrapped at Coltishall Feb 1992*
	WE188	EE Canberra T4	Solway Aviation Society, Carlisle
	WE402	DH Venom FB50 (G-VENI/J-1523)	Vintage Aircraft Team, Cranfield
	WE569	Auster T7 (G-ASAJ)	Privately owned, Middle Wallop
	WE600	Auster T7 (mod) (7602M)	RAF Cosford Aerospace Museum
	WE925	Gloster Meteor F8	Classic Jet Aircraft Collection, Loughborough
	WE982	Slingsby Prefect TX1 (8781M)	RAF Cosford Aerospace Museum
	WE990	Slingsby Prefect TX1 (BGA 2583)	Privately owned, RAF Swanton Morley
	WF118	Percival Sea Prince T1 (G-DACA)	Privately owned, Charlwood, Surrey
	WF122	Percival Sea Prince T1 (A2673) [575/CU]	Flambards Village Theme Park, Helston
	WF125	Percival Sea Prince T1 (A2674) [576]	RN Predannack Fire School
	WF128	Percival Sea Prince T1 (8611M) [CU]	Norfolk & Suffolk Aviation Museum, Flixton
	WF137	Percival Sea Prince C1	Second World War Aircraft Preservation Society, Lasham
	WF225	Hawker Sea Hawk F1 (A2645) [CU]	RNAS Culdrose, at main gate
	WF259	Hawker Sea Hawk F2 (A2483) [171/A]	Royal Scottish Museum of Flight, East Fortune
	WF369	Vickers Varsity T1 [F]	Newark Air Museum, Winthorpe
	WF372	Vickers Varsity T1 [T]	Brooklands Aviation Museum
	WF376	Vickers Varsity T1	Bristol Airport Fire Section
	WF408	Vickers Varsity T1 (8395M)	RAF Northolt, for ground instruction
	WF410	Vickers Varsity T1 [F]	Brunel Technical College, Lulsgate
	WF425	Vickers Varsity T1	Imperial War Museum, Duxford
	WF643	Gloster Meteor F8 [X]	Norfolk & Suffolk Aviation Museum, Flixton
	WF714	Gloster Meteor F8 (Really WK914)	Privately owned, Duxford
	WF784	Gloster Meteor T7 (7895M)	RAF Quedgeley, at main gate
	WF825	Gloster Meteor T7 (8359M) [A]	Avon Aviation Museum, Yatesbury
	WF877	Gloster Meteor T7 (G-BPOA)	Aces High Ltd, North Weald
	WF890	EE Canberra T17	*Scrapped at Wyton 1992*
	WF911	EE Canberra B2 (nose only)	Pennine Aviation Museum, store
	WF916	EE Canberra T17 [EL]	RAF No 360 Sqn, Wyton
	WF922	EE Canberra PR3	Midland Air Museum, Coventry
	WG300	DH Chipmunk T10 PAX	RAFGSA, Bicester
	WG303	DH Chipmunk T10 PAX (8208M)	RAFGSA, Bicester
	WG307	DH Chipmunk T10 (G-BCYJ)	Privately owned, Shempston Farm, Lossiemouth, Scotland
	WG308	DH Chipmunk T10 [71]	RAF No 7 AEF, Newton
	WG316	DH Chipmunk T10 (G-BCAH)	Privately owned, Shoreham
	WG321	DH Chipmunk T10 [G]	AAC BFWF, Middle Wallop
	WG323	DH Chipmunk T10 [F]	AAC BFWF, Middle Wallop
	WG348	DH Chipmunk T10 (G-BBMV)	Privately owned, Moulton St Mary
	WG350	DH Chipmunk T10 (G-BPAL)	Privately owned, Old Sarum
	WG362	DH Chipmunk T10 PAX (8437M/8630M)	RAF Swinderby, instructional use
	WG403	DH Chipmunk T10 [O]	AAC BFWF, Middle Wallop
	WG407	DH Chipmunk T10 [67]	RAF No 9 AEF, Finningley
	WG418	DH Chipmunk T10 PAX (8209M/G-ATDY)	RAF No 10 AEF, Woodvale
	WG419	DH Chipmunk T10 PAX (8206M)	No 1053 Sqn ATC, Armthorpe
	WG422	DH Chipmunk T10 (G-BFAX/8394M) [16]	Privately owned, St Just
	WG430	DH Chipmunk T10 [3]	RAF No 1 AEF, Manston
	WG432	DH Chipmunk T10 [L]	AAC BFWF, Middle Wallop
	WG458	DH Chipmunk T10 [B]	RAF No 4 AEF, Exeter
	WG463	DH Chipmunk T10 PAX (8363M/G-ATDX)	No 188 Sqn ATC, Ipswich
	WG464	DH Chipmunk T10 PAX (8364M/G-ATEA)	
	WG465	DH Chipmunk T10 (G-BCEY)	Privately owned, White Waltham
	WG466	DH Chipmunk T10	RAF Gatow Station Flight, Berlin
	WG469	DH Chipmunk T10 [72]	RAF No 7 AEF, Newton
	WG471	DH Chipmunk T10 PAX (8210M)	No 301 Sqn ATC, Bury St Edmunds

Serial	Type	Owner or Operator	Notes
WG472	DH Chipmunk T10 (G-AOTY)	Privately owned, Netherthorpe	
WG477	DH Chipmunk T10 PAX (8362M/G-ATDI/G-ATDP)	No 281 Sqn ATC, Birkdale	
WG478	DH Chipmunk T10 [L]	RAF EFTS, Swinderby	
WG479	DH Chipmunk T10 [K]	RAF EFTS, Swinderby	
WG480	DH Chipmunk T10 [D]	RAF EFTS, Swinderby	
WG486	DH Chipmunk T10	RAF Gatow Station Flight, Berlin	
WG511	Avro Shackleton T4 (nose)	Flambards Village Theme Park, Helston	
WG718	WS51 Dragonfly HR3 (A2531) [934/-]	Wales Aircraft Museum, Cardiff	
WG719	WS51 Dragonfly HR5 (G-BRMA) [902]	International Helicopter Museum, Weston-super-Mare	
WG724	WS51 Dragonfly HR5 [932]	North-East Aircraft Museum, Usworth	
WG751	WS51 Dragonfly HR5	Privately owned, Ramsgreave, Lancs	
WG754	WS51 Dragonfly HR3 [912/CU] (really WG725) (7703M)	Flambards Village Theme Park, Helston	
WG760	EE P1A (7755M)	RAF Cosford Aerospace Museum	
WG763	EE P1A (7816M)	Greater Manchester Museum of Science and Industry	
WG768	Short SB5 (8005M)	RAF Cosford Aerospace Museum	
WG774	BAC 221	Science Museum, RNAS Yeovilton	
WG777	Fairey FD2 (7986M)	RAF Cosford Aerospace Museum	
WG789	EE Canberra B2/6 (cockpit only)	Booker Aircraft Museum	
WH132	Gloster Meteor T7 (7906M) [J]	No 276 Sqn ATC, Chelmsford	
WH166	Gloster Meteor T7 (8052M)	Privately owned, Birlingham, Worcs	
WH291	Gloster Meteor F8	Second World War Aircraft Preservation Society, Lasham	
WH301	Gloster Meteor F8 (7930M) [T]	RAF Museum, Hendon	
WH364	Gloster Meteor F8 (8169M)	Avon Aviation Museum, stored Wroughton	
WH453	Gloster Meteor D16 [L]	MoD(PE) T & EE Llanbedr	
WH589	Hawker Fury FB 10 [115/NW] (G-BTTA) (really 243 Iraq AF)	*Repainted as R Neth AF 115*	
WH646	EE Canberra T17A [EG]	RAF No 360 Sqn, Wyton	
WH657	EE Canberra B2	Brenzett Aeronautical Collection	
WH664	EE Canberra T17 [EH]	*Scrapped at Wyton 1992*	
WH665	EE Canberra T17 (8763M) [J]	BAe Filton, Fire Section	
WH670	EE Canberra B2 [CB]	*Scrapped at Wyton 1992*	
WH699	EE Canberra B2T (8755M) (really WJ637)	RAFC Cranwell, Trenchard Hall on display	
WH703	EE Canberra B2 (8490M) [S]	*To Pendine, January 1992*	
WH718	EE Canberra TT18 [CW]	*Scrapped at Wyton 1992*	
WH724	EE Canberra T19 (nose only)	RAF Shawbury Fire Section	
WH725	EE Canberra B2	Imperial War Museum, Duxford	
WH734	EE Canberra TT18	MoD(PE) DRA, Llanbedr	
WH740	EE Canberra T17 (8762M)	East Midlands Airport Aero Park	
WH773	EE Canberra PR7 (8696M)	Privately owned, Charlwood, Surrey	
WH775	EE Canberra PR7 [O] (8128M/8868M)	RAF No 2 SoTT, Cosford	
WH779	EE Canberra PR7 [BP]	RAF CTSF, Wyton	
WH791	EE Canberra PR7 (8165M/8176M/8187M)	RAF Cottesmore, at main gate	
WH840	EE Canberra T4 (8350M)(nose)	Staffordshire Aviation Museum, Seighford	
WH846	EE Canberra T4	Yorkshire Air Museum, Elvington	
WH849	EE Canberra T4 [BE]	RAF St Athan, stored	
WH850	EE Canberra T4	Macclesfield Historical Aviation Society, Chelford, Cheshire	
WH854	EE Canberra T4 (nose only)	Martin Baker Aircraft, Chalgrove	
WH863	EE Canberra T17 (8693M) (nose only)	Newark Air Museum, Winthorpe	
WH869	EE Canberra B2 (8515M)	RAF Abingdon, BDRF	
WH887	EE Canberra TT18 [847]	RN stored, St Athan	
WH902	EE Canberra T17 [EK]	RAF No 360 Sqn, Wyton	
WH903	EE Canberra B2 (nose only) (5854M)	Privately owned, Charlwood, Surrey	
WH904	EE Canberra T19 [04]	Newark Air Museum, Winthorpe	
WH914	EE Canberra B2 (G-27-373) [U]	*Scrapped at Samlesbury*	
WH946	EE Canberra B6 (Mod) (8185M) (nose only)	Privately owned, Tetney, Grimsby	
WH952	EE Canberra B6	Royal Arsenal, Woolwich	
WH953	EE Canberra B6 (Mod)	MoD(PE) DRA Bedford	
WH957	EE Canberra E15 (8869M) (nose)	Privately owned, Bruntingthorpe	

Notes	Serial	Type	Owner or Operator
	WH960	EE Canberra B15 (8344M) (nose)	Privately owned, Bruntingthorpe
	WH964	EE Canberra E15 (8870M) (nose)	Privately owned, Bruntingthorpe
	WH981	EE Canberra E15 [CN]	*Scrapped at Wyton 1992*
	WH983	EE Canberra E15 [CP]	*Scrapped at Wyton December 1991*
	WH984	EE Canberra B15 (8101M) (nose only)	Privately owned, Bruntingthorpe
	WH991	WS51 Dragonfly HR3	Privately owned, Storwood, East Yorks
	WJ231	Hawker Sea Fury FB11 [115/O] (really WE726)	FAA Museum, stored Lee-on-Solent
	WJ237	WAR Sea Fury Replica (G-BLTG) [113/O]	Privately owned, Little Gransden
	WJ350	Percival Sea Prince C2	Guernsey Airport Fire Section
	WJ358	Auster AOP6 (G-ARYD)	Museum of Army Flying, Middle Wallop
	WJ565	EE Canberra T17 (8871M) (nose)	Privately owned, Bruntingthorpe
	WJ567	EE Canberra B2 [CC]	*Scrapped at Wyton December 1991*
	WJ574	EE Canberra TT18 [844]	RN stored, St Athan
	WJ576	EE Canberra T17	Wales Aircraft Museum, Cardiff
	WJ607	EE Canberra T17A [EB]	RAF No 360 Sqn, Wyton
	WJ614	EE Canberra TT18 [846]	RN stored, St Athan
	WJ629	EE Canberra TT18 (8747M) [845]	RAF Chivenor, BDRT
	WJ630	EE Canberra T17 [ED]	RAF No 360 Sqn, Wyton
	WJ633	EE Canberra T17A [EF]	RAF No 360 Sqn, Wyton
	WJ636	EE Canberra TT18 [CX]	*Scrapped at Wyton 1992*
	WJ639	EE Canberra TT18 [39]	North-East Aircraft Museum, Usworth
	WJ640	EE Canberra B2 (8722M) [L]	*Scrapped at Cosford, August 1991*
	WJ676	EE Canberra B2 (7796M) (nose)	Privately owned, Bruntingthorpe
	WJ677	EE Canberra B2 (nose only)	RNAS Yeovilton Fire Section
	WJ678	EE Canberra B2 (8864M) [CF]	RAF Abingdon, BDRF
	WJ680	EE Canberra TT18 [CT] (G-BURM)	Privately owned, Wyton
	WJ682	EE Canberra TT18 [CU]	*Scrapped at Wyton 1992*
	WJ715	EE Canberra TT18 [CV]	*Scrapped at Wyton 1992*
	WJ717	EE Canberra TT18 [841] (9052M)	RAF CTTS, St Athan
	WJ721	EE Canberra TT18 [21]	Pennine Aviation Museum, Bacup
	WJ731	EE Canberra B2T [BK]	RAF CTSF, Wyton
	WJ756	EE Canberra E15 [BB]	*Scrapped at Wyton 1992*
	WJ775	EE Canberra B6 (8581M) [Z]	CSDE, RAF Swanton Morley
	WJ815	EE Canberra PR7 (8729M)	RAF Coningsby Fire Section
	WJ817	EE Canberra PR7 (8695M) [FU2]	RAF Wyton, Fire Section
	WJ821	EE Canberra PR7 (8668M)	Bassingbourn, on display
	WJ861	EE Canberra T4 [BF]	RAF St Athan, Fire Section
	WJ863	EE Canberra T4 (nose only)	Cambridge Airport Fire Section
	WJ865	EE Canberra T4	DRA Farnborough Apprentice School
	WJ866	EE Canberra T4 [BL]	RAF CTSF, Wyton
	WJ870	EE Canberra T4 (8683M)	*Scrapped at St Mawgan, March 1992*
	WJ872	EE Canberra T4 (8492M) (nose only)	No 327 Sqn ATC, Kilmarnock
	WJ874	EE Canberra T4 [BM]	RAF CTSF, Wyton
	WJ876	EE Canberra T4 (nose only)	RAF Exhibition Flight, Abingdon
	WJ879	EE Canberra T4 [BH]	*Scrapped at Wyton 1992*
	WJ880	EE Canberra T4 (8491M) [39] (nose only)	South Yorkshire Aircraft Preservation Society, Firbeck
	WJ893	Vickers Varsity T1	DRA Aberporth Fire Section
	WJ903	Vickers Varsity T1 [C] (nose)	Dumfries & Galloway Aviation Museum, Tinwald Downs
	WJ944	Vickers Varsity T1	Wales Aircraft Museum, Cardiff
	WJ945	Vickers Varsity T1 (G-BEDV) [21]	Duxford Aviation Society
	WJ975	EE Canberra T19 [S]	Bomber County Aviation Museum, Hemswell
	WJ981	EE Canberra T17A [EN]	RAF No 360 Sqn, Wyton
	WJ986	EE Canberra T17 [EP]	RAF No 360 Sqn, Wyton
	WJ992	EE Canberra T4	MoD(PE) DRA Bedford
	WK102	EE Canberra T17 [EQ] (8780M) (nose)	Privately owned, Bruntingthorpe
	WK111	EE Canberra T17 [EA]	RAF No 360 Sqn, Wyton
	WK118	EE Canberra TT18 [CQ]	*Scrapped at Wyton 1992*
	WK119	EE Canberra B2 (nose)	RAF Wyton, fire section
	WK122	EE Canberra TT18 [22]	Flambards Village Theme Park, Helston
	WK123	EE Canberra TT18 [CY]	*To Germany 8 Jan 1992*
	WK124	EE Canberra TT18 [CR] (9093M)	FSCTE, RAF Manston
	WK126	EE Canberra TT18 [843]	RN, stored St Athan
	WK127	EE Canberra TT18 [FO] (8985M)	RAF Wyton, BDRT

Serial	Type	Owner or Operator	Notes
WK128	EE Canberra B2	MoD(PE) DRA, Llanbedr	
WK142	EE Canberra TT18 [848]	RN, stored St Athan	
WK143	EE Canberra B2	DRA Llanbedr Fire Section	
WK162	EE Canberra B2 [CA] (8887M)	RAF Wyton Fire Section	
WK163	EE Canberra B6	MoD(PE) DRA Bedford	
WK198	VS Swift F4 (7428M)	North-East Aircraft Museum, Usworth	
WK275	VS Swift F4	Privately owned, Upper Hill, nr Leominster	
WK277	VS Swift FR5 (7719M) [N]	Newark Air Museum, Winthorpe	
WK281	VS Swift FR5 (7712M) [S]	RAF Museum, Hendon	
WK511	DH Chipmunk T10 [905]	RN, stored Shawbury	
WK512	DH Chipmunk T10 [A]	AAC BFWF, Middle Wallop	
WK517	DH Chipmunk T10 [84]	RAF No 11 AEF, Leeming	
WK518	DH Chipmunk T10	RAF BBMF, Coningsby	
WK522	DH Chipmunk T10 (G-BCOU)	Privately owned, High Easter	
WK549	DH Chipmunk T10 [Y] (G-BTWF)	Privately owned, Rufforth	
WK550	DH Chipmunk T10 [J]	RAF EFTS, Swinderby	
WK554	DH Chipmunk T10 [4]	RAF No 1 AEF, Manston	
WK558	DH Chipmunk T10 (G-ARMG)	Privately owned, Wellesbourne Mountford	
WK559	DH Chipmunk T10 [M]	AAC BFWF, Middle Wallop	
WK562	DH Chipmunk T10 [91]	RAF No 10 AEF, Woodvale	
WK570	DH Chipmunk T10 PAX (8211M)	No 424 Sqn ATC, Southampton	
WK572	DH Chipmunk T10 [92]	RAF No 10 AEF, Woodvale	
WK574	DH Chipmunk T10	RNAS Yeovilton Station Flight	
WK576	DH Chipmunk T10 PAX (8357M)	No 1206 Sqn ATC, Lichfield	
WK585	DH Chipmunk T10	RAF No 12 AEF, Turnhouse	
WK586	DH Chipmunk T10 [X]	RAF No 8 Sqn, Lossiemouth	
WK587	DH Chipmunk T10 PAX (8212M)		
WK589	DH Chipmunk T10 [C]	RAF No 6 AEF, Abingdon	
WK590	DH Chipmunk T10 [69]	RAF No 9 AEF, Finningley	
WK608	DH Chipmunk T10 [906]	RN Flying Grading Flt, Plymouth	
WK609	DH Chipmunk T10 [93]	RAF No 10 AEF, Woodvale	
WK611	DH Chipmunk T10 (G-ARWB)	Privately owned, Shoreham	
WK613	DH Chipmunk T10 [P]	Pennine Aviation Museum, Bacup	
WK620	DH Chipmunk T10 [T]	AAC BFWF, Middle Wallop	
WK622	DH Chipmunk T10 (G-BCZH)	Privately owned, Norwich	
WK624	DH Chipmunk T10 [N]	RAF No 3 AEF, Colerne	
WK626	DH Chipmunk T10 PAX (8213M)	No 358 Sqn ATC, Welling, Kent	
WK628	DH Chipmunk T10 (G-BBMW)	Privately owned, Shoreham	
WK630	DH Chipmunk T10 [11]	RAF No 2 AEF, Bournemouth	
WK633	DH Chipmunk T10 [B]	RAF EFTS, Swinderby	
WK634	DH Chipmunk T10 [902]	RN Flying Grading Flt, Plymouth	
WK635	DH Chipmunk T10	RNAS Yeovilton Station Flight	
WK638	DH Chipmunk T10 [83]	RAF No 11 AEF, Leeming	
WK639	DH Chipmunk T10 [L]	RAF No 3 AEF, Colerne	
WK640	DH Chipmunk T10 [C]	RAF EFTS, Swinderby	
WK642	DH Chipmunk T10 [94]	RAF No 10 AEF, Woodvale	
WK643	DH Chipmunk T10 [G]	RAF EFTS, Swinderby	
WK654	Gloster Meteor F8 (8092M) [X]	RAF Neatishead, at main gate	
WK714	Gloster Meteor F8 (really WK914)	Privately owned, Duxford	
WK800	Gloster Meteor D16 [Z]	MoD(PE) T&EE Llanbedr	
WK864	Gloster Meteor F8 (really WL168) [C] (7750M)	*Repainted as WL168*	
WK935	Gloster Meteor Prone Pilot (7869M)	RAF Cosford Aerospace Museum	
WK991	Gloster Meteor F8 (7825M)	Imperial War Museum, Duxford	
WL131	Gloster Meteor F8 (nose only) (7751M)	4th Guernsey (Forest) Air Scouts, Guernsey Airport	
WL181	Gloster Meteor F8 [X]	North-East Aircraft Museum, Usworth	
WL332	Gloster Meteor T7	Stratford Aircraft Coll'n, Long Marston	
WL345	Gloster Meteor T7	Privately owned, Hollington, E Sussex	
WL349	Gloster Meteor T7 [Z]	Gloucestershire Airport, on display	
WL360	Gloster Meteor T7 (7920M) [G]	Gloucestershire Aviation Collection, Hucclecote	
WL375	Gloster Meteor T7	Dumfries & Galloway Aviation Museum, Tinwald Downs	
WL405	Gloster Meteor T7	Martin Baker Aircraft, Chalgrove	
WL419	Gloster Meteor T7	Martin Baker Aircraft, Chalgrove	
WL505	DH Vampire FB9 (7705M) (G-FBIX)	Privately owned, Cranfield	
WL626	Vickers Varsity T1 (G-BHDD) [P]	East Midlands Airport Aeropark	
WL627	Vickers Varsity T1 (8488M) [D] (nose)	Privately owned, Preston, Humberside	

Notes	Serial	Type	Owner or Operator
	WL635	Vickers Varsity T1	RAF Machrihanish Police School
	WL679	Vickers Varsity T1	RAF Cosford Aerospace Museum
	WL732	BP Sea Balliol T21	RAF Cosford Aerospace Museum, stored
	WL756	Avro Shackleton AEW2(9101M)	RAF St Mawgan, Fire Section
	WL790	Avro Shackleton AEW2	Shackleton Preservation Trust, Coventry
	WL795	Avro Shackleton MR2C (8753M) [T]	RAF St Mawgan, on display
	WL798	Avro Shackleton MR2C (8114M) (nose only)	Privately owned, Elgin
	WL925	Slingsby Cadet TX3 (really WV925) (cockpit only)	RAF No 633 VGS, Cosford, instructional use
	WM145	AW Meteor NF11 (nose only)	N. Yorks Recovery Group, Chop Gate
	WM167	AW Meteor NF11 (G-LOSM)	Jet Heritage, Bournemouth
	WM223	AW Meteor TT20	Second World War Aircraft Preservation Society, Lasham
	WM267	Gloster Meteor NF11 (nose)	South Yorkshire Aviation Museum, Firbeck
	WM292	AW Meteor TT20 [841]	Wales Aircraft Museum, Cardiff
	WM311	AW Meteor TT20 (8177M) (really WM224)	Privately owned, North Weald
	WM366	AW Meteor NF13 (4X-FNA)	Second World War Aircraft Preservation Society, Lasham
	WM367	AW Meteor NF13	*Scrapped at Powick, March 1991*
	WM571	DH Sea Venom FAW 21 [VL]	Southampton Hall of Aviation
	WM729	DH Vampire NF10 (pod only) [A]	Privately owned, Ruislip
	WM913	Hawker Sea Hawk FB5 [456-J] (A2510/8162M)	Newark Air Museum, Winthorpe
	WM961	Hawker Sea Hawk FB5 [J] (A2517)	Air World, Caernarfon
	WM969	Hawker Sea Hawk FB5 (A2530)	Imperial War Museum, Duxford
	WM993	Hawker Sea Hawk FB5 (A2522) [034]	Privately owned, Peasedown St John, Avon
	WM994	Hawker Sea Hawk FB5 (A2503/G-SEAH)	Jet Heritage, Bournemouth
	WN105	Hawker Sea Hawk FB3 (A2662/ A2509/8164M) (really WF299)	Privately owned, Birlingham
	WN108	Hawker Sea Hawk FB5 [033]	Ulster Aviation Society, Langford Lodge
	WN149	BP Balliol T2 (nose only)	Pennine Aviation Museum, Bacup
	WN493	WS51 Dragonfly HR5	FAA Museum, RNAS Yeovilton
	WN499	WS51 Dragonfly HR5 [Y]	Caernarfon Air World
	WN516	BP Balliol T2 (nose only)	North-East Aircraft Museum, Usworth
	WN534	BP Balliol T2 (nose only)	Pennine Aviation Museum, Bacup
	WN890	Hawker Hunter F2 (nose only)	Privately owned, Chandlers Ford, Hants
	WN904	Hawker Hunter F2 (7544M) [3]	RE 39 Regt, Waterbeach, on display
	WN907	Hawker Hunter F2 (7416M) (cockpit only)	Blyth Valley Aviation Collect'n, Walpole
	WP180	Hawker Hunter F5 (7582M/ 8473M) [K] (really WP190)	RAF Stanbridge, at main gate
	WP185	Hawker Hunter F5 (7583M)	
	WP250	DH Vampire NF10 (nose only)	Privately owned, North Weald
	WP255	DH Vampire NF10 (pod)	South Yorkshire Air Museum, Firbeck
	WP270	EoN Eton TX1 (8598M)	Greater Manchester Museum of Science and Industry
	WP271	EoN Eton TX1	Privately owned, stored, Keevil
	WP309	Percival Sea Prince T1 [570/CU]	RNAS Yeovilton Fire Section
	WP313	Percival Sea Prince T1 [568/CU]	FAA Museum, stored Wroughton
	WP314	Percival Sea Prince T1 (8634M) [573/CU]	Privately owned, Preston, Humberside
	WP321	Percival Sea Prince T1 (G-BRFC) [750/CU]	Privately owned, Bourn
	WP503	WS51 Dragonfly HR3 [901]	Privately owned, Storwood, East Yorks
	WP515	EE Canberra B2 [CD]	*Scrapped at St Athan 1992*
	WP772	DH Chipmunk T10 [Q]	AAC, stored St Athan
	WP776	DH Chipmunk T10 [817/CU]	RN No 771 Sqn, Culdrose
	WP784	DH Chipmunk T10 PAX	Privately owned, Boston
	WP786	DH Chipmunk T10 [G]	RAF No 6 AEF, Abingdon
	WP788	DH Chipmunk T10 (G-BCHL)	Privately owned, Sleap
	WP790	DH Chipmunk T10 (G-BBNC) [T]	Mosquito Aircraft Museum, London Colney
	WP795	DH Chipmunk T10 [901]	RN Flying Grading Flt, Plymouth
	WP800	DH Chipmunk T10 [2] (G-BCXN)	Privately owned, Halton
	WP801	DH Chipmunk T10 [911]	RN Flying Grading Flt, Plymouth

Serial	Type	Owner or Operator	Notes
WP803	DH Chipmunk T10	RAF No 4 AEF, Exeter	
WP805	DH Chipmunk T10 [D]	RAF No 6 AEF, Abingdon	
WP808	DH Chipmunk T10 (G-BDEU)	Privately owned, Binham	
WP809	DH Chipmunk T10 [778]	RN No 771 Sqn, Culdrose	
WP833	DH Chipmunk T10 [A]	RAF No 4 AEF, Exeter	
WP835	DH Chipmunk T10 (G-BDCB)	Privately owned, Booker	
WP837	DH Chipmunk T10 [L]	RAF No 5 AEF, Cambridge	
WP839	DH Chipmunk T10 [A]	RAF No 8 AEF, Shawbury	
WP840	DH Chipmunk T10 [9]	RAF No 2 AEF, Bournemouth	
WP843	DH Chipmunk T10 (G-BDBP) [F]	Privately owned, Tollerton	
WP844	DH Chipmunk T10 [85]	RAF No 11 AEF, Leeming	
WP845	DH Chipmunk T10 PAX	No 1329 Sqn ATC, Stroud	
WP855	DH Chipmunk T10 [5]	RAF No 1 AEF, Manston	
WP856	DH Chipmunk T10 [904]	RN Flying Grading Flt, Plymouth	
WP857	DH Chipmunk T10 (G-BDRJ)[24]	Privately owned, Elstree	
WP859	DH Chipmunk T10 [E]	RAF No 8 AEF, Shawbury	
WP860	DH Chipmunk T10	*Written off*	
WP863	DH Chipmunk T10 PAX (8360M/G-ATJI)	No 2293 Sqn ATC, Marlborough	
WP864	DH Chipmunk T10 PAX (8214M)	RAF No 7AEF, Newton	
WP869	DH Chipmunk T10 PAX (8215M)	RAF	
WP871	DH Chipmunk T10 [W]	AAC BFWF, Middle Wallop	
WP872	DH Chipmunk T10	RAF No 12 AEF, Turnhouse	
WP896	DH Chipmunk T10 [M]	RAF No 3 AEF, Colerne	
WP900	DH Chipmunk T10 [V]	RAF No 3 AEF, Colerne	
WP901	DH Chipmunk T10 [B]	RAF No 6 AEF, Abingdon	
WP903	DH Chipmunk T10 (G-BCGC)	RN Gliding Club, Culdrose	
WP904	DH Chipmunk T10 [909]	RN Flying Grading Flt, Plymouth	
WP906	DH Chipmunk T10 [816/CU]	RN No 771 Sqn, Culdrose	
WP907	DH Chipmunk T10 PAX (7970M)	Privately owned, Reading	
WP912	DH Chipmunk T10 (8467M)	RAF Cosford Aerospace Museum	
WP914	DH Chipmunk T10 [E]	RAF No 6 AEF, Abingdon	
WP920	DH Chipmunk T10 [10]	RAF No 2 AEF, Bournemouth	
WP921	DH Chipmunk T10 PAX	No 1924 Sqn ATC, Croydon	
WP925	DH Chipmunk T10 [C]	AAC BFWF, Middle Wallop	
WP927	DH Chipmunk T10 PAX (8216M/G-ATJK)	No 247 Sqn ATC, Ashton-under-Lyne	
WP928	DH Chipmunk T10 [D]	AAC BFWF, Middle Wallop	
WP929	DH Chipmunk T10 [F]	RAF No 8 AEF, Shawbury	
WP930	DH Chipmunk T10 [J]	AAC BFWF, Middle Wallop	
WP962	DH Chipmunk T10 [95]	RAF No 10 AEF, Woodvale	
WP964	DH Chipmunk T10 [Y]	AAC BFWF, Middle Wallop	
WP967	DH Chipmunk T10	RAF No 12 AEF, Turnhouse	
WP970	DH Chipmunk T10 [T]	RAF No 5 AEF, Cambridge	
WP971	DH Chipmunk T10 (G-ATHD)	Privately owned, White Waltham	
WP972	DH Chipmunk T10 PAX(8667M)	CSDE, RAF Swanton Morley	
WP974	DH Chipmunk T10 [96]	RAF No 10 AEF, Woodvale	
WP977	DH Chipmunk T10 (G-BHRD) [N]	Privately owned, Kidlington	
WP978	DH Chipmunk T10 PAX (7467M)	RAF No 2 AEF, Lee-on-Solent	
WP979	DH Chipmunk T10 [J]	CSDE, RAF Swanton Morley	
WP980	DH Chipmunk T10 [E]	RAF EFTS, Swinderby	
WP981	DH Chipmunk T10 [D]	RAF No 5 AEF, Cambridge	
WP983	DH Chipmunk T10 [B]	AAC BFWF, Middle Wallop	
WP984	DH Chipmunk T10 [73]	RAF No 7 AEF, Newton	
WR410	DH Venom FB54 (really J1790/ G-BLKA) [N]	Vintage Aircraft Team, Cranfield	
WR539	DH Venom FB4 (8399M) [F]	Wales Aircraft Museum, Cardiff	
WR960	Avro Shackleton AEW2 (8772M)	Greater Manchester Museum of Science and Industry	
WR963	Avro Shackleton AEW2	Shackleton Preservation Trust, Coventry	
WR971	Avro Shackleton MR3 (8119M) [Q]	Privately owned, Narborough, Norfolk	
WR974	Avro Shackleton MR3 (8117M) [K]	Privately owned, Charlwood, Surrey	
WR977	Avro Shackleton MR3 (8186M) [B]	Newark Air Museum, Winthorpe	
WR982	Avro Shackleton MR3 (8106M) [J]	Privately owned, Charlwood, Surrey	
WR985	Avro Shackleton MR3 (8103M)	Stratford Aircraft Collect'n, Long Marston	
WS103	Gloster Meteor T7 [709/VL]	FAA Museum, stored Wroughton	
WS692	Gloster Meteor NF12 (7605M) [C]	Newark Air Museum, Winthorpe	

Notes	Serial	Type	Owner or Operator
	WS726	Gloster Meteor NF14 (7960M) [G]	No 1855 Sqn ATC, Royton
	WS739	Gloster Meteor NF14 (7961M)	Newark Air Museum, Winthorpe
	WS760	Gloster Meteor NF14 (7964M)	Classic Jet A/c Collect'n, Loughborough
	WS774	Gloster Meteor NF14 (7959M)	RAF Hospital, Ely, at main gate
	WS776	Gloster Meteor NF14 (7716M) [K]	RAF North Luffenham, at main gate
	WS788	Gloster Meteor NF14 (7967M) [Z]	Yorkshire Air Museum, Elvington
	WS792	Gloster Meteor NF14 (7965M) [K]	Privately owned, Brighouse Bay, D&G
	WS807	Gloster Meteor NF14 (7973M) [N]	Privately owned
	WS832	Gloster Meteor NF14 [W]	Solway Aviation Society, Carlisle Airport
	WS838	Gloster Meteor NF14	Midland Air Museum, Coventry
	WS840	Gloster Meteor NF14 (7969M)	*Scrapped at Aldegrove, November 1991*
	WS843	Gloster Meteor NF14 (7937M) [Y]	RAF Museum, Hendon
	WT121	Douglas Skyraider AEW1 [415/CU] (really WT983)	FAA Museum, RNAS Yeovilton
	WT212	EE Canberra B2	Lovaux Ltd, Macclesfield
	WT301	EE Canberra B6 (Mod)	Defence School, Chattenden
	WT308	EE Canberra B(I)6	RN Predannack Fire School
	WT309	EE Canberra B(I)6	MoD(PE) A&AEE, Boscombe Down Apprentice School
	WT327	EE Canberra B(I)8	MoD(PE) DRA Bedford
	WT333	EE Canberra B(I)8	MoD(PE) DRA Bedford
	WT339	EE Canberra B(I)8 (8198M)	RAF Barkston Heath Fire Section
	WT346	EE Canberra B(I)8 (8197M)	RAF Cosford Aerospace Museum, stored
	WT478	EE Canberra T4 [BA]	*Scrapped at Wyton July 1992*
	WT480	EE Canberra T4 [BC]	RAF CTSF, Wyton
	WT482	EE Canberra T4 (nose)	Stratford Aircraft Collect'n, Long Marston
	WT483	EE Canberra T4 [83]	Stratford Aircraft Collect'n, Long Marston
	WT486	EE Canberra T4 (8102M) [C]	Aldergrove Fire Section
	WT488	EE Canberra T4	BAe Dunsfold Fire Section
	WT507	EE Canberra PR7 (8131M/ 8548M) [44] (nose only)	No 384 Sqn ATC, Mansfield
	WT509	EE Canberra PR7 [CG]	RAF No 39(1 PRU) Sqn, Wyton
	WT518	EE Canberra PR7 (8133M/ 8691M) (rear fuselage only)	Cardiff Airport Fire Section
	WT519	EE Canberra PR7 [CH]	*Scrapped at Wyton 1992*
	WT520	EE Canberra PR7 (8094M/ 8184M) (nose only)	Privately owned, Burntwood, Staffs
	WT532	EE Canberra PR7 (8728M/ 8890M) (fuselage only)	Lovaux, Bournemouth, derelict
	WT534	EE Canberra PR7 (8549M) [43] (nose only)	No 492 Sqn ATC, Shirley, W. Mids
	WT536	EE Canberra PR7 (8063M) (nose)	Privately owned, Bruntingthorpe
	WT537	EE Canberra PR7	BAe Samlesbury, on display
	WT538	EE Canberra PR7 [CJ]	*Scrapped at Wyton 1992*
	WT555	Hawker Hunter F1 (7499M)	Privately owned, Greenford, London
	WT569	Hawker Hunter F1 (7491M)	No 2117 Sqn ATC, Kenfig Hill, Mid-Glamorgan
	WT589	Supermarine 544 (nose only)	Brooklands Aviation Museum
	WT612	Hawker Hunter F1 (7496M)	RAF Henlow on display
	WT619	Hawker Hunter F1 (7525M)	Greater Manchester Museum of Science and Industry
	WT648	Hawker Hunter F1 (7530M)(nose)	Wiltshire Historic Aviation Grp, Salisbury
	WT651	Hawker Hunter F1 [C]	Newark Air Museum
	WT660	Hawker Hunter F1 (7421M) [C]	RAF,stored Carlisle
	WT680	Hawker Hunter F1 (7533M) [Z]	No 1429 Sqn ATC at T & EE Aberporth
	WT684	Hawker Hunter F1 (7422M)	RAF Brize Norton Fire Section
	WT694	Hawker Hunter F1 (7510M)	RAF Newton, at main gate
	WT711	Hawker Hunter GA11 [833/DD]	RNAS Culdrose, SAH
	WT720	Hawker Hunter F51 [B] (8565M) (really E-408)	RAF Sealand, on display
	WT722	Hawker Hunter T8C [878/VL]	RN FRADU, Yeovilton
	WT723	Hawker Hunter PR11 [866/VL]	RN FRADU Yeovilton
	WT744	Hawker Hunter GA11 [868/VL]	RN FRADU, Yeovilton
	WT745	Hawker Hunter T8C (parts XL565) (fuselage only)	Jet Heritage, Bournemouth
	WT746	Hawker Hunter F4 (7770M) [A]	Army, Saighton, Chester
	WT799	Hawker Hunter T8C [879]	RN, stored Shawbury
	WT804	Hawker Hunter GA11 [831/DD]	RNAS Culdrose, SAH
	WT806	Hawker Hunter GA11	RAF, stored Shawbury
	WT859	Supermarine 544 (cockpit only)	Brooklands Museum
	WT867	Slingsby Cadet TX3	Privately owned, Eaglescott

Serial	Type	Owner or Operator	Notes
WT899	Slingsby Cadet TX3	Privately owned, Rush Green	
WT902	Slingsby Cadet TX3 (BGA 3147)	Privately owned, Lleweni Parc, Clwyd	
WT905	Slingsby Cadet TX3	Aberdeen Technical College	
WT910	Slingsby Cadet TX3	Privately owned, Clapham	
WT933	Bristol Sycamore 3 (G-ALSW/7709M)	Newark Air Museum, Winthorpe	
WV106	Douglas Skyraider AEW1 [427/C]	Flambards Village Theme Park, Helston	
WV198	S55 Whirlwind HAR21 (G-BJWY/A2576) [K]	South Yorkshire Air Museum, Firbeck	
WV256	Hawker Hunter GA11 [862/VL]	RN FRADU, Yeovilton	
WV267	Hawker Hunter GA11 [836/DD]	RNAS Culdrose, SAH	
WV276	Hawker Hunter F4 (7847M) [D]	MoD(PE) DRA, Farnborough	
WV309	Hawker Hunter F51 (really E-409)	Wales Air Museum, Cardiff	
WV318	Hawker Hunter T7B [A]	RAF No 12 Sqn, Lossiemouth	
WV322	Hawker Hunter T8C (9096M)	SIF, RAFC Cranwell	
WV332	Hawker Hunter F4 (7673M) (nose only)	No 1254 Sqn ATC, Godalming	
WV363	Hawker Hunter T8C [872/VL]	*Crashed off Stornaway 15 Feb 1992*	
WV372	Hawker Hunter T7 [877/VL]	RN FRADU, Yeovilton	
WV381	Hawker Hunter GA11 [732/VL]	UKAEA, Culham, Oxon	
WV383	Hawker Hunter T7	MoD(PE) DRA Farnborough	
WV396	Hawker Hunter T8C [879/VL]	RN FRADU, Yeovilton	
WV483	Percival Provost T1 (7693M) [N-E]	Privately owned	
WV486	Percival Provost T1 (7694M) [N-D]	Privately owned, Grazeley, Berks	
WV493	Percival Provost T1 (G-BDYG/7696M) [29]	Royal Scottish Museum of Flight, East Fortune	
WV495	Percival Provost T1 (7697M) [P-C]	Vintage Aircraft Team, Cranfield	
WV499	Percival Provost T1 (7698M) [P-G]	Privately owned, North Weald	
WV562	Percival Provost T1 (7606M) [P-C]	RAF Cosford Aerospace Museum, stored	
WV605	Percival Provost T1 [T-B]	Norfolk & Suffolk Aviation Museum, Flixton	
WV606	Percival Provost T1 (7622M) [P-B]	Newark Air Museum, Winthorpe	
WV666	Percival Provost T1 (7925M/G-BTDH) [O-D]	Privately owned, Shoreham	
WV679	Percival Provost T1 (7615M) [O-J]	Privately owned, Wellesbourne Mountford	
WV686	Percival Provost T1 (7621M) (G-BLFT) [O-P]	Privately owned, Cranfield	
WV703	Percival Pembroke C1 (8108M) (*G-IIIM*)	Privately owned, Tattershall Thorpe	
WV705	Percival Pembroke C1 (nose only)	Southampton Hall of Aviation, stored	
WV740	Percival Pembroke C1 (G-BNPH)	Privately owned, Shoreham	
WV746	Percival Pembroke C1 (8938M)	RAF Cosford Aerospace Museum	
WV753	Percival Pembroke C1 (8113M)	Wales Aircraft Museum, Cardiff	
WV781	Bristol Sycamore HR12 (G-ALTD/7839M)	Caernarfon Air World	
WV783	Bristol Sycamore HR12 (G-ALSP/7841M)	RNAY Fleetlands Museum	
WV787	EE Canberra B2/8 (8799M)	Newark Air Museum, Winthorpe	
WV795	Hawker Sea Hawk FGA6 (A2661/8151M)	*Sold to Cyprus, November 1992*	
WV797	Hawker Sea Hawk FGA6 [491/J] (A2637/8155M)	Midland Air Museum, Coventry	
WV798	Hawker Sea Hawk FGA6 (A2557) [028/CU]	Second World War Aircraft Preservation Society, Lasham	
WV826	Hawker Sea Hawk FGA6 (A2532) [147/Z]	Wales Aircraft Museum, Cardiff	
WV856	Hawker Sea Hawk FGA6 [163]	FAA Museum, RNAS Yeovilton	
WV903	Hawker Sea Hawk FGA4 (A2632/8153M) [128/C]	RN, stored Lee-on-Solent	
WV908	Hawker Sea Hawk FGA6 (A2660/8154M) [188/A]	RN Historic Flight, RNAS Yeovilton, stored	
WV911	Hawker Sea Hawk FGA4 (A2526) [115/C]	RN Lee-on-Solent stored	
WW138	DH Sea Venom FAW22 [227/O]	FAA Museum, RNAS Yeovilton	

Notes	Serial	Type	Owner or Operator
	WW145	DH Sea Venom FAW22 [680/LM]	Royal Scottish Museum of Flight, East Fortune
	WW217	DH Sea Venom FAW22 [736]	Newark Air Museum, Winthorpe
	WW388	Percival Provost T1 (7616M)[O-F]	Wales Aircraft Museum, Cardiff
	WW397	Percival Provost T1 (8060M/ G-BKHP) [N-E]	Privately owned, RAF Lyneham
	WW421	Percival Provost T1 (7688M) [O]	Lincolnshire Aviation Museum, East Kirkby
	WW442	Percival Provost T1 (7618M) [N]	Privately owned, Kings Langley
	WW444	Percival Provost T1 [D]	Privately owned
	WW447	Percival Provost T1	Privately owned, Grazeley, Berks
	WW453	Percival Provost T1 (G-TMKI) [W-S]	Privately owned, Newbury
	WW654	Hawker Hunter GA11 [834/DD]	RNAS Culdrose, SAH
	WX660	Hover-Air HA-5 Hoverhawk III (really XW660)	Privately owned, Cheltenham
	WX788	DH Venom NF3	Wales Aircraft Museum, Cardiff
	WX853	DH Venom NF3 (7443M)	Mosquito Aircraft Museum, London Colney
	WX905	DH Venom NF3 (7458M)	Newark Air Museum, Winthorpe
	WZ304	DH Vampire FB50 [A-T]	Vintage Aircraft Team, Cranfield
	WZ415	DH Vampire T11 [72]	No 2 Sqn ATC, Leavesden
	WZ425	DH Vampire T11	Wales Aircraft Museum, Cardiff
	WZ450	DH Vampire T11 (pod only)	Privately owned North Weald
	WZ458	DH Vampire T11 (7728M) [31] (cockpit only)	Blyth Valley Aviation Collect'n, Walpole
	WZ464	DH Vampire T11 (N62430) [40]	Vintage Aircraft Team, Cranfield
	WZ476	DH Vampire T11 (really XE985)	Mosquito Aircraft Museum, stored Hatfield
	WZ507	DH Vampire T11 (G-VTII)	Vintage Aircraft Team, Cranfield
	WZ514	DH Vampire T11	Privately owned, Meols, Merseyside
	WZ515	DH Vampire T11 [60]	Solway Aviation Society, Carlisle
	WZ518	DH Vampire T11	North East Aircraft Museum, Usworth
	WZ549	DH Vampire T11 [F] (8118M)	Ulster Heritage Centre, Langford Lodge
	WZ550	DH Vampire T11 (7902M) [R]	Booker Aircraft Museum
	WZ553	DH Vampire T11 [40]	Vintage Aircraft Team, Cranfield
	WZ557	DH Vampire T11	NYARC, Chop Gate
	WZ559	DH Vampire T11 (7736M) [45] (pod only)	RAF Halton Fire Section
	WZ581	DH Vampire T11 [77]	Privately owned, Ruislip
	WZ584	DH Vampire T11 [K]	St Albans College of FE
	WZ589	DH Vampire T11 [19]	Lashenden Air Warfare Museum, Headcorn
	WZ590	DH Vampire T11 [19]	Imperial War Museum, Duxford
	WZ608	DH Vampire T11 [56] (pod only)	Privately owned, Romford
	WZ616	DH Vampire T11 [60]	*Sold to USA*
	WZ620	DH Vampire T11 [68]	Privately owned, Keevil
	WZ662	Auster AOP9 (G-BKVK)	Privately owned, Cranfield
	WZ679	Auster AOP9 (7863M)	Privately owned, Gransden, Cambs
	WZ706	Auster AOP9 (7851M) (G-BURR)	Privately owned, Netheravon
	WZ711	Auster 9/Beagle E3 (G-AVHT)	Privately owned, Middle Wallop
	WZ721	Auster AOP9	Museum of Army Flying, Middle Wallop
	WZ724	Auster AOP9 (7432M)	AAC Middle Wallop, at main gate
	WZ736	Avro 707A (7868M)	Greater Manchester Museum of Science and Industry
	WZ744	Avro 707C (7932M)	RAF Cosford Aerospace Museum
	WZ753	Slingsby Grasshopper TX1	Southampton Hall of Aviation
	WZ767	Slingsby Grasshopper TX1	North-East Aircraft Museum, Usworth
	WZ779	Slingsby Grasshopper TX1	Stratford Aircraft Collect'n, Long Marston
	WZ791	Slingsby Grasshopper TX1 (8944M)	RAF Syerston, store
	WZ796	Slingsby Grasshopper TX1	Privately owned, Nympsfield, stored
	WZ819	Slingsby Grasshopper TX1 (BGA 3498)	Privately owned, West Malling
	WZ822	Slingsby Grasshopper TX1	Robertsbridge Aviation Society, Mayfield
	WZ826	Vickers Valiant B(K)1 (7872M) (nose only) (really XD826)	Wales Aircraft Museum, Cardiff
	WZ831	Slingsby Grasshopper TX1	Privately owned, Nympsfield, stored
	WZ845	DH Chipmunk T10 [6]	RAF No 1 AEF, Manston
	WZ846	DH Chipmunk T10 PAX (G-BCSC/8439M)	No 1404 Sqn ATC, Chatham
	WZ847	DH Chipmunk T10 [F]	RAF No 6 AEF, Abingdon
	WZ856	DH Chipmunk T10 [74]	RAF No 7 AEF, Newton

Serial	Type	Owner or Operator	Notes
WZ862	DH Chipmunk T10 [M]	RAF EFTS, Swinderby	
WZ866	DH Chipmunk T10 PAX (8217M) (G-ATEB)	No 2296 Sqn ATC, Dunoon, Strathclyde	
WZ868	DH Chipmunk T10 (G-BCIW) [H]	*Written off 26 November 1991*	
WZ869	DH Chipmunk T10 PAX (8019M) [R]	No 391 Sqn ATC, Handforth	
WZ872	DH Chipmunk T10 [E]	RAF No 5 AEF, Cambridge	
WZ876	DH Chipmunk T10 (G-BBWN)	Privately owned, Netherthorpe	
WZ877	DH Chipmunk T10 [75]	RAF No 7 AEF, Newton	
WZ878	DH Chipmunk T10 [86]	RAF No 11 AEF, Leeming	
WZ879	DH Chipmunk T10 [X]	RAF EFTS, Swinderby	
WZ882	DH Chipmunk T10 [K]	AAC BFWF, Middle Wallop	
WZ884	DH Chipmunk T10 [P]	AAC BFWF, Middle Wallop	
XA109	DH Sea Vampire T22	Royal Scottish Museum of Flt, East Fortune	
XA127	DH Sea Vampire T22 (nose)	FAA Museum, RNAS Yeovilton	
XA129	DH Sea Vampire T22	FAA Museum, stored	
XA231	Slingsby Grasshopper TX1 (8888M)	E. Cheshire & S. Manchester Wing ATC HQ, RAF Sealand	
XA243	Slingsby Grasshopper TX1 (8886M)	RAF, stored St Athan	
XA244	Slingsby Grasshopper TX1	RAF, stored Cosford	
XA282	Slingsby Cadet TX3	Caernarfon Air World	
XA289	Slingsby Cadet TX3	Privately owned, Eaglescott	
XA292	Slingsby Cadet TX3 (BGA3350)	Brooklands Aviation Museum*	
XA293	Slingsby Cadet TX3	Stratford Aircraft Collect'n, Long Marston	
XA302	Slingsby Cadet TX3 (BGA3786)	Privately owned	
XA454	Fairey Gannet COD4	RNAS Yeovilton Fire Section	
XA459	Fairey Gannet ECM6 (A2608) [E/-]	Privately owned, Cirencester	
XA460	Fairey Gannet ECM6 [768/BY]	North-East Wales Institute of HE, Connah's Quay	
XA466	Fairey Gannet COD4 [777/LM]	FAA Museum, stored	
XA508	Fairey Gannet T2 (A2472) [627/GN]	Midland Air Museum, Coventry	
XA553	Gloster Javelin FAW1 (7470M)	RAF Stanmore Park, on display	
XA564	Gloster Javelin FAW1 (7464M)	RAF Cosford Aerospace Museum	
XA571	Gloster Javelin FAW1 (nose) (7663M/7722M)	Booker Aircraft Museum	
XA634	Gloster Javelin FAW4 (7641M) [L]	RAF Leeming, on display	
XA699	Gloster Javelin FAW5 (7809M)	Midland Air Museum, Coventry	
XA801	Gloster Javelin FAW2 (7739M)	RAF Stafford, at main gate	
XA847	EE P1B (8371M)	Privately owned, Southampton Docks	
XA862	WS55 Whirlwind HAR1 (A2542/ G-AMJT) [9]	IHM, Weston-super-Mare	
XA864	WS55 Whirlwind HAR1	FAA Museum, stored Wroughton	
XA868	WS55 Whirlwind HAR1	IHM, Weston-super-Mare	
XA870	WS55 Whirlwind HAR1 (A2543)	Flambards Village Theme Park, Helston	
XA880	DH Devon C2	MoD(PE) T&EE, Llanbedr	
XA893	Avro Vulcan B1 (8591M) (nose)	RAF Cosford Aerospace Museum	
XA903	Avro Vulcan B1 (nose)	Wales Aircraft Museum, Cardiff	
XA909	Avro Vulcan B1 (nose only)	Lincolnshire Aviation Museum, East Kirkby	
XA917	HP Victor B1 (7827M) (nose)	RAF Marham, Fire Section	
XB259	Blackburn Beverley C1 (G-AOAI)	Museum of Army Transport, Beverley	
XB261	Blackburn Beverley C1 (cockpit)	Imperial War Museum, Duxford	
XB285	Blackburn Beverley C1	*Scrapped at Worminghall, Bucks*	
XB288	Blackburn Beverley C1	*Scrapped at Worminghall, Bucks*	
XB446	Grumman Avenger ECM6B [992/C]	FAA Museum, RNAS Yeovilton	
XB480	Hiller HT1 (A2577) [537]	FAA Museum, RNAS Yeovilton	
XB733	Canadair Sabre 4 (G-ATBF)	Privately owned	
XD145	Saro SR53	RAF Cosford Aerospace Museum	
XD163	WS55 Whirlwind HAR10 (8645M) [X]	International Helicopter Museum, Weston-super-Mare	
XD165	WS55 Whirlwind HAR10 (8673M) [B]	AAC Netheravon, instructional use	
XD186	WS55 Whirlwind HAR10 (8730M)	RAF Chivenor, on display	
XD215	VS Scimitar F1 (A2573) (cockpit only)	Privately owned, Ottershaw	
XD219	VS Scimitar F1 (fuselage)	RNAS Yeovilton, Fire Section	
XD234	VS Scimitar F1 [834]	DRA, derelict Farnborough	

Notes	Serial	Type	Owner or Operator
	XD235	VS Scimitar F1 (cockpit only)	Privately owned, Ottershaw
	XD244	VS Scimitar F1 (cockpit only)	Privately owned, Ottershaw
	XD317	VS Scimitar F1 [112/R]	FAA Museum, RNAS Yeovilton
	XD332	VS Scimitar F1 (A2574) [194/C]	Flambards Village Theme Park, Helston
	XD375	DH Vampire T11 (7887M) [72]	City of Norwich Aviation Museum
	XD377	DH Vampire T11 (pod)(8203M) [A]	Yorkshire Aircraft Museum, Elvington, stored
	XD382	DH Vampire T11 (8033M)	Privately owned, Ripley, Derbys
	XD425	DH Vampire T11 [16]	Dumfries & Galloway Aviation Museum, Tinwald Downs
	XD434	DH Vampire T11 [25]	Fenland Air Museum, Wisbech
	XD435	DH Vampire T11 [26] (pod)	Privately owned, Lapworth, Warwicks
	XD445	DH Vampire T11 [51]	Bomber County Aviation Museum, Hemswell
	XD447	DH Vampire T11 [50]	Vampire Preservation Group, Long Marston
	XD452	DH Vampire T11 (7990M) [66]	Privately owned
	XD453	DH Vampire T11 (7890M) [64]	No 58 Sqn ATC, Yorkshire Air Museum, Elvington
	XD459	DH Vampire T11 [63] (pod only)	Privately owned, Bruntingthorpe
	XD463	DH Vampire T11 (8023M)	No 1360 Sqn ATC, Stapleford, Notts
	XD506	DH Vampire T11 (7983M)	RAF Swinderby, on display
	XD515	DH Vampire T11 (7998M/*XM515*)	Newark Air Museum, Winthorpe
	XD525	DH Vampire T11 (7882M) (pod only)	Campbell College CCF, Belfast
	XD528	DH Vampire T11 (8159M)	Privately owned, Ottershaw
	XD534	DH Vampire T11 [41]	Military Aircraft Preservation Group, Hadfield, Derbys
	XD535	DH Vampire T11 (pod)	Aircraft Restoration Flt, North Weald
	XD536	DH Vampire T11 (7734M) [H]	Privately owned
	XD542	DH Vampire T11 (7604M/*XD429*) [28]	RAF Edzell, Scotland, on display
	XD547	DH Vampire T11 [Z] (pod only)	Scotland West Aircraft Investigation Group, stored Aberfoyle
	XD593	DH Vampire T11 [50]	Newark Air Museum, Winthorpe
	XD595	DH Vampire T11 (pod only)	Privately owned, Altrincham
	XD596	DH Vampire T11 (7939M)	Southampton Hall of Aviation
	XD599	DH Vampire T11 [A]	Caernarfon Air World
	XD602	DH Vampire T11 (7737M) (pod)	Privately owned, Brands Hatch
	XD616	DH Vampire T11 [56]	No 1239 Sqn ATC, Hoddesdon, Herts
	XD622	DH Vampire T11 (8160M)	No 2214 Sqn ATC, Usworth
	XD624	DH Vampire T11 [O]	Macclesfield Technical College
	XD626	DH Vampire T11 [Q]	Midland Air Museum, Coventry
	XD674	Hunting Jet Provost T1 (7570M)[T]	RAF Cosford Aerospace Museum
	XD816	Vickers Valiant B(K)1 (nose)	Brooklands Museum, Weybridge
	XD818	Vickers Valiant B(K)1 (7894M)	RAF Museum, Hendon
	XD857	Vickers Valiant B(K) 1 (nose)	Privately owned, Romford
	XD875	Vickers Valiant B(K)1 (nose)	South Yorks Aviation Museum, Firbeck
	XE317	Bristol Sycamore HR14 (G-AMWO) [S-N]	Newark Air Museum, Winthorpe
	XE327	Hawker Sea Hawk FGA6 (A2556) [644/LH]	Privately owned, Kings Langley, Herts
	XE339	Hawker Sea Hawk FGA6 (8156M/A2635) [149/E]	RNAS, stored Lee-on-Solent
	XE340	Hawker Sea Hawk FGA6 [131/Z]	Privately owned, Strathallan
	XE368	Hawker Sea Hawk FGA6 (A2534) [200/J]	Flambards Village Theme Park, Helston
	XE369	Hawker Sea Hawk FGA6 [5] (A2580/8158M/A2633)	RNAS Yeovilton Fire Section
	XE489	Hawker Sea Hawk FGA6 (G-JETH)	Privately owned, Charlwood, Surrey
	XE521	Fairey Rotodyne Y (parts)	International Helicopter Museum, Weston-super-Mare
	XE584	Hawker Hunter FGA9 (front fuselage)	Macclesfield Historical Av Soc, Chelford, Cheshire
	XE587	Hawker Hunter F6 [7]	*To USA as N587XE March 1992*
	XE597	Hawker Hunter FGA9 (8874M) (nose)	RAF Halton Fire Section
	XE601	Hawker Hunter FGA9	MoD(PE) A&AEE Boscombe Down
	XE624	Hawker Hunter FGA9 (8875M) [G]	RAF Brawdy, on display
	XE627	Hawker Hunter F6A [T]	Imperial War Museum, Duxford

Serial	Type	Owner or Operator	Notes
XE643	Hawker Hunter FGA9 (8586M) (nose)	RAF Exhibition Flight, St Athan	
XE650	Hawker Hunter FGA9 (G-9-449) (nose)	South Yorkshire Aviation Museum, Firbeck	
XE653	Hawker Hunter F6A (8829M) [S]	RAF TMTS, Scampton	
XE656	Hawker Hunter F6 (8678M)	RAF No 1 SoTT, Halton	
XE665	Hawker Hunter T8C [876/VL]	RN FRADU, Yeovilton	
XE668	Hawker Hunter GA11 [832/DD]	RNAS Culdrose, SAH	
XE670	Hawker Hunter F4 (7762M/ 8585M) (nose)	RAF Exhibition Flight, St Athan	
XE677	Hawker Hunter F4 (G-HHUN)	Jet Heritage, Bournemouth	
XE682	Hawker Hunter GA11	RNAS Culdrose Fire Section	
XE685	Hawker Hunter GA11 [861/VL]	RN FRADU, Yeovilton	
XE689	Hawker Hunter GA11 [864/VL]	RN FRADU, Yeovilton	
XE707	Hawker Hunter GA11 [865/VL]	RN FRADU, Yeovilton	
XE712	Hawker Hunter GA11 [708]	RN Predannack Fire School	
XE793	Slingsby Cadet TX3 (8666M)	RAF St Athan, instructional use	
XE799	Slingsby Cadet TX3 [R] (8943M)	RAF ACCGS Syerston, preserved	
XE802	Slingsby Cadet TX3	Privately owned, Cupar, Fife	
XE849	DH Vampire T11 (7928M) [V3]	Avon Aviation Museum, Calne, Wilts	
XE852	DH Vampire T11 [H]	No 2247 Sqn ATC, Hawarden	
XE855	DH Vampire T11 (pod)	Midland Air Museum, Coventry	
XE856	DH Vampire T11	Stratford Aircraft Collect'n, Long Marston	
XE864	DH Vampire T11 (composite with XD435)	Privately owned, Stretton, Cheshire	
XE872	DH Vampire T11 [62]	Midland Air Museum, Coventry	
XE874	DH Vampire T11 (8582M) [61]	Privately owned, Aston juxta Mondrum	
XE897	DH Vampire T11 (really XD403)	Privately owned, RAF Leuchars	
XE920	DH Vampire T11 (8196M) [D]	RAF, stored Scampton	
XE921	DH Vampire T11 [64]	Privately owned, Keevil	
XE935	DH Vampire T11 [30]	South Yorks Air Museum, Firbeck	
XE946	DH Vampire T11 (7473M) (pod only)	RAF Museum Restoration Centre, Cardington	
XE956	DH Vampire T11	St Albans College of FE	
XE979	DH Vampire T11 [54]	Privately owned, Birlingham, Worcs	
XE982	DH Vampire T11 (7564M)	Privately owned, Dunkeswell	
XE993	DH Vampire T11 (8161M)	Privately owned, Cosford	
XE995	DH Vampire T11 [53]	Privately owned, Marden, Kent	
XE998	DH Vampire T11 [36]	Privately owned, Wisbech	
XF113	VS Swift F7 (nose) [19]	Wiltshire Historic Aircraft Group, Salisbury	
XF114	VS Swift F7 (G-SWIF)	Jet Heritage, Bournemouth	
XF289	Hawker Hunter T8C [875/VL]	RN FRADU, Yeovilton	
XF300	Hawker Hunter GA11 [860/VL]	RN FRADU, Yeovilton	
XF301	Hawker Hunter GA11 [834/VL]	RN FRADU, Yeovilton	
XF310	Hawker Hunter T8C [869/VL]	RN FRADU, Yeovilton	
XF314	Hawker Hunter F51 [N] (really E-412)	Tangmere Military Aviation Museum	
XF319	Hawker Hunter F4 (7849M) [B] (G-BTCY)	*Sold to USA*	
XF321	Hawker Hunter T7	RNEC Manadon	
XF357	Hawker Hunter T8C [871/VL]	RN FRADU, Yeovilton	
XF358	Hawker Hunter T8C [870/VL]	RN FRADU, Yeovilton	
XF368	Hawker Hunter GA11 [863/VL]	RN FRADU, Yeovilton	
XF375	Hawker Hunter F6) (8736M/G-BUEZ [05]	Privately owned, Cranwell	
XF382	Hawker Hunter F6A [15]	Midland Air Museum, Coventry	
XF383	Hawker Hunter F6 (8706M) (wears *8506M*) [V]	RAF Wittering, BDRT	
XF383	Hawker Hunter F51 [71] (really E-409)	*Repainted as WV369*	
XF445	Hawker Hunter FGA9 (8715M) [T] (really XG264)	RAF Brawdy Fire Section	
XF509	Hawker Hunter F6 (8708M)	RAF Chivenor, at main gate	
XF515	Hawker Hunter F6A (8830M) [C]	RAF TMTS, Scampton	
XF516	Hawker Hunter F6A (8685M) [66]	SIF, RAFC Cranwell	
XF519	Hawker Hunter FGA9 [J]	*Burnt by June 1992*	
XF522	Hawker Hunter F6 (nose)	No 1365 Sqn ATC, Aylesbury	
XF526	Hawker Hunter F6 (8679M) [78/E]	RAF St Athan Fire Section	
XF527	Hawker Hunter F6 (8680M)	RAF Halton, on display	
XF545	Percival Provost T1 (7957M) [O-K]	RAF Linton-on-Ouse, on display	

Notes	Serial	Type	Owner or Operator
	XF597	Percival Provost T1 (G-BKFW) [A-H]	Privately owned, Aldermaston
	XF603	Percival Provost T1 [H]	Rolls-Royce Tech Coll, Filton
	XF690	Percival Provost T1 (G-BGKA/G-MOOS) (8041M)	Privately owned, Newton
	XF708	Avro Shackleton MR3 [203/C]	Imperial War Museum, Duxford
	XF785	Bristol 173 (G-ALBN/7648M	RAF Cosford Aerospace Museum
	XF836	Percival Provost T1 (8043M/ G-AWRY) [JG]	Privately owned, Thatcham
	XF844	Percival Provost T1 [70]	DRA Farnborough Apprentice School
	XF877	Percival Provost T1 (G-AWVF) [JX]	Privately owned, Norwich
	XF926	Bristol 188 (8368M)	RAF Cosford Aerospace Museum
	XF967	Hawker Hunter T8C [B]	RAF Lossiemouth
	XF974	Hawker Hunter F4 (7949M/ G-BTCX) [C]	*Sold to USA*
	XF994	Hawker Hunter T8C [873/VL]	RN FRADU, Yeovilton
	XF995	Hawker Hunter T8B [W]	RAF No 208 Sqn, Lossiemouth
	XG154	Hawker Hunter FGA9 (8863M) [54]	RAF Museum, Hendon
	XG160	Hawker Hunter F6A [U] (8831M)	RAF TMTS, Scampton
	XG164	Hawker Hunter F6 (8681M)	RAF No 1 SoTT, Halton
	XG172	Hawker Hunter F6A [A] (8832M)	RAF TMTS, Scampton
	XG194	Hawker Hunter FGA9 [55] (8839M)	RAF North Luffenham Training Area
	XG195	Hawker Hunter FGA9	Bomber County Air Museum, Hemswell (composite with XG297)
	XG196	Hawker Hunter F6A [31] (8702M)	RAF Bracknell, on display
	XG209	Hawker Hunter F6 [69] (8709M)	SIF RAFC, Cranwell
	XG210	Hawker Hunter F6	DRA Apprentice School, Bedford
	XG225	Hawker Hunter F6A (8713M) [S]	RAF Cosford on display
	XG226	Hawker Hunter F6A (8800M)	Stratford Aircraft Collect'n, Long Marston
	XG226	Hawker Hunter F6A (8800M) [28] (nose)	Privately owned, Faygate
	XG252	Hawker Hunter FGA9 (8840M) [U]	RAF Credenhill, on display
	XG254	Hawker Hunter FGA9 (8881M)	RAF Coltishall
	XG274	Hawker Hunter F6 [71] (8710M)	RAF No 1 SoTT, Halton
	XG290	Hawker Hunter F6 [74] (8711M) (fuselage)	Jet Heritage, Bournemouth
	XG297	Hawker Hunter FGA9 (nose)	Pennine Aviation Museum, Bacup
	XG325	EE Lightning F1 (nose)	No 1312 Sqn ATC, Southend, Essex
	XG327	EE Lightning F1 (8188M)	*Burnt by June 1992*
	XG329	EE Lightning F1 (8050M)	RAF Swinderby
	XG331	EE Lightning F1 (nose)	Stratford Aircraft Collect'n, Long Marston
	XG337	EE Lightning F1 (8056M) [M]	RAF Cosford Aerospace Museum
	XG452	Bristol Belvedere HC1 (G-BRMB/7997M)	IHM, Weston-super-Mare
	XG454	Bristol Belvedere HC1 (8366M)	Greater Manchester Museum of Science and Industry
	XG474	Bristol Belvedere HC1 (8367M)	RAF Museum, Hendon
	XG502	Bristol Sycamore HR14	Museum of Army Flying, Middle Wallop
	XG506	Bristol Sycamore HR14 (7852M)	Bomber County Aviation Museum, Hemswell
	XG518	Bristol Sycamore HR14 (8009M) [S-E]	North-East Aircraft Museum, Usworth
	XG540	Bristol Sycamore HR14 (7899M/ 8345M) [Y-S]	Privately owned, Preston, Lancs
	XG544	Bristol Sycamore HR14	Privately owned, Lower Tremar
	XG547	Bristol Sycamore HR14 (G-HAPR/8010M) [S-T]	IHM, Weston-super-Mare
	XG573	WS55 Whirlwind HAR3	CDE, Porton Down, Wilts
	XG574	WS55 Whirlwind HAR3 (A2575) [752/PO]	FAA Museum, RNAS Yeovilton
	XG577	WS55 Whirlwind HAR3	RAF Leconfield Crash Rescue Training (A2571/9050M)
	XG592	WS55 Whirlwind HAS7 [54]	
	XG594	WS55 Whirlwind HAS7 [517/PO]	Royal Scottish Museum of Flight, East Fortune
	XG596	WS55 Whirlwind HAS7 (A2651) [66]	IHM, Weston-super-Mare
	XG613	DH Sea Venom FAW21	Imperial War Museum, Duxford
	XG629	DH Sea Venom FAW22 (pod)	Stratford Aircraft Collect'n, Long Marston

Serial	Type	Owner or Operator	Notes
XG680	DH Sea Venom FAW22 [735/VL]	North-East Aircraft Museum, Usworth	
XG691	DH Sea Venom FAW22 [93/J]	Flambards Village Theme Park, Helston	
XG692	DH Sea Venom FAW22 [668/LM]	Midland Warplane Museum, Hatton	
XG730	DH Sea Venom FAW22 [499/A]	Mosquito Aircraft Museum, London Colney	
XG736	DH Sea Venom FAW22	Ulster Aviation Society, Newtownards	
XG737	DH Sea Venom FAW22 [220/Z]	Vampire Preservation Group, Long Marston	
XG743	DH Sea Vampire T22 [597/LM]	Imperial War Museum, Duxford	
XG797	Fairey Gannet ECM6 [766/BY]	Imperial War Museum, Duxford	
XG831	Fairey Gannet ECM6 (A2539) [396]	Flambards Village Theme Park, Helston	
XG882	Fairey Gannet T5 (8754M) [771/LM]	Privately owned, Errol	
XG883	Fairey Gannet T5 [773/BY]	Wales Aircraft Museum, Cardiff	
XG888	Fairey Gannet T5 [-/LM]	RNAS Lee-on-Solent, stored	
XG900	Short SC1	FAA Museum, RNAS Yeovilton	
XG905	Short SC1	Ulster Folk & Transport Museum, Holywood, County Down	
XH131	EE Canberra PR9 [AF]	RAF No 39(1 PRU) Sqn, Wyton	
XH132	Short SC9 Canberra (8915M)	*Scrapped at St Mawgan 1992*	
XH133	EE Canberra PR9	RAF St Athan, Fire Section	
XH134	EE Canberra PR9 [AA]	RAF No 39(1 PRU) Sqn, Wyton	
XH135	EE Canberra PR9 [AG]	RAF No 39(1 PRU) Sqn, Wyton	
XH136	EE Canberra PR9 (8782M) [W]	RAF No 2 SoTT, Cosford	
XH165	EE Canberra PR9	RAF St Athan, Fire Section	
XH168	EE Canberra PR9 [AB]	RAF No 39(1 PRU) Sqn, Wyton	
XH169	EE Canberra PR9 [AC]	RAF No 39(1 PRU) Sqn, Wyton	
XH170	EE Canberra PR9 (8739M)	RAF Wyton, on display	
XH171	EE Canberra PR9 (8746M) [U]	RAF Cosford Aerospace Museum	
XH174	EE Canberra PR9 (cockpit only)	RAF, stored St Athan	
XH175	EE Canberra PR9	*Scrapped at St Athan 1992*	
XH177	EE Canberra PR9 (nose)	Wales Aircraft Museum, Cardiff	
XH278	DH Vampire T11 (8595M/7866M)	No 2482 Sqn ATC, RAF Henlow	
XH312	DH Vampire T11 [18]	Privately owned, Chester	
XH313	DH Vampire T11 [E]	St Albans College of FE	
XH328	DH Vampire T11 [66] (dismantled)	Jet Heritage, Bournemouth	
XH330	DH Vampire T11 [73]	Privately owned, Bridgnorth	
XH537	Avro Vulcan B2MRR (8749M) (cockpit only)	Privately owned, Camberley	
XH558	Avro Vulcan B2	RAF Vulcan Display Flight, Waddington	
XH560	Avro Vulcan K2 (nose)	Privately owned, Rayleigh, Essex	
XH567	EE Canberra B6(mod)	MoD(PE) DRA Bedford	
XH568	EE Canberra B6(mod)	MoD(PE) DRA Bedford	
XH583	EE Canberra T4 (G-27-374) (nose)	North-East Aircraft Museum, Usworth	
XH590	HP Victor K1A	*Burnt by June 1992*	
XH592	HP Victor K1A (8429M) [L]	RAF Cosford Aerospace Museum	
XH593	HP Victor K1A (8428M) [T]	*Scrapped at Cosford*	
XH616	HP Victor K1A	*Burnt by June 1992*	
XH648	HP Victor K1A	Imperial War Museum, Duxford	
XH669	HP Victor K2 (9092M)	RAF Waddington, wfu	
XH670	HP Victor SR2 (nose)	Privately owned, Rayleigh, Essex	
XH671	HP Victor K2	RAF No 55 Sqn, Marham	
XH672	HP Victor K2	RAF No 55 Sqn, Marham	
XH673	HP Victor K2 (8911M)	RAF Marham, on display	
XH675	HP Victor K2	RAF Marham Fire Section	
XH767	Gloster Javelin FAW9 (7955M) [A]	City of Norwich Aviation Museum	
XH837	Gloster Javelin FAW7 (8032M) (nose)	Caernarfon Air World	
XH892	Gloster Javelin FAW9 (7982M) [J]	Norfolk & Suffolk Aviation Museum, Flixton	
XH897	Gloster Javelin FAW9	Imperial War Museum, Duxford	
XH903	Gloster Javelin FAW9 (7938M)	Gloucestershire Aviation Collection, Hucclecote	
XH980	Gloster Javelin FAW8 (7867M) [A]	RAF West Raynham, at main gate	
XH992	Gloster Javelin FAW8 (7829M) [P]	Newark Air Museum, Winthorpe	
XJ314	RR Thrust Measuring Rig	FAA Museum, RNAS Yeovilton	
XJ380	Bristol Sycamore HR14 (8628M)	Privately owned, North Byth, Grampian	
XJ389	Fairey Jet Gyrodyne (XD759/ G-AJJP)	RAF Cosford Aerospace Museum, stored	
XJ393	WS55 Whirlwind HAR3 (A2538)	Royal Marines Museum, Eastney	
XJ396	WS55 Whirlwind HAR10	DRA Farnborough Fire Section	

Notes	Serial	Type	Owner or Operator
	XJ409	WS55 Whirlwind HAR10	
	XJ430	WS55 Whirlwind HAR10	*Burnt by June 1992*
	XJ435	WS55 Whirlwind HAR10 (8671M) [V]	AAC Netheravon, instructional use
	XJ445	WS55 Whirlwind HAR5	CDE, Porton Down, Wilts
	XJ476	DH Sea Vixen FAW1 (nose section)	No 424 Sqn ATC, Southampton Hall of Aviation
	XJ481	DH Sea Vixen FAW1 [VL]	RNAY Fleetlands Museum
	XJ482	DH Sea Vixen FAW1 (A2598) [713/VL]	Norfolk & Suffolk Aviation Museum, Flixton
	XJ494	DH Sea Vixen FAW2	Privately owned, Kings Langley, Herts
	XJ560	DH Sea Vixen FAW2 (8142M) [242]	Newark Air Museum, Winthorpe
	XJ565	DH Sea Vixen FAW2 [127/E]	Mosquito Aircraft Museum, London Colney
	XJ571	DH Sea Vixen FAW2 (8140M) [242/R]	Privately owned, Dunsfold
	XJ575	DH Sea Vixen FAW2 (A2611) (nose)	Wellesbourne Wartime Museum
	XJ580	DH Sea Vixen FAW2 [131/E]	Christchurch Memorial Group
	XJ582	DH Sea Vixen FAW2 (8139M) [702]	Privately owned, Chelmsford
	XJ607	DH Sea Vixen FAW2 (8171M) [701/VL]	Privately owned, Dunsfold
	XJ608	DH Sea Vixen FAW2 (8802M)	*Scrapped 1992*
	XJ634	Hawker Hunter F6A (8684M) [29]	RAF SIF, Cranwell
	XJ639	Hawker Hunter F6A (8687M) [31]	RAF SIF, Cranwell
	XJ676	Hawker Hunter F6A (8844M)	Privately owned, Duxford
	XJ690	Hawker Hunter FGA9 (composite with XG195)	Staffordshire Aviation Museum, Seighford
	XJ723	WS55 Whirlwind HAR10	No 2288 Sqn ATC, Montrose
	XJ726	WS55 Whirlwind HAR10 [F]	Caernarfon Air World
	XJ727	WS55 Whirlwind HAR10 (8661M) [L]	RAF No 1 SoTT, Halton
	XJ729	WS55 Whirlwind HAR10 (8732M)	RAF Finningley, BDRT
	XJ758	WS55 Whirlwind HAR10 (8464M)	Privately owned, Oswestry
	XJ763	WS55 Whirlwind HAR10 (G-BKHA) [P]	Privately owned, Tarrant Keynston, Dorset
	XJ772	DH Vampire T11 [H]	Brooklands Technical College
	XJ823	Avro Vulcan B2A	Solway Aviation Society, Carlisle Airport
	XJ824	Avro Vulcan B2A	Imperial War Museum, Duxford
	XJ825	Avro Vulcan K2	*Scrapped at Waddington, Jan 1992*
	XJ917	Bristol Sycamore HR14 [H-S]	Flambards Village Theme Park, Helston
	XJ918	Bristol Sycamore HR14 (8190M)	RAF Cosford Aerospace Museum
	XK149	Hawker Hunter F6A [34] (8714M)	RAF SIF, Cranwell
	XK378	Auster AOP9 (TAD200)	Privately owned
	XK416	Auster AOP9 (G-AYUA/7855M)	Vintage Aircraft Team, Cranfield
	XK417	Auster AOP9 (G-AVXY)	Privately owned, Leicester East
	XK418	Auster AOP9 (7976M)	Second World War Aircraft Preservation Society, Lasham
	XK421	Auster AOP9 (8365M)	Stratford Aircraft Collect'n, Long Marston
	XK482	Saro Skeeter AOP12 (7840M/ G-BJWC) [C]	Helicopter Museum of GB, Squires Gate
	XK488	Blackburn Buccaneer S1	FAA Museum, RNAS Yeovilton
	XK526	Blackburn Buccaneer S2 (8648M)	RAF Honington, at main gate
	XK527	Blackburn Buccaneer S2D (nose) (8818M)	Privately owned, New Milton, Hants
	XK530	Blackburn Buccaneer S1	DRA Bedford Fire Section
	XK531	Blackburn Buccaneer S1 (8403M) [LM]	Defence School, Winterbourne Gunner
	XK532	Blackburn Buccaneer S1 (8867M/A2581) [632/LM]	The Fresson Trust, Inverness Airport
	XK533	Blackburn Buccaneer S1 (nose)	Royal Scottish Museum of Flight, East Fortune
	XK590	DH Vampire T11 [V]	Wellesbourne Wartime Museum
	XK623	DH Vampire T11 [56] (*G-VAMP*)	Caernarfon Air World
	XK624	DH Vampire T11 [32]	Norfolk & Suffolk Aviation Museum, Flixton
	XK625	DH Vampire T11 [12]	Brenzett Aviation Museum
	XK627	DH Vampire T11	Pennine Aviation Museum, Bacup
	XK632	DH Vampire T11 [67]	No 2 Sqn ATC, Leavesden
	XK637	DH Vampire T11 [56]	No 1855 Sqn ATC, Royton, Greater Manchester

Serial	Type	Owner or Operator	Notes
XK655	DH Comet C2R (G-AMXA) (nose)	Privately owned, Carlisle	
XK659	DH Comet C2R (G-AMXC) (nose)	Privately owned, Elland, W. Yorks	
XK695	DH Comet 2R (G-AMXH) (fuselage only)	RAF Newton, instructional use	
XK699	DH Comet C2 (7971M)	RAF Lyneham on display	
XK724	Folland Gnat F1 (7715M)	RAF Cosford Aerospace Museum	
XK740	Folland Gnat F1 (8396M)	Southampton Hall of Aviation	
XK741	Folland Gnat F1	Midland Air Museum, Coventry	
XK776	ML Utility 1	Museum of Army Flying, Middle Wallop	
XK819	Slingsby Grasshopper TX1	The Aviation Collection, Warmingham, Cheshire	
XK822	Slingsby Grasshopper TX1	Privately owned, West Malling	
XK824	Slingsby Grasshopper TX1	Privately owned, Narborough, Norfolk	
XK895	DH Sea Devon C20 [19/CU] (G-SDEV)	Privately owned, North Weald	
XK896	DH Sea Devon C20 (G-RNAS)	Privately owned, Staverton	
XK906	WS55 Whirlwind HAS7	*Scrapped at Netheravon*	
XK907	WS55 Whirlwind HAS7 [U]	Midland Air Museum, Coventry	
XK911	WS55 Whirlwind HAS7 (A2603) [519/PO]	Privately owned, Ipswich	
XK912	WS55 Whirlwind HAS7 [60/CU]	*Scrapped by April 1991*	
XK936	WS55 Whirlwind HAS7 [62]	Imperial War Museum, Duxford	
XK943	WS55 Whirlwind HAS7	*Scrapped at Abingdon, Jan 1992*	
XK944	WS55 Whirlwind HAS7 (A2607)	No 617 Sqn ATC, Malpas School, Cheshire	
XK968	WS55 Whirlwind HAR10 (8445M) [E]	FSCTE, RAF Manston	
XK969	WS55 Whirlwind HAR10 (8646M)	*Burnt by June 1992*	
XK970	WS55 Whirlwind HAR10 (8789M)	*Scrapped at Odiham*	
XK986	WS55 Whirlwind HAR10 (8790M)	*Scrapped at Odiham*	
XK987	WS55 Whirlwind HAR10 (8393M)	MoD Swynnerton, Staffs	
XK988	WS55 Whirlwind HAR10 (A2646) [D]	Museum of Army Flying, Middle Wallop	
XL149	Blackburn Beverley C1 (nose) (7988M)	Newark Air Museum, Winthorpe	
XL158	HP Victor K2	RAF, stored Marham	
XL160	HP Victor K2 (8910M)	RAF Marham, BDRT	
XL161	HP Victor K2	RAF No 55 Sqn, Marham	
XL162	HP Victor K2 (9114M)	FSCTE RAF Manston	
XL164	HP Victor K2	RAF No 55 Sqn, Marham	
XL188	HP Victor K2 (9100M)	RAF Kinloss Fire Section	
XL190	HP Victor K2	RAF, stored Marham	
XL192	HP Victor K2 (9024M)	RAF Marham Fire Section	
XL231	HP Victor K2	RAF No 55 Sqn, Marham	
XL318	Avro Vulcan B2 (8733M)	RAF Museum, Hendon	
XL319	Avro Vulcan B2	North-East Aircraft Museum, Usworth	
XL360	Avro Vulcan B2A	Midland Air Museum, Coventry	
XL384	Avro Vulcan B2 (8505M/8670M)	*Burnt at RAF Scampton*	
XL386	Avro Vulcan B2A (8760M)	FSCTE, RAF Manston	
XL388	Avro Vulcan B2 (nose)	Blyth Valley Aviation Collection, Walpole	
XL391	Avro Vulcan B2	Privately owned, Blackpool	
XL392	Avro Vulcan B2 (8745M)	RAF Valley Fire Section	
XL426	Avro Vulcan B2 (G-VJET)	Privately owned, Southend	
XL427	Avro Vulcan B2 (8756M)	RAF Machrihanish Fire Section	
XL445	Avro Vulcan K2 (8811M) (cockpit only)	Blyth Valley Aviation Collection, Walpole	
XL449	Fairey Gannet AEW3	Wales Aircraft Museum, Cardiff	
XL472	Fairey Gannet AEW3 [044/R]	Privately owned, Charlwood, Surrey	
XL497	Fairey Gannet AEW3 [041/R]	RN, Prestwick, on display	
XL500	Fairey Gannet AEW3 (A2701) [LM]	RN, stored Lee-on-Solent	
XL502	Fairey Gannet AEW3 (8610M) (G-BMYP)	Privately owned, Carlisle	
XL503	Fairey Gannet AEW3 [070/E]	FAA Museum, RNAS Yeovilton	
XL511	HP Victor K2	*Burnt by June 1992*	
XL512	HP Victor K2	RAF, stored Marham	
XL563	Hawker Hunter T7	MoD(PE) IAM Farnborough	
XL564	Hawker Hunter T7 [4]	MoD(PE) ETPS Boscombe Down	
XL565	Hawker Hunter T7 (composite with WT745)	RAF, stored Shawbury	

Notes	Serial	Type	Owner or Operator
	XL567	Hawker Hunter T7 (8723M) [84] (fuselage only)	Privately owned, Exeter
	XL568	Hawker Hunter T7A [C]	RAF No 208 Sqn, Lossiemouth
	XL569	Hawker Hunter T7 (8833M) [80]	RAF (for disposal)
	XL573	Hawker Hunter T7	RAF, stored Shawbury
	XL577	Hawker Hunter T7 (8676M) [01]	RAF SIF, Cranwell
	XL578	Hawker Hunter T7	Privately owned, Cranfield
	XL580	Hawker Hunter T8M [723]	RN No 899 Sqn, Yeovilton
	XL586	Hawker Hunter T7	RAF, stored Shawbury
	XL587	Hawker Hunter T7 (8807M) [Z]	RAF TMTS, Scampton
	XL591	Hawker Hunter T7	RAF, stored Shawbury
	XL592	Hawker Hunter T7 (8836M) [Y]	RAF TMTS, Scampton
	XL595	Hawker Hunter T7 (G-BTYL)	Privately owned, Coltishall
	XL598	Hawker Hunter T8C [880/VL]	RN FRADU, Yeovilton
	XL600	Hawker Hunter T7 [Y/FL]	*Sold to USA*
	XL601	Hawker Hunter T7 [874/VL]	RN FRADU, Yeovilton
	XL602	Hawker Hunter T8M	MoD(PE) BAe Dunsfold
	XL603	Hawker Hunter T8M [724]	RN No 899 Sqn, Yeovilton
	XL609	Hawker Hunter T7 (8866M) [YF]	RAF Lossiemouth, BDRT
	XL612	Hawker Hunter T7 [2]	MoD(PE) ETPS, Boscombe Down
	XL613	Hawker Hunter T7A	RAF, stored Shawbury
	XL614	Hawker Hunter T7	RAF No 12 Sqn, Lossiemouth
	XL616	Hawker Hunter T7 [D]	RAF Lossiemouth
	XL618	Hawker Hunter T7 (8892M) [05]	RAF Cottesmore Fire Section
	XL623	Hawker Hunter T7 [90] (8770M)	RAF Newton
	XL629	EE Lightning T4	A&AEE Boscombe Down, at main gate
	XL703	SAL Pioneer CC1 (8034M)	RAF Cosford Aerospace Museum, stored
	XL728	WS58 Wessex HAS1	RAF Brawdy, Fire Section
	XL735	Saro Skeeter AOP12	Privately owned, Tattershall Thorpe
	XL738	Saro Skeeter AOP12 (7860M)	Museum of Army Flying, Middle Wallop
	XL762	Saro Skeeter AOP12 (8017M)	Royal Scottish Museum of Flight, East Fortune
	XL763	Saro Skeeter AOP12	Southall Technical College
	XL764	Saro Skeeter AOP12 (7940M)	Newark Air Museum, Winthorpe
	XL765	Saro Skeeter AOP12	Privately owned, Pimlico
	XL770	Saro Skeeter AOP12 (8046M)	Southampton Hall of Aviation
	XL809	Saro Skeeter AOP12 (G-BLIX/ PH-HOF)	Privately owned, Clapham, Beds
	XL811	Saro Skeeter AOP12 [157]	Privately owned, Stoke-On-Trent
	XL812	Saro Skeeter AOP12 (G-SARO)	Privately owned, Old Buckenham
	XL813	Saro Skeeter AOP12	Museum of Army Flying, Middle Wallop
	XL814	Saro Skeeter AOP12	AAC Historic Aircraft Flt, Middle Wallop
	XL824	Bristol Sycamore HR14 (8021M)	Greater Manchester Museum of Science and Industry
	XL829	Bristol Sycamore HR14	Bristol Industrial Museum
	XL836	WS55 Whirlwind HAS7 (A2642) [65]	*Burnt at RN Predannack Fire School*
	XL840	WS55 Whirlwind HAS7	Stratford Aircraft Collect'n, Long Marston
	XL846	WS55 Whirlwind HAS7 (A2625) [85]	*Burnt by October 1992*
	XL847	WS55 Whirlwind HAS7 (A2626) [83]	AAC Middle Wallop, Fire Section
	XL853	WS55 Whirlwind HAS7 [LS] (A2630)	RNAY Fleetlands Museum
	XL875	WS55 Whirlwind HAR9	Air Service Training, Perth
	XL880	WS55 Whirlwind HAR9 (A2714) [35]	RN Lee-on-Solent, Fire Service
	XL898	WS55 Whirlwind HAR9 (8654M) [30/ED]	Privately owned, New Byth, Grampian
	XL929	Percival Pembroke C1 (G-BNPU)	Northbrook College, Shoreham Airport
	XL954	Percival Pembroke C1 (N4234C)	Privately owned, White Waltham
	XL993	SAL Twin Pioneer CC1 (8388M)	RAF Cosford Aerospace Museum
	XM135	BAC Lightning F1 [13	Imperial War Museum, Duxford
	XM144	BAC Lightning F1 (8417M) [J]	Privately owned, Burntwood, Staffs
	XM169	BAC Lightning F1A (8422M) (nose)	N. Yorkshire Recovery Group, Chop Gate
	XM172	BAC Lightning F1A (8427M) [B]	RAF Coltishall, on display
	XM173	BAC Lightning F1A (8414M) [A]	RAF Bentley Priory, at main gate
	XM191	BAC Lightning F1A (7854M/ 8590M) (nose)	RAF Exhibition Flight, Abingdon
	XM192	BAC Lightning F1A (8413M) [K]	RAF Wattisham, at main gate
	XM223	DH Devon C2 [J]	MoD(PE) DRA West Freugh

Serial	Type	Owner or Operator	Notes
XM279	EE Canberra B(I)8 (nose)	S Yorks Air Museum, Firbeck	
XM296	DH Heron C4	FAA Museum, stored Yeovilton	
XM300	WS58 Wessex HAS1	Wales Industrial and Maritime Museum, Cardiff	
XM327	WS58 Wessex HAS3 [401/KE]	College of Nautical Studies, Warsash	
XM328	WS58 Wessex HAS3	RNAS Culdrose, SAH	
XM329	WS58 Wessex HAS1 (A2609) [33/PO]	RN Predannack Fire School	
XM330	WS58 Wessex HAS1	MoD(PE) DRA Farnborough, stored	
XM349	Hunting Jet Provost T3A [T] (9046M)	RAF No 2 SoTT, Cosford	
XM350	Hunting Jet Provost T3A [89] (9036M)	RAF Church Fenton, Fire Section	
XM351	Hunting Jet Provost T3 [Y] (8078M)	RAF Aerospace Museum, Cosford	
XM352	Hunting Jet Provost T3A [21]	RAF No 1 FTS, Linton-on-Ouse	
XM355	Hunting Jet Provost T3 (8229M) [D]	Privately owned, Bruntingthorpe	
XM357	Hunting Jet Provost T3A [45]	RAF No 1 FTS, Linton-on-Ouse	
XM358	Hunting Jet Provost T3A [53] (8987M)	Privately owned, Bruntingthorpe	
XM362	Hunting Jet Provost T3 (8230M)	RAF No 1 SoTT, Halton	
XM365	Hunting Jet Provost T3A [37]	RAF No 1 FTS, Linton-on-Ouse	
XM367	Hunting Jet Provost T3 [Z] (8083M)	RAF No 2 SoTT, Cosford	
XM369	Hunting Jet Provost T3 (8084M) [C]	Privately owned, North Byth, Grampian	
XM370	Hunting Jet Provost T3A [10]	RAF No 1 FTS, Linton-on-Ouse	
XM371	Hunting Jet Provost T3A [31/K] (8962M)	RAF Halton (for disposal)	
XM372	Hunting Jet Provost T3A [55] (8917M)	RAF Linton-on-Ouse Fire Section	
XM374	Hunting Jet Provost T3A [18]	RAF No 1 FTS, Linton-on-Ouse	
XM375	Hunting Jet Provost T3 (8231M) [B]	RAF Cottesmore Fire Section	
XM376	Hunting Jet Provost T3A [27]	RAF No 1 FTS, Linton-on-Ouse	
XM378	Hunting Jet Provost T3A [34]	RAF No 1 FTS, Linton-on-Ouse	
XM379	Hunting Jet Provost T3	Army Apprentice College, Arborfield	
XM383	Hunting Jet Provost T3A [90]	Privately owned, Crowland, Lincs	
XM386	Hunting Jet Provost T3 (8076M) [08]	RAF No 1 SoTT, Halton	
XM387	Hunting Jet Provost T3A [I]	RAF, stored Shawbury	
XM401	Hunting Jet Provost T3A [17]	RAF, Linton-on-Ouse (for disposal)	
XM402	Hunting Jet Provost T3 (8055AM) [J]	Privately owned, Narborough, Norfolk	
XM403	Hunting Jet Provost T3A [V] (9048M)	RAF No 2 SoTT, Cosford	
XM404	Hunting Jet Provost T3 (8055BM)	RAF No 1 SoTT, Halton	
XM405	Hunting Jet Provost T3A [42] (G-TORE)	Privately owned, Cranfield	
XM408	Hunting Jet Provost T3 (8233M) [D]	Privately owned, Bruntingthorpe	
XM409	Hunting Jet Provost T3 (8082M) [A]	*Scrapped, 1992*	
XM410	Hunting Jet Provost T3 (8054AM) [B]	RAF North Luffenham Training Area	
XM411	Hunting Jet Provost T3 (8434M) [L	RAF Halton (for disposal)	
XM412	Hunting Jet Provost T3A [41/T] (9011M)	RAF No 1 SoTT, Halton	
XM413	Hunting Jet Provost T3	Army Apprentice College, Arborfield	
XM414	Hunting Jet Provost T3A [101/H] (8996M)	RAF No 1 SoTT, Halton	
XM417	Hunting Jet Provost T3 (8054BM)	RAF No 1 SoTT, Halton	
XM419	Hunting Jet Provost T3A [102] (8990M)	RAF CTTS, St Athan	
XM424	Hunting Jet Provost T3A	RAF No 1 FTS, Linton-on-Ouse	
XM425	Hunting Jet Provost T3A [88] (8995M)	RAF No 1 SoTT, Halton	
XM426	Hunting Jet Provost T3 [64] (nose)	Robertsbridge Aviation Museum	
XM455	Hunting Jet Provost T3A [K] (8960M)	RAF No 2 SoTT, Cosford	

Notes	Serial	Type	Owner or Operator
	XM459	Hunting Jet Provost T3A [F]	RAF, stored Shawbury
	XM461	Hunting Jet Provost T3A [11]	RAF stored, Linton-on-Ouse
	XM463	Hunting Jet Provost T3A [38]	RAF Museum, Hendon
	XM464	Hunting Jet Provost T3A [23]	RAF No 1 FTS, Linton-on-Ouse
	XM465	Hunting Jet Provost T3A [55]	RAF St Athan
	XM466	Hunting Jet Provost T3A	RAF No 1 FTS, Linton-on-Ouse
	XM467	Hunting Jet Provost T3 (8085M)	
	XM468	Hunting Jet Provost T3 (8081M)	RAF CTTS, St Athan
	XM470	Hunting Jet Provost T3A [12]	RAF No 1 FTS, Linton-on-Ouse
	XM471	Hunting Jet Provost T3A [L/93] (8968M)	RAF No 2 SoTT, Cosford
	XM472	Hunting Jet Provost T3A (nose) (9051M)	No 1005 Sqn ATC
	XM473	Hunting Jet Provost T3A [33] (fuselage) (8974M)	Privately owned, Norwich
	XM474	Hunting Jet Provost T3 (8121M)	No 1330 Sqn ATC, Warrington
	XM475	Hunting Jet Provost T3A (9112M) [44]	RAF FSCTE, Manston
	XM478	Hunting Jet Provost T3A [33] (8983M)	RAF No 1 SoTT, Halton
	XM479	Hunting Jet Provost T3A [54]	RAF No 1 FTS, Linton-on-Ouse
	XM480	Hunting Jet Provost T3 [02] (8080M)	RAF No 1 SoTT, Halton
	XM515	DH Vampire T11 (7998M) (really XD515)	Solway Aviation Society, Carlisle
	XM529	Saro Skeeter AOP12 (7979M/ G-BDNS)	Privately owned, Handforth
	XM553	Saro Skeeter AOP12 (G-AWSV)	Privately owned, Middle Wallop
	XM555	Saro Skeeter AOP12 (8027M)	RAF Cosford Aerospace Museum, stored
	XM556	Saro Skeeter AOP12 (G-HELI/7870M) [V]	International Helicopter Museum, Weston-super-Mare
	XM561	Saro Skeeter AOP12 (7980M)	Lincolnshire Aviation Museum, East Kirkby, stored
	XM564	Saro Skeeter AOP12	Royal Armoured Corps Museum, Bovington
	XM569	Avro Vulcan B2	Wales Aircraft Museum, Cardiff
	XM575	Avro Vulcan B2A (G-BLMC)	East Midlands Airport Aeropark
	XM594	Avro Vulcan B2	Newark Air Museum, Winthorpe
	XM597	Avro Vulcan B2	Royal Scottish Museum of Flight, East Fortune
	XM598	Avro Vulcan B2 (8778M)	RAF Cosford Aerospace Museum
	XM602	Avro Vulcan B2 (8771M)	Privately owned
	XM603	Avro Vulcan B2	Avro Aircraft Restoration Society, BAe Woodford
	XM607	Avro Vulcan B2 (8779M)	RAF Waddington, on display
	XM612	Avro Vulcan B2	City of Norwich Aviation Museum
	XM652	Avro Vulcan B2 (nose)	Privately owned, Burntwood, Staffs
	XM655	Avro Vulcan B2 (G-VULC/ N655AV)	Privately owned, Wellesbourne Mountford
	XM656	Avro Vulcan B2 (8757M) (nose)	Privately owned, Romford
	XM657	Avro Vulcan B2A (8734M)	*Burnt by June 1992*
	XM660	WS55 Whirlwind HAS7 [78]	North-East Aircraft Museum, Usworth
	XM665	WS55 Whirlwind HAS7	Booker Aircraft Museum
	XM685	WS55 Whirlwind HAS7 (G-AYZJ) [513/PO]	Newark Air Museum, Winthorpe
	XM693	HS Gnat T1 (7891M)	BAe Hamble on display
	XM694	HS Gnat T1	DRA Bedford Apprentice School
	XM697	HS Gnat T1 (G-NAAT)	Jet Heritage, Bournemouth
	XM706	HS Gnat T1 (8572M) [12]	RAF Swinderby Fire Section
	XM708	HS Gnat T1 (8573M)	RAF Locking, on display
	XM709	HS Gnat T1 (8617M) [67]	Privately owned
	XM715	HP Victor K2	RAF No 55 Sqn, Marham
	XM717	HP Victor K2	RAF No 55 Sqn, Marham
	XM833	WS58 Wessex HAS3	Stratford Aircraft Collection, Long Marston
	XM836	WS58 Wessex HAS3	*Scrapped December 1991*
	XM838	WS58 Wessex HAS3 [05]	RN Predannack Fire School
	XM843	WS58 Wessex HAS1 (A2693) [527/LS]	RNAS Lee-on-Solent on display
	XM845	WS58 Wessex HAS1 (A2682) [530/PO]	*Scrapped RNAS Yeovilton*
	XM868	WS58 Wessex HAS1 (A2706) [517]	RN Predannack Fire School

Serial	Type	Owner or Operator	Notes
XM870	WS58 Wessex HAS3 [PO]	RN NACDS, Culdrose	
XM874	WS58 Wessex HAS1 (A2689) [521/CU]	RN Predannack Fire School	
XM916	WS58 Wessex HAS3 [666/PO]	*Burnt by July 1990*	
XM917	WS58 Wessex HAS1 (A2692) [528/PO]	*Scrapped at RNAS Lee-on-Solent*	
XM923	WS58 Wessex HAS 3	RNAY Fleetlands, Fire Section	
XM927	WS58 Wessex HAS3 (8814M) [660/PO]	RAF Shawbury, Fire Section	
XM987	BAC Lightning T4	RAF Coningsby, Fire Section	
XN126	WS55 Whirlwind HAR10 (8655M) [S]	RAF Benson BDRT	
XN132	Sud Alouette AH2 (G-BUIV)	*Repainted as G-BUIV*	
XN185	Slingsby Sedbergh TX1 (8942M)	RAF ACCGS Syerston, preserved	
XN198	Slingsby Cadet TX3	Privately owned, Challock Lees	
XN239	Slingsby Cadet TX3 [G] (8889M)	Imperial War Museum, Duxford	
XN243	Slingsby Cadet TX3 (BGA 3145)	RAFGSA, Bicester	
XN246	Slingsby Cadet TX3	Southampton Hall of Aviation	
XN258	WS55 Whirlwind HAR9 [589/CU]	Flambards Village Theme Park, Helston	
XN259	WS55 Whirlwind HAS7	London City Airport, Fire Section	
XN297	WS55 Whirlwind HAR9 [12] (really XN311/A2643)	Privately owned, Hull	
XN298	WS55 Whirlwind HAR9 [810/LS]	International Fire Training Centre, Chorley	
XN299	WS55 Whirlwind HAS7 [ZZ]	Royal Marines' Museum, Portsmouth	
XN302	WS55 Whirlwind HAS7 (A2654/ 9037M) [LS]	RAF Finningley, Fire Section	
XN304	WS55 Whirlwind HAS7 [64]	Norfolk & Suffolk Aviation Museum, Flixton	
XN308	WS55 Whirlwind HAS7 (A2605) [510/PO]	RNAS Yeovilton Fire Section	
XN332	Saro P531 (G-APNV/A2579) [759]	IHM, Weston-super-Mare	
XN334	Saro P531 (A2525)	IHM, Crawley College of Technology	
XN341	Saro Skeeter AOP12 (8022M)	Privately owned, Luton Airport	
XN344	Saro Skeeter AOP12 (8018M)	Science Museum, South Kensington	
XN351	Saro Skeeter AOP12 (G-BKSC)	Privately owned, Inverness Airport	
XN359	WS55 Whirlwind HAR9 (A2712) [34/ED]	RNAS Lee-on-Solent, Fire Section	
XN380	WS55 Whirlwind HAS7 [67]	Lashenden Air Warfare Museum, Headcorn	
XN385	WS55 Whirlwind HAS7	RN, stored	
XN386	WS55 Whirlwind HAR9 [435/ED] (A2713)	Privately owned, Blackpool	
XN412	Auster AOP9	Cotswold Aircraft Restoration Group, RAF Innsworth	
XN435	Auster AOP9 (G-BGBU)	Privately owned, Egham	
XN437	Auster AOP9 (G-AXWA)	Privately owned, Welling, Kent	
XN441	Auster AOP9 (G-BGKT)	Privately owned, Reymerston Hall	
XN453	DH Comet 2e	*Scrapped at Farnborough*	
XN459	Hunting Jet Provost T3A [N]	RAF, stored Shawbury	
XN461	Hunting Jet Provost T3A [28]	RAF No 1 FTS, Linton-on-Ouse	
XN462	Hunting Jet Provost T3A [17]	RAF No 1 FTS, Linton-on-Ouse	
XN466	Hunting Jet Provost T3A [29]	RAF No 1 FTS, Linton-on-Ouse	
XN467	Hunting Jet Provost T4 (8559M) [F]	RAF No 1 SoTT, Halton	
XN470	Hunting Jet Provost T3A [41]	RAF No 1 FTS, Linton-on-Ouse	
XN471	Hunting Jet Provost T3A [24]	RAF No 1 FTS, Linton-on-Ouse	
XN472	Hunting Jet Provost T3A [5/86] (8959M)	RAF No 2 SoTT, Cosford	
XN473	Hunting Jet Provost T3A (8862M) [98] (cockpit only)	RAF Church Fenton Fire Section	
XN492	Hunting Jet Provost T3 (8079M) [M])	RAF No 2 SoTT, Cosford	
XN493	Hunting Jet Provost T3	Privately owned, Ottershaw	
XN494	Hunting Jet Provost T3A [43/R] (9012M)	RAF No 1 SoTT, Halton	
XN495	Hunting Jet Provost T3A [102] (8786M)	RAF Finningley, Fire Section	
XN497	Hunting Jet Provost T3A [52]	RAF St Athan	
XN498	Hunting Jet Provost T3A [16]	RAF No 1 FTS, Linton-on-Ouse	
XN499	Hunting Jet Provost T3A [L]	RAF, stored Shawbury	
XN500	Hunting Jet Provost T3A [48]	RAF No 1 FTS, Linton-on-Ouse	
XN501	Hunting Jet Provost T3A [G] (8958M)	RAF No 2 SoTT, Cosford	

Notes	Serial	Type	Owner or Operator
	XN502	Hunting Jet Provost T3A [D]	RAF, stored Shawbury
	XN505	Hunting Jet Provost T3A [25]	RAF No 1 FTS, Linton-on-Ouse
	XN506	Hunting Jet Provost T3A [19]	RAF, stored Shawbury
	XN508	Hunting Jet Provost T3A [47]	RAF No 1 FTS, Linton-on-Ouse
	XN509	Hunting Jet Provost T3A [50]	RAF No 1 FTS, Linton-on-Ouse
	XN510	Hunting Jet Provost T3A [40]	RAF No 1 FTS, Linton-on-Ouse
	XN511	Hunting Jet Provost T3 [21] (nose only)	Newark Air Museum, Winthorpe
	XN512	Hunting Jet Provost T3 (8435M)	RAF No 1 SoTT, Halton
	XN549	Hunting Jet Provost T3 [32] (8235M) [P]	RAF Shawbury Fire Section
	XN551	Hunting Jet Provost T3A [100] (8984M)	RAF CTTS, St Athan
	XN552	Hunting Jet Provost T3A [32]	RAF No 1 FTS, Linton-on-Ouse
	XN553	Hunting Jet Provost T3A	RAF, stored Shawbury
	XN554	Hunting Jet Provost T3 (8436M) [K]	RAF North Luffenham Training Area
	XN577	Hunting Jet Provost T3A [89/F] (8956M)	RAF No 2 SoTT, Cosford
	XN579	Hunting Jet Provost T3A [14]	RAF North Luffenham Training Area
	XN581	Hunting Jet Provost T3A [C]	*Scrapped at Scampton 1991*
	XN582	Hunting Jet Provost T3A [95/H] (8957M)	RAF No 2 SoTT, Cosford
	XN584	Hunting Jet Provost T3A [E] (9014M)	RAF No 1 SoTT, Halton
	XN586	Hunting Jet Provost T3A [91/S] (9039M)	RAF No 2 SoTT, Cosford
	XN589	Hunting Jet Provost T3A [46]	RAF No 1 FTS, Linton-on-Ouse
	XN592	Hunting Jet Provost T3 (nose only)	No 1105 Sqn ATC, Winchester
	XN593	Hunting Jet Provost T3A [97/Q] (8988M)	RAF No 2 SoTT, Cosford
	XN594	Hunting Jet Provost T3 [W] (8077M)	RAF No 2 SoTT, Cosford
	XN595	Hunting Jet Provost T3A [43]	RAF, stored Shawbury
	XN597	Hunting Jet Provost T3 (nose)	North East Aircraft Museum, Usworth
	XN600	Hunting Jet Provost T3A (nose only)	N. Yorks Recovery Group, Chop Gate
	XN602	Hunting Jet Provost T3 (8088M)	FSCTE, RAF Manston
	XN606	Hunting Jet Provost T3A [51] (9121M)	RAF Brawdy, BDRT
	XN629	Hunting Jet Provost T3A [49]	RAF, stored Shawbury
	XN632	Hunting Jet Provost T3 (8352M)	RAF Chivenor, crash rescue training
	XN634	Hunting Jet Provost T3A [53]	RAF No 1 FTS, Linton-on-Ouse
	XN636	Hunting Jet Provost T3A [15] (9045M)	RAF No 2 SoTT, Cosford
	XN637	Hunting Jet Provost T3 (G-BKOU)	Vintage Aircraft Team, Cranfield
	XN640	Hunting Jet Provost T3A [99/R] (9016M)	RAF No 2 SoTT, Cosford
	XN641	Hunting Jet Provost T3A (8865M) [47]	RAF Newton Fire Section
	XN647	DH Sea Vixen FAW2 (A2610) [707/VL]	Flambards Village Theme Park, Helston
	XN649	DH Sea Vixen FAW2 [126]	MoD(PE), stored DRA Farnborough
	XN650	DH Sea Vixen FAW2 (A2612/ A2620/A2639) [VL]	Wales Aircraft Museum, Cardiff
	XN651	DH Sea Vixen FAW2 (A2616) (nose only)	Privately owned, Portsmouth
	XN652	DH Sea Vixen FAW2 (8817M)	*Scrapped at Catterick*
	XN657	DH Sea Vixen D3 [TR-1]	T&EE Llanbedr Fire Section
	XN685	DH Sea Vixen FAW2 (8173M) [P] [03/VL]	Midland Air Museum, Coventry
	XN688	DH Sea Vixen FAW2 (8141M) [511]	DRA Farnborough Fire Section
	XN691	DH Sea Vixen FAW2 [N] [247/H] (8143M)	Aces High, North Weald
	XN692	DH Sea Vixen FAW2 (A2624) [125/E]	RNAS Yeovilton, on display
	XN694	DH Sea Vixen FAW2	MoD(PE) DRA, Llanbedr
	XN696	DH Sea Vixen FAW2 (cockpit only)	Blyth Valley Aviation Collection, Walpole
	XN699	DH Sea Vixen FAW2 [752] (8224M)	*Scrapped 1992 at North Luffenham*
	XN714	Hunting H126	RAF Cosford Aerospace Museum

Serial	Type	Owner or Operator	Notes
XN724	EE Lightning F2A [F] (8513M)	Privately owned, Newcastle-on-Tyne	
XN728	EE Lightning F2A (8546M) [V]	Privately owned, Balderton, Notts	
XN734	EE Lightning F3A (8346M/ G-27-239/G-BNCA)	Vintage Aircraft Team, Cranfield	
XN769	EE Lightning F2 (8402M) [Z]	London ATCC, West Drayton	
XN774	EE Lightning F2A (8551M) [F]	RAF Coningsby, Fire Section	
XN776	EE Lightning F2A [C] (8535M)	Royal Scottish Museum of Flight, East Fortune	
XN817	AW Argosy C1	MoD(PE) DRA West Freugh Fire Section	
XN819	AW Argosy C1 (8205M) (nose only)	Newark Air Museum, Winthorpe	
XN855	AW Argosy E1 (8556M)	FSCTE, RAF Manston	
XN923	HS Buccaneer S1 [13]	Privately owned, Charlwood, Surrey	
XN928	HS Buccaneer S1 (8179M)	Wales Aircraft Museum, Cardiff	
XN929	HS Buccaneer S1 (8051M) (nose only)	*Scrapped at RAF Cranwell*	
XN930	HS Buccaneer S1 (8180M) [632/LM] (nose only)	Privately owned, Chelmsford	
XN934	HS Buccaneer S1 (A2600) [631]	*Scrappped at RN Predannack*	
XN953	HS Buccaneer S1 (A2655/8182M)	RN Predannack Fire School	
XN957	HS Buccaneer S1 [630/LM]	FAA Museum, RNAS Yeovilton	
XN964	HS Buccaneer S1 [613/LM]	Newark Air Museum, Winthorpe	
XN967	HS Buccaneer S1 (A2627) [103/E]	Flambards Village Theme Park, Helston	
XN972	HS Buccaneer S1 (8183M) (nose only) (really XN962)	RAF Exhibition Flight, St Athan	
XN974	HS Buccaneer S2A	Yorkshire Air Museum, Elvington	
XN976	HS Buccaneer S2B	*Crashed into North Sea 9 July 1992*	
XN979	HS Buccaneer S2 (nose only)	ATC, RAF Stanbridge	
XN981	HS Buccaneer S2B	RAF No 12 Sqn, Lossiemouth	
XN982	HS Buccaneer S2A	MoD(PE) BAe Brough	
XN983	HS Buccaneer S2B	RAF, stored Shawbury	
XP110	WS58 Wessex HAS3 [55/FL]	RNAY Fleetlands Apprentice School	
XP116	WS58 Wessex HAS3 (A2618) [520]	RN AES, Lee-on-Solent	
XP137	WS58 Wessex HAS3 [CU]	RN NACDS, Culdrose	
XP140	WS58 Wessex HAS3 (8806M) [653/PO]	RAF Chilmark, BDRT	
XP142	WS58 Wessex HAS3	FAA Museum, RNAS Yeovilton	
XP150	WS58 Wessex HAS3 [406/AN]	RN AES, Lee-on-Solent	
XP151	WS58 Wessex HAS1 (A2684) [R]	RN Lee-on-Solent Fire Section	
XP157	WS58 Wessex HAS1 (A2680)	RN AES, Lee-on-Solent	
XP158	WS58 Wessex HAS1 (A2688) [522/CU]	RN AES, Lee-on-Solent	
XP159	WS58 Wessex HAS1 (8877M) [047/R]	Privately owned, Brands Hatch	
XP160	WS58 Wessex HAS1 (A2650) [521/CU]	RNAS Lee-on-Solent Fire Section	
XP165	WS Scout AH1	IHM, Weston-super-Mare	
XP166	WS Scout AH1 (G-APVL)	DRA Farnborough, Apprentice School	
XP190	WS Scout AH1	South Yorkshire Air Museum, Firbeck	
XP191	WS Scout AH1	*Scrapped at Middle Wallop*	
XP226	Fairey Gannet AEW3 (A2667) [073/E]	Newark Air Museum, Winthorpe	
XP241	Auster AOP9	Rebel Air Museum, Andrewsfield	
XP242	Auster AOP9 (G-BUCI)	Privately owned, Middle Wallop	
XP244	Auster AOP9 (7864M) [M7922]	Army Apprentice College, Arborfield	
XP248	Auster AOP9 (7822M)	Privately owned, Little Gransden	
XP279	Auster AOP9 (G-BWKK)	Privately owned, Goodwood	
XP280	Auster AOP9	Leicester Technology Museum, Coalville	
XP281	Auster AOP9	Imperial War Museum, Duxford	
XP282	Auster AOP9 (G-BGTC)	Privately owned, Tollerton	
XP283	Auster AOP9 (7859M)	Privately owned, Lichfield	
XP299	WS55 Whirlwind HAR10 (8726M)	RAF Cosford Aerospace Museum	
XP329	WS55 Whirlwind HAR10 [V] (8791M) [UN]	Privately owned, Tattershall Thorpe	
XP330	WS55 Whirlwind HAR10	CAA Fire School, Teesside Airport	
XP333	WS55 Whirlwind HAR10 (8650M) [G]	*Burnt by June 1992*	
XP338	WS55 Whirlwind HAR10 (8647M) [N]	RAF No 2 SoTT, Cosford	

Notes	Serial	Type	Owner or Operator
	XP344	WS55 Whirlwind HAR10 (8764M) [X]	RAF North Luffenham
	XP345	WS55 Whirlwind HAR10 [UN] (8792M)	Privately owned, Storwood, East Yorks
	XP346	WS55 Whirlwind HAR10 (8793M)	Stratford Aircraft Collection, Long Marston
	XP350	WS55 Whirlwind HAR10	Flambards Village Theme Park, Helston
	XP351	WS55 Whirlwind HAR10 (8672M) [Z]	RAF Shawbury, on display
	XP353	WS55 Whirlwind HAR10 (8720M)	Privately owned, Brands Hatch
	XP354	WS55 Whirlwind HAR10 (8721M)	RAF No 1 SoTT, Halton
	XP355	WS55 Whirlwind HAR10 (8463M/ (G-BEBC) [A]	City of Norwich Aviation Museum
	XP357	WS55 Whirlwind HAR10 (8499M)	*Burnt by June 1992*
	XP359	WS55 Whirlwind HAR10 (8447M)	RAF Stafford, Fire Section
	XP360	WS55 Whirlwind HAR10 [V]	North East Aircraft Museum, Usworth
	XP361	WS55 Whirlwind HAR10 (8731M)	RAF Coltishall, Fire Section
	XP393	WS55 Whirlwind HAR10 [U]	DRA Farnborough, Fire Section
	XP395	WS55 Whirlwind HAR10 [A] (8674M)	Privately owned, Tattershall Thorpe
	XP398	WS55 Whirlwind HAR10 (8794M)	Privately owned, Charlwood, Surrey
	XP399	WS55 Whirlwind HAR10	Privately owned, Raunds, Northants
	XP400	WS55 Whirlwind HAR10 [N] (8444M)	*Burnt by June 1992*
	XP404	WS55 Whirlwind HAR10 (8682M)	International Helicopter Museum, Weston-super-Mare
	XP405	WS55 Whirlwind HAR10 [Y] (8656M)	Junior Infantry Reg't, Shorncliffe, Kent
	XP411	AW Argosy C1 (8442M) [C]	RAF Cosford Aerospace Museum
	XP458	Slingsby Grasshopper TX1	City of Norwich Aviation Museum
	XP488	Slingsby Grasshopper TX1	Fenland Aircraft Preservation Society, Wisbech
	XP502	HS Gnat T1 (8576M) [02]	RAF St Athan, CTTS
	XP503	HS Gnat T1 (8568M) [73]	Privately owned, Bruntingthorpe
	XP504	HS Gnat T1 (8618M) (G-TIMM) [04]	Privately owned, Cranfield
	XP505	HS Gnat T1 [05]	Science Museum, Wroughton
	XP511	HS Gnat T1 (8619M) [65]	*To USA 1992*
	XP516	HS Gnat T1 (8580M) [16]	MoD(PE) DRA Farnborough
	XP530	HS Gnat T1 (8606M) [60]	RAF No 1 SoTT, Halton
	XP532	HS Gnat T1 (8577M/8615M) [32]	Privately owned, Colchester
	XP534	HS Gnat T1 (8620M) [64]	RAF No 1 SoTT, Halton
	XP540	HS Gnat T1 (8608M) [62]	Privately owned, Bruntingthorpe
	XP542	HS Gnat T1 (8575M) [42]	RAF St Athan, CTTS
	XP547	Hunting Jet Provost T4 [N/03] (8992M)	RAF No 2 SoTT, Cosford
	XP556	Hunting Jet Provost T4 [B] (9027M)	RAF No 1 SoTT, Halton
	XP557	Hunting Jet Provost T4 (8494M)	Privately owned, Bruntingthorpe
	XP558	Hunting Jet Provost T4 (8627M/ A2628) [20]	RAF St Athan Fire Section
	XP563	Hunting Jet Provost T4 [C] (9028M)	RAF No 1 SoTT, Halton
	XP567	Hunting Jet Provost T4 (8510M) [23]	RAF No 1 SoTT, Halton
	XP568	Hunting Jet Provost T4	Stratford Aircraft Collection, Long Marston
	XP573	Hunting Jet Provost T4 (8236M) [19]	RAF No 1 SoTT, Halton
	XP585	Hunting Jet Provost T4 (8407M) [24]	RAF No 1 SoTT, Halton
	XP627	Hunting Jet Provost T4	North-East Aircraft Museum, Usworth
	XP629	Hunting Jet Provost T4 [P] (9026M)	RAF North Luffenham Training Area
	XP638	Hunting Jet Provost T4 [A] (9034M)	RAF No 1 SoTT, Halton
	XP640	Hunting Jet Provost T4 (8501M) [D]	RAF No 1 SoTT, Halton
	XP672	Hunting Jet Provost T4 (8458M) [27]	RAF No 1 SoTT, Halton
	XP677	Hunting Jet Provost T4 (8587M) (nose only)	No 2530 Sqn ATC, Headley Court, Uckfield, East Sussex
	XP680	Hunting Jet Provost T4 (8460M)	RAF St Athan, Fire Section
	XP686	Hunting Jet Provost T4 (8401M/ 8502M) [G]	RAF North Luffenham Training Area

Serial	Type	Owner or Operator	Notes
XP688	Hunting Jet Provost T4 [E] (9031M)	RAF No 1 SoTT, Halton	
XP693	BAC Lightning F6	Privately ownd, Exeter	
XP701	BAC Lightning F3 (8924M)(nose)	Robertsbridge Aviation Society	
XP703	BAC Lightning F3 (nose only)	Lightning Preservation Group, Bruntingthorpe	
XP706	BAC Lightning F3 (8925M)	Lincolnshire Lightning Preservation Society, Strubby	
XP741	BAC Lightning F3 [AR] (8939M)	FSCTE, RAF Manston	
XP745	BAC Lightning F3 (8453M) [H]	RAF Boulmer, at main gate	
XP749	BAC Lightning F3 (8926M)	*Scrapped 1988*	
XP750	BAC Lightning F3 (8927M)	*Scrapped 1988*	
XP764	BAC Lightning F3 [DC] (8929M)	*Scrapped 1988*	
XP772	DHC Beaver AL1 (G-BUCJ)	AAC Historic Aircraft Flight, Middle Wallop	
XP775	DHC Beaver AL1	Privately owned	
XP806	DHC Beaver AL1	Museum of Army Flying, Middle Wallop,	
XP820	DHC Beaver AL1	AAC Historic Aircraft Flight, Middle Wallop	
XP821	DHC Beaver AL1 [MCO]	Museum of Army Flying, Middle Wallop	
XP822	DHC Beaver AL1	Museum of Army Flying, Middle Wallop	
XP831	Hawker P1127 (8406M)	Science Museum, South Kensington	
XP841	Handley-Page HP115	FAA/Concorde Museum, RNAS Yeovilton	
XP846	WS Scout AH1 [B, H] (fuselage)	Army, No 39 Engineer Reg't, Waterbeach	
XP847	WS Scout AH1	Museum of Army Flying, Middle Wallop	
XP848	WS Scout AH1	AAC SAE, Middle Wallop	
XP849	WS Scout AH1	MoD(PE) ETPS, Boscombe Down	
XP850	WS Scout AH1 (fuselage)	AAC, stored Dishforth	
XP853	WS Scout AH1	AAC SAE, Middle Wallop	
XP854	WS Scout AH1 (7898M/TAD043)	AAC SAE, Middle Wallop	
XP855	WS Scout AH1	Army Apprentice College, Arborfield	
XP856	WS Scout AH1	AAC Middle Wallop, BDRT	
XP857	WS Scout AH1	AAC Middle Wallop Fire Section	
XP883	WS Scout AH1	AAC No 658 Sqn, Netheravon	
XP884	WS Scout AH1	AAC SAE, Middle Wallop	
XP886	WS Scout AH1	Army Apprentice College, Arborfield	
XP888	WS Scout AH1	AAC SAE, Middle Wallop	
XP890	WS Scout AH1 [G] (fuselage)	AAC, stored Perth	
XP891	WS Scout AH1 [S]	AAC No 666 (TA) Sqn, Netheravon	
XP893	WS Scout AH1	AAC, stored Middle Wallop	
XP899	WS Scout AH1 [D]	Army Apprentice College, Arborfield	
XP902	WS Scout AH1	AAC, Netheravon, BDRT	
XP905	WS Scout AH1	AAC SAE, Middle Wallop	
XP907	WS Scout AH1	AAC, stored Fleetlands	
XP908	WS Scout AH1 [Y]	AAC No 660 Sqn, Sek Kong	
XP909	WS Scout AH1	*Written off, 19 Feb 1991*	
XP910	WS Scout AH1	AAC, SAE, Middle Wallop	
XP919	DH Sea Vixen FAW2 (8163M) [706/VL]	City of Norwich Aviation Museum	
XP921	DH Sea Vixen FAW2 (8226M) [753]	*Scrapped at North Luffenham 1992*	
XP924	DH Sea Vixen D3	MoD(PE) T&EE Llanbedr	
XP925	DH Sea Vixen FAW2 (nose only) [752]	DRA Farnborough	
XP956	DH Sea Vixen FAW2	Privately owned, Dunsfold	
XP980	Hawker P.1127 (A2700)	FAA Museum, RNAS Yeovilton	
XP984	Hawker P.1127 (A2658)	RNEC Manadon, for instruction	
XR137	AW Argosy E1	Caernarfon Air World	
XR140	AW Argosy E1 (8579M) (fuselage only)	RAF Halton, Fire Section	
XR220	BAC TSR2 (7933M)	RAF Cosford Aerospace Museum	
XR222	BAC TSR2	Imperial War Museum, Duxford	
XR232	Sud Alouette AH2 (F-WEIP)	Museum of Army Flying, Middle Wallop	
XR240	Auster AOP9 (G-BDFH)	*Written off 3 May 1992*	
XR241	Auster AOP9 (G-AXRR)	Museum of Army Flying, Middle Wallop	
XR243	Auster AOP9 (8057M)		
XR244	Auster AOP9	AAC Historic Aircraft Flight, Middle Wallop	
XR246	Auster AOP9 (7862M/G-AZBU)	Privately owned, Reymerston Hall	
XR267	Auster AOP9 (G-BJXR)	Cotswold Aircraft Restoration Group, RAF Innsworth	
XR271	Auster AOP9	Museum of Artillery, Woolwich	
XR363	SC5 Belfast C1 (G-OHCA)	Privately owned, Southend	
XR371	SC5 Belfast C1	RAF Cosford Aerospace Museum	
XR379	Sud Alouette AH2	AAC, Perth, stored	

Notes	Serial	Type	Owner or Operator
	XR396	DH Comet 4C (8882M) (G-BDIU)	RAF Kinloss BDRT
	XR436	Saro Scout AH1	AAC Middle Wallop, BDRT
	XR443	DH Sea Heron C1 (G-ODLG)	Privately owned, Booker, stored
	XR453	WS55 Whirlwind HAR10 (8873M) [A]	RAF Odiham, on gate
	XR458	WS55 Whirlwind HAR10 (8662M) [H]	Museum of Army Flying, Middle Wallop
	XR478	WS55 Whirlwind HAR10	Defence School, Winterbourne Gunner
	XR482	WS55 Whirlwind HAR10 [G]	Defence School, Winterbourne Gunner
	XR485	WS55 Whirlwind HAR10 [Q]	Norfolk & Suffolk Aviation Museum, Flixton
	XR486	WS55 Whirlwind HCC12 (8727M)	Privately owned, Tattershall Thorpe (G-RWWW)
	XR497	WS58 Wessex HC2 [F]	RAF No 72 Sqn, Aldergrove
	XR498	WS58 Wessex HC2 [X]	RAF No 72 Sqn, Aldergrove
	XR499	WS58 Wessex HC2 [W]	RAF No 72 Sqn, Aldergrove
	XR501	WS58 Wessex HC2	RAF No 22 Sqn, St Mawgam
	XR502	WS58 Wessex HC2 [Z]	RAF No 72 Sqn, Aldergrove
	XR503	WS58 Wessex HC2	MoD(PE), stored DRA Bedford
	XR504	WS58 Wessex HC2	RAF SARTU, Valley
	XR505	WS58 Wessex HC2 [WA]	RAF No 2 FTS, Shawbury
	XR506	WS58 Wessex HC2 [V]	RAF No 72 Sqn, Aldergrove
	XR507	WS58 Wessex HC2	RAF No 22 Sqn, Chivenor
	XR508	WS58 Wessex HC2 [D]	RAF No 28 Sqn, Sek Kong
	XR509	WS58 Wessex HC2 (8752M)	RAF Benson, BDRT
	XR511	WS58 Wessex HC2 [L]	RAF No 60 Sqn, Benson
	XR515	WS58 Wessex HC2 [B]	RAF No 28 Sqn, Sek Kong
	XR516	WS58 Wessex HC2 [WB]	RAF No 2 FTS, Shawbury
	XR517	WS58 Wessex HC2 [N]	RAF No 60 Sqn, Benson
	XR518	WS58 Wessex HC2	RAF No 22 Sqn, B Flt, Leuchars
	XR519	WS58 Wessex HC2 [WC]	RAF No 2 FTS, Shawbury
	XR520	WS58 Wessex HC2	RAF No 22 Sqn, St Mawgam
	XR521	WS58 Wessex HC2 [WD]	RAF No 2 FTS, Shawbury
	XR522	WS58 Wessex HC2 [A]	RAF No 28 Sqn, Sek Kong
	XR523	WS58 Wessex HC2 [M]	RAF No 60 Sqn, Benson
	XR524	WS58 Wessex HC2	RAF SARTU, Valley
	XR525	WS58 Wessex HC2 [G]	RAF No 60 Sqn, Benson
	XR526	WS58 Wessex HC2 (8147M)	*Scrapped*
	XR527	WS58 Wessex HC2 [K]	RAF No 72 Sqn, Aldergrove
	XR528	WS58 Wessex HC2 [T]	RAF No 72 Sqn, Aldergrove
	XR529	WS58 Wessex HC2 [E]	RAF No 72 Sqn, Aldergrove
	XR534	HS Gnat T1 (8578M) [65]	RAF Valley on display
	XR535	HS Gnat T1 (8569M) [05]	RAF No 1 SoTT, Halton
	XR537	HS Gnat T1 (8642M) [T] (G-NATY)	Privately owned, Bournemouth
	XR538	HS Gnat T1 (8621M) [69]	RAF No 1 SoTT, Halton
	XR569	HS Gnat T1 (8560M) [08]	RAF Linton-on-Ouse Fire Section
	XR571	HS Gnat T1 (8493M)	RAF *Red Arrows*, Scampton, on display
	XR574	HS Gnat T1 (8631M) [72]	RAF No 1 SoTT, Halton
	XR588	WS58 Wessex HC2	RAF SARTU, Valley
	XR595	WS Scout AH1 [M]	AAC, Fleetlands
	XR597	WS Scout AH1	AAC SAE, Middle Wallop
	XR600	WS Scout AH1 (fuselage)	AAC Aldergrove, BDRT
	XR601	WS Scout AH1	Army Apprentice College, Arborfield
	XR602	WS Scout AH1	AAC, Netheravon, Fire Section
	XR627	WS Scout AH1	AAC, SAE, Middle Wallop
	XR628	WS Scout AH1	AAC, stored Perth
	XR629	WS Scout AH1 (fuselage)	AAC, stored RNAW Almondbank, Perth
	XR630	WS Scout AH1 [U]	AAC Middle Wallop BDRT
	XR632	WS Scout AH1	AAC, No 658 Sqn, Netheravon
	XR635	WS Scout AH1	AAC SAE, Middle Wallop
	XR639	WS Scout AH1 [X] (fuselage)	AAC, stored RNAW Almondbank, Perth
	XR643	Hunting Jet Provost T4 (8516M) [26]	RAF Halton (for disposal)
	XR650	Hunting Jet Provost T4 (8459M) [28]	RAF No 1 SoTT, Halton
	XR651	Hunting Jet Provost T4 (8431M) [A]	RAF No 1 SoTT, Halton
	XR653	Hunting Jet Provost T4 [H] (9035M)	RAF Halton (for disposal)
	XR654	Hunting Jet Provost T4 (fuselage)	Macclesfield Historical Av Soc, Chelford
	XR658	Hunting Jet Provost T4 (8192M)	North Wales Institute of Higher Education, Connah's Quay

Serial	*Type*	*Owner or Operator*	*Notes*
XR662	Hunting Jet Provost T4 (8410M) [25]	RAF Halton (for disposal)	
XR669	Hunting Jet Provost T4 (8062M) [02] (nose only)	RAF No 1 SoTT, Halton	
XR670	Hunting Jet Provost T4 (8498M)	RAF Odiham, instructional use	
XR672	Hunting Jet Provost T4 [50] (8495M)	RAF Halton, Fire Section	
XR673	Hunting Jet Provost T4 [L] (9032M)	RAF No 1 SoTT, Halton	
XR674	Hunting Jet Provost T4 [D] (9030M)	RAF No 1 SoTT, Halton	
XR679	Hunting Jet Provost T4 [M/04] (8991M)	RAF No 2 SoTT, Cosford	
XR681	Hunting Jet Provost T4 (8588M) (nose only)	No 1349 Sqn ATC, Odiham	
XR700	Hunting Jet Provost T4 (8589M) (nose only)	RAF Exhibition Flight, Aldergrove	
XR701	Hunting Jet Provost T4 [K/21] (9025M)	RAF No 1 SoTT, Halton	
XR704	Hunting Jet Provost T4 (8506M) [30]	RAF No 1 SoTT, Halton	
XR713	BAC Lightning F3 (8935M) [C]	RAF Leuchars	
XR716	BAC Lightning F3 (8940M)	RAF Cottesmore Fire Section	
XR718	BAC Lightning F6 (8932M)	*Scrapped at Wattisham 1992*	
XR724	BAC Lightning F6 (G-BTSY)	Privately owned, Binbrook	
XR725	BAC Lightning F6	Privately owned, Binbrook	
XR726	BAC Lightning F6 (cockpit)	Privately owned, Harrogate	
XR728	BAC Lightning F6 [JS]	Lightning Preservation Group, Bruntingthorpe	
XR747	BAC Lightning F6 (nose only)	Privately owned, Plymouth	
XR749	BAC Lightning F3 (8934M) (nose)	South Yorkshire Aircraft Preservation Society, Firbeck	
XR751	BAC Lightning F3	Privately owned, Lower Tremar, Cornwall	
XR753	BAC Lightning F6 [BP] (8969M)	RAF Leeming on display	
XR754	BAC Lightning F6 [BC] (8972M)	RAF Honington, BDRT	
XR755	BAC Lightning F6	Privately owned, Callington, Cornwall	
XR757	BAC Lightning F6 (nose only)	Privately owned, New Waltham	
XR759	BAC Lightning F6	Privately owned, Rossington	
XR770	BAC Lightning F6 [JS]	Privately owned, New Waltham	
XR771	BAC Lightning F6 [BM]	Midland Air Museum, Coventry	
XR773	BAC Lightning F6 [BR]	MoD(PE) BAe Warton	
XR777	WS Scout AH1 (really XT625)	St George's Barracks, Sutton Coldfield	
XR806	BAC VC10 C1	RAF No 10 Sqn, Brize Norton	
XR807	BAC VC10 C1	RAF No 10 Sqn, Brize Norton	
XR808	BAC VC10 C1	RAF No 10 Sqn, Brize Norton	
XR810	BAC VC10 C1	RAF No 10 Sqn, Brize Norton	
XR944	Wallis WA116 (G-ATTB)	RAF Museum, Hendon	
XR953	HS Gnat T1 (8609M) [63]	RAF No 1 SoTT, Halton	
XR954	HS Gnat T1 (8570M) [30]	RAF No 1 SoTT, Halton	
XR955	HS Gnat T1 (A2678) [SAH-2]	Privately owned, Leavesden	
XR977	HS Gnat T1 (8640M)	RAF Cosford Aerospace Museum	
XR980	HS Gnat T1 (8622M) [70]	RAF No 1 SoTT, Halton	
XR985	HS Gnat T1 (7886M)	Vintage Aircraft Team, Cranfield	
XR991	HS Gnat T1 (XS102/G-MOUR)	Intrepid Aviation Co, North Weald	
XR998	HS Gnat T1 (8623M) [71]	RAF No 1 SoTT, Halton	
XS100	HS Gnat T1 (8561M) [57]	*Sold to USA 1992 as N7CV*	
XS101	HS Gnat T1 (8638M) (G-GNAT)	Privately owned, Cranfield	
XS109	HS Gnat T1 (8626M) [75]	*To USA 1992*	
XS119	WS58 Wessex HAS3 [55]	*Burnt by October 1992*	
XS122	WS58 Wessex HAS3 (A2707) [655/PO]	RNEC Manadon, for instruction	
XS125	WS58 Wessex HAS1 (A2648) [I7]	*Burnt by October 1992*	
XS128	WS58 Wessex HAS1 (A2670) [37]	RNAS Yeovilton, BDRT	
XS149	WS58 Wessex HAS3 [661/GL]	IHM, Weston-super-Mare	
XS153	WS58 Wessex HAS3 [662/PO]	RN Lee-on-Solent, BDRF	
XS176	Hunting Jet Provost T4 (8514M) [N]	RAF No 1 SoTT, Halton	
XS177	Hunting Jet Provost T4 [N] (9044M)	RAF Valley Fire Section	
XS178	Hunting Jet Provost T4 [P/05] (8994M)	RAF No 2 SoTT, Cosford	

Notes	Serial	Type	Owner or Operator
	XS179	Hunting Jet Provost T4 (8237M) [20]	RAF No 1 SoTT, Halton
	XS180	Hunting Jet Provost T4 (8238M) [21]	RAF Halton (for disposal)
	XS181	Hunting Jet Provost T4 [F] (9033M)	RAF No 1 SoTT, Halton
	XS186	Hunting Jet Provost T4 (8408M) [M]	RAF North Luffenham Training Area
	XS209	Hunting Jet Provost T4 (8409M) [29]	RAF No 1 SoTT, Halton
	XS210	Hunting Jet Provost T4 (8239M) [22]	RAF Halton (for disposal)
	XS215	Hunting Jet Provost T4 (8507M) [17]	RAF Halton (for disposal)
	XS216	Hunting Jet Provost T4 (nose only)	RAF Finningley Fire Section
	XS217	Hunting Jet Provost T4 [O] (9029M)	RAF No 1 SoTT, Halton
	XS218	Hunting Jet Provost T4 (8508M) [18] (nose)	Museum of Berkshire Aviation, Woodley
	XS219	Hunting Jet Provost T4 [O/06] (8993M)	RAF No 2 SoTT, Cosford
	XS230	BAC Jet Provost T5P	MoD(PE) ETPS, Boscombe Down
	XS231	BAC Jet Provost T5 (G-ATAJ)	Privately owned, Bruntingthorpe
	XS235	DH Comet 4C	MoD(PE) A&AEE Boscombe Down
	XS241	WS58 Wessex HU5	RAF, Benson for spares
	XS416	BAC Lightning T5	Privately owned, New Waltham
	XS417	BAC Lightning T5	Newark Air Museum, Winthorpe
	XS419	BAC Lightning T5	Privately owned, Rossington
	XS420	BAC Lightning T5	Privately owned, Narborough, Norfolk
	XS422	BAC Lightning T5	Privately owned, Southampton Docks
	XS451	BAC Lightning T5 (8503M) (G-LTNG)	Privately owned, Plymouth
	XS452	BAC Lightning T5 [BT] (G-BPFE)	Privately owned, Bruntingthorpe
	XS456	BAC Lightning T5	Privately owned, Wainfleet
	XS457	BAC Lightning T5 (nose only)	Privately owned, New Waltham
	XS458	BAC Lightning T5 [DY]	Privately owned, Cranfield
	XS459	BAC Lightning T5	Privately owned, Narborough, Norfolk
	XS463	WS Wasp HAS1 (really XT431)	IHM, Weston-super-Mare
	XS463	WS Wasp HAS1 (A2647)	RN Predannack Fire School
	XS479	WS58 Wessex HU5 [XF] (8819M)	JATE, RAF Brize Norton
	XS481	WS58 Wessex HU5	AAC, Dishforth for BDRT
	XS482	WS58 Wessex HU5 [A-D]	DRA Farnborough Apprentice School
	XS483	WS58 Wessex HU5 [T]	RN AES, Lee-on-Solent
	XS484	WS58 Wessex HU5 [821/CU]	RAF Finningley, Fire Section
	XS485	WS58 Wessex HC5C (*Hearts*)	RAF No 84 Sqn, Akrotiri
	XS486	WS58 Wessex HU5 [524]	RN Recruiting Team, Lee-on-Solent
	XS488	WS58 Wessex HU5 [XK] (9056M)	RAF No 1 SoTT, Halton
	XS489	WS58 Wessex HU5 [R]	RAF Odiham, instructional use
	XS491	WS58 Wessex HU5 [XM]	RAF Stafford
	XS492	WS58 Wessex HU5 [623]	RN, stored
	XS493	WS58 Wessex HU5	RN, stored Fleetlands
	XS496	WS58 Wessex HU5 [625/PO]	RN AES, Lee-on-Solent
	XS498	WS58 Wessex HC5C (*Joker*)	RAF No 84 Sqn, Akrotiri
	XS506	WS58 Wessex HU5 [XE]	RAF Shawbury instructional use
	XS507	WS58 Wessex HU5 [627/PO]	RN AES, Lee-on-Solent
	XS508	WS58 Wessex HU5	RN, stored Lee-on-Solent
	XS509	WS58 Wessex HU5 (A2597)	MoD(PE) ETPS, Boscombe Down
	XS510	WS58 Wessex HU5 [626/PO]	RN AES, Lee-on-Solent
	XS511	WS58 Wessex HU5 [M]	RN AES, Lee-on-Solent
	XS513	WS58 Wessex HU5 [419]	RN AES, Lee-on-Solent
	XS514	WS58 Wessex HU5 [L]	RN AES, Lee-on-Solent
	XS515	WS58 Wessex HU5 [N]	RN AES, Lee-on-Solent
	XS516	WS58 Wessex HU5 [Q]	RN AES, Lee-on-Solent
	XS517	WS58 Wessex HC5C (*Diamonds*)	RAF No 84 Sqn, Akrotiri
	XS520	WS58 Wessex HU5 [F]	RN AES, Lee-on-Solent
	XS522	WS58 Wessex HU5 [ZL/VL]	RN AES, Lee-on-Solent
	XS523	WS58 Wessex HU5 [824/CU]	RNAY Fleetlands
	XS527	WS Wasp HAS1	FAA Museum, RNAS Yeovilton
	XS529	WS Wasp HAS1 [461]	RN AES, Lee-on-Solent
	XS535	WS Wasp HAS1 [432]	RAOC, West Moors, Dorset
	XS538	WS Wasp HAS1 [451] (A2725)	RNAS Lee-on-Solent, Fire Section
	XS539	WS Wasp HAS1 [435]	RN, stored Fleetlands

Serial	Type	Owner or Operator	Notes
XS541	WS Wasp HAS1 [602]	RN, stored Fleetlands	
XS545	WS Wasp HAS1 (A2702) [635]	RN AES, Lee-on-Solent	
XS562	WS Wasp HAS1 [605]	RN, stored Fleetlands	
XS567	WS Wasp HAS1 [434/E]	Imperial War Museum, Duxford	
XS568	WS Wasp HAS1 [441]	RNAY Fleetlands Apprentice School	
XS569	WS Wasp HAS1	RNAY Fleetlands Apprentice School	
XS570	WS Wasp HAS1 (A2699) [445/P]	Warship Preservation Trust, Birkenhead	
XS572	WS Wasp HAS1 (8845M) [414]	RAF Stafford Fire Section (No 16 MU)	
XS576	DH Sea Vixen FAW2 [125/E]	Imperial War Museum, Duxford	
XS577	DH Sea Vixen D3	MoD(PE) T&EE Llanbedr	
XS587	DH Sea Vixen FAW(TT)2 (8828M) (G-VIXN)	Privately owned, Charlwood, Surrey	
XS590	DH Sea Vixen FAW2 [131/E]	FAA Museum, RNAS Yeovilton	
XS596	HS Andover C1(PR)	MoD(PE), A&AEE Boscombe Down	
XS597	HS Andover C1	RAF, stored Shawbury	
XS598	HS Andover C1 (fuselage only)	RAF AMS, Brize Norton	
XS603	HS Andover E3	RAF No 115 Sqn, Benson	
XS605	HS Andover E3	RAF No 115 Sqn, Benson	
XS606	HS Andover C1	MoD(PE) ETPS, Boscombe Down	
XS607	HS Andover C1	MoD(PE) DRA Farnborough	
XS610	HS Andover E3	RAF No 115 Sqn, Benson	
XS637	HS Andover C1	RAF, stored Shawbury	
XS639	HS Andover E3A	RAF No 32 Sqn, Northolt	
XS640	HS Andover E3	RAF No 115 Sqn, Benson	
XS641	HS Andover C1(PR)	RAF, stored Shawbury	
XS642	HS Andover C1 [C] (8785M)	RAF Benson Fire Section	
XS643	HS Andover E3A	RAF No 32 Sqn, Northolt	
XS644	HS Andover E3A	RAF No 32 Sqn, Northolt	
XS646	HS Andover C1 (mod)	MoD(PE) DRA Farnborough	
XS650	Slingsby Swallow TX1 (8801M)	*To BGA3823*	
XS674	WS58 Wessex HC2 [R]	RAF No 60 Sqn, Benson	
XS675	WS58 Wessex HC2	RAF No 22 Sqn, B Flt, Leuchars	
XS676	WS58 Wessex HC2 [WJ]	RAF No 2 FTS, Shawbury	
XS677	WS58 Wessex HC2 [WK]	RAF No 2 FTS, Shawbury	
XS679	WS58 Wessex HC2 [WG]	RAF No 2 FTS, Shawbury	
XS695	HS Kestrel FGA1 (A2619) [SAH-6]	RN, stored Culdrose	
XS709	HS Dominie T1 [M]	RAF No 6 FTS, Finningley	
XS710	HS Dominie T1 [O]	RAF No 6 FTS, Finningley	
XS711	HS Dominie T1 [L]	RAF No 6 FTS, Finningley	
XS712	HS Dominie T1 [A]	RAF No 6 FTS, Finningley	
XS713	HS Dominie T1 [C]	RAF No 6 FTS, Finningley	
XS714	HS Dominie T1 [P]	RAF No 6 FTS, Finningley	
XS726	HS Dominie T1 [T]	RAF No 6 FTS, Finningley	
XS727	HS Dominie T1 [D]	RAF No 6 FTS, Finningley	
XS728	HS Dominie T1 [E]	RAF No 6 FTS, Finningley	
XS729	HS Dominie T1 [G]	RAF No 6 FTS, Finningley	
XS730	HS Dominie T1 [H]	RAF, stored St Athan	
XS731	HS Dominie T1 [J]	RAF No 6 FTS, Finningley	
XS732	HS Dominie T1 [B] (Fuselage)	DRA, Fort Halstead, Kent	
XS733	HS Dominie T1 [Q]	RAF No 6 FTS, Finningley	
XS734	HS Dominie T1 [N]	RAF No 6 FTS, Finningley	
XS735	HS Dominie T1 [R]	RAF No 6 FTS, Finningley	
XS736	HS Dominie T1 [S]	RAF No 6 FTs, Finningley	
XS737	HS Dominie T1 [K]	RAF, stored Finningley	
XS738	HS Dominie T1 [U]	RAF No 6 FTS, Finningley	
XS739	HS Dominie T1 [F]	RAF, stored St Athan	
XS743	Beagle Basset CC1	MoD(PE) ETPS, Boscombe Down	
XS770	Beagle Basset CC1 (G-HRHI)	Privately owned, Cranfield	
XS789	HS Andover CC2	RAF No 32 Sqn, Northolt	
XS790	HS Andover CC2	MoD(PE) DRA Farnborough	
XS791	HS Andover CC2	RAF No 32 Sqn, Northolt	
XS792	HS Andover CC2	RAF No 32 Sqn, Northolt	
XS793	HS Andover CC2	RAF, stored, Northolt	
XS794	HS Andover CC2	RAF No 32 Sqn, Northolt	
XS862	WS58 Wessex HAS3 [650]	RN AES Lee-on-Solent	
XS863	WS58 Wessex HAS1	Imperial War Museum, Duxford	
XS865	WS58 Wessex HAS1 (A2694) [529/CU]	RNAS Lee-on-Solent Fire Section	
XS866	WS58 Wessex HAS1 (A2705) [520/CU]	RN SAH, Culdrose	
XS868	WS58 Wessex HAS1 (A2691)	RNAY Fleetlands, on gate	
XS870	WS58 Wessex HAS1 (A2697) [-/PO]	RN AES Lee-on-Solent	

Notes	Serial	Type	Owner or Operator
	XS871	WS58 Wessex HAS1 (8457M) [AI]	RAF Odiham Fire Section
	XS872	WS58 Wessex HAS1 (A2666) [572/CU]	RNAY Fleetlands Apprentice School
	XS873	WS58 Wessex HAS1 (A2686) [525/CU]	RN Predannack Fire School
	XS876	WS58 Wessex HAS1 (A2695) [523]	RN SAH, Culdrose
	XS877	WS58 Wessex HAS1 (A2687) [16/PO]	RN, stored Culdrose
	XS878	WS58 Wessex HAS1 (A2683)	RN, Lee-on-Solent, Fire Section
	XS881	WS58 Wessex HAS1 (A2675) [046/CU]	FAA Museum, stored
	XS885	WS58 Wessex HAS1 (A2668) [12/CU]	RNAS Culdrose, SAH
	XS886	WS58 Wessex HAS1 (A2685) [527/CU]	Sea Scouts, Evesham, Worcs
	XS887	WS58 Wessex HAS1 [403/FI] [403/FI]	Flambards Village Theme Park, Helston
	XS888	WS58 Wessex HAS1 [521]	RNAS Lee-on-Solent, Fire Section
	XS897	BAC Lightning F6	Privately owned, Rossington
	XS898	BAC Lightning F6 [BD]	Privately owned, Cranfield
	XS899	BAC Lightning F6 [BL]	Privately owned, Cranfield
	XS903	BAC Lightning F6 [BA]	Yorkshire Air Museum, Elvington
	XS904	BAC Lightning F6	Privately owned, Bruntingthorpe
	XS919	BAC Lightning F6	Privately owned, Liskeard, Cornwall
	XS922	BAC Lightning F6 [BJ] (8973M)	RAF Wattisham Fire Section
	XS923	BAC Lightning F6 [BE]	Privately owned, Cranfield
	XS925	BAC Lightning F6 [BA] (8961M)	RAF Museum, Hendon
	XS928	BAC Lightning F6	BAe Warton, on display
	XS932	BAC Lightning F6 (cockpit only)	Privately owned, Bruntingthorpe
	XS933	BAC Lightning F6 (cockpit only)	Privately owned, Narborough, Norfolk
	XS935	BAC Lightning F6	Privately owned, Rossington
	XS936	BAC Lightning F6	Privately owned, Liskeard, Cornwall
	XT108	Agusta-Bell Sioux AH1 [U]	Museum of Army Flying, Middle Wallop
	XT131	Agusta-Bell Sioux AH1 [B]	AAC Historic Aircraft Flight, Middle Wallop
	XT133	Agusta-Bell Sioux AH1 (7923M)	Royal Engineers' Museum, Chatham
	XT140	Agusta-Bell Sioux AH1	Air Service Training, Perth
	XT141	Agusta-Bell Sioux AH1 (8509M)	
	XT148	Agusta-Bell Sioux AH1	Privately owned, Panshanger
	XT150	Agusta-Bell Sioux AH1 (7883M) [R]	AAC Netheravon, on display
	XT151	WS Sioux AH1 [W]	Museum of Army Flying store, Middle Wallop
	XT175	WS Sioux AH1 (TAD175)	CSE Oxford for ground instruction
	XT176	WS Sioux AH1 [U]	FAA Museum, RNAS Yeovilton
	XT190	WS Sioux AH1	Museum of Army Flying, Middle Wallop
	XT200	WS Sioux AH1 [F]	Newark Air Museum, Winthorpe
	XT236	WS Sioux AH1 (frame only)	North-East Air Museum, Usworth
	XT242	WS Sioux AH1 (composite)	The Aeroplane Collection, Warmingham, Cheshire
	XT255	WS58 Wessex HAS3 (8751M)	RAF No 14 MU, Carlisle, BDRT
	XT257	WS58 Wessex HAS3 (8719M)	RAF No 1 SoTT, Halton
	XT271	HS Buccaneer S2A	*Scrapped, July 1991*
	XT272	HS Buccaneer S2	DRA Farnborough Fire Section
	XT273	HS Buccaneer S2A	*Scrapped, July 1991*
	XT274	HS Buccaneer S2A (8856M) [E]	*Scrapped, 1992*
	XT277	HS Buccaneer S2A (8853M) [M]	RAF No 2 SoTT, Cosford
	XT279	HS Buccaneer S2B	*Scrapped, July 1991*
	XT280	HS Buccaneer S2B	RAF No 208 Sqn, Lossiemouth
	XT281	HS Buccaneer S2B (8705M) [ET]	RAF Lossiemouth, ground instruction
	XT283	HS Buccaneer S2A	RAF Lossiemouth, Fire Section
	XT284	HS Buccaneer S2A (8855M) [T]	*Scrapped at Abingdon 1992*
	XT286	HS Buccaneer S2A	RAF, stored Shawbury
	XT287	HS Buccaneer S2B	*Scrapped, May 1992*
	XT288	HS Buccaneer S2B (9134M)	RAF Lossiemouth, WLT
	XT415	WS Wasp HAS1 [FIR3]	Airwork Ltd, Bournemouth
	XT420	WS Wasp HAS1 [606]	RN, stored Fleetlands
	XT422	WS Wasp HAS1 [324]	Privately owned, Burgess Hill
	XT426	WS Wasp HAS1 [FIR2]	*To Malaysia, Aug 1992*
	XT427	WS Wasp HAS1 [606]	Flambards Village Theme Park, Helston
	XT429	WS Wasp HAS1 [445/PLY]	*To Malaysia, Aug 1992*

Serial	Type	Owner or Operator	Notes
XT430	WS Wasp HAS1 [444]	Defence School, Winterbourne Gunner	
XT434	WS Wasp HAS1 [455]	RNAY Fleetlands Apprentice School	
XT437	WS Wasp HAS1 [423]	RN AES, Lee-on-Solent	
XT439	WS Wasp HAS1 [605]	Cranfield Institute of Technology	
XT443	WS Wasp HAS1 [422/AU]		
XT449	WS58 Wessex HU5 [C]	RN AES, Lee-on-Solent	
XT450	WS58 Wessex HU5 [V]	RN Predannack Fire School	
XT451	WS58 Wessex HU5 [XN]	RAF Shawbury, instructional use	
XT453	WS58 Wessex HU5 [A]	RN AES, Lee-on-Solent	
XT455	WS58 Wessex HU5 [U]	RN AES, Lee-on-Solent	
XT456	WS58 Wessex HU5 [XZ] (8941M)	RAF Aldergrove, BDRT	
XT458	WS58 Wessex HU5 [622]	RN AES, Lee-on-Solent	
XT459	WS58 Wessex HU5 [D]	Privately owned, Faygate	
XT460	WS58 Wessex HU5 [K]	RN AES, Lee-on-Solent	
XT463	WS58 Wessex HC5C	RAF No 84 Sqn, Akrotiri	
XT466	WS58 Wessex HU5 [XV] (8921M)	RAF No 2 SoTT, Cosford	
XT468	WS58 Wessex HU5 [628/PO]	RN AES, Lee-on-Solent	
XT469	WS58 Wessex HU5 (8920M)	RAF Stafford ground instruction	
XT470	WS58 Wessex HU5 [A]	AAC, Netheravon, Fire Section	
XT471	WS58 Wessex HU5	AAC, Dishforth for BDRT	
XT472	WS58 Wessex HU5 [XC]	IHM, Weston-super-Mare	
XT475	WS58 Wessex HU5 [624] (9108M)	FSCTE, RAF Manston	
XT479	WS58 Wessex HC5C (*Spades*)	RAF No 84 Sqn, Akrotiri	
XT480	WS58 Wessex HU5 [XQ]	RN Fleetlands	
XT481	WS58 Wessex HU5 [XF]	RNAS, Culdrose, Fire Section	
XT482	WS58 Wessex HU5 [ZM/VL]	RN AES, Lee-on-Solent	
XT484	WS58 Wessex HU5 [H]	RN AES, Lee-on-Solent	
XT485	WS58 Wessex HU5 [621]	RN AES, Lee-on-Solent	
XT486	WS58 Wessex HU5 [XR] (8919M)	RAF JATE, Brize Norton, preserved	
XT487	WS58 Wessex HU5 (A2723) [815] [815/LS]	RNAS Lee-on-Solent Fire Section	
XT575	Vickers Viscount 837 (OE-LAG)	MoD(PE) DRA Bedford	
XT595	McD Phantom FG1 (fuselage) (8550M)	RAF Wattisham, BDRT	
XT595	McD Phantom FG1 (nose only) (8851M)	RAF Exhibition Flight, St Athan	
XT596	McD Phantom FG1	FAA Museum, RNAS Yeovilton	
XT597	McD Phantom FG1	MoD(PE) A&AEE Boscombe Down	
XT601	WS58 Wessex HC2	RAF No 22 Sqn, St Mawgan	
XT602	WS58 Wessex HC2	RAF No 22 Sqn, St Mawgan	
XT603	WS58 Wessex HC2 [WF]	RAF No 2 FTS, Shawbury	
XT604	WS58 Wessex HC2	RAF SARTU, Valley	
XT605	WS58 Wessex HC2 [E]	RAF No 28 Sqn, Sek Kong	
XT606	WS58 Wessex HC2 [WL]	RAF No 2 FTS, Shawbury	
XT607	WS58 Wessex HC2 [P]	RAF No 72 Sqn, Aldergrove	
XT614	WS Scout AH1 [C]	AAC No 660 Sqn, Sek Kong	
XT616	WS Scout AH1 (fuselage)	AAC, stored RNAW Almondbank, Perth	
XT617	WS Scout AH1	AAC, stored RNAW Almondbank, Perth	
XT620	WS Scout AH1	AAC Aldergrove, BDRT	
XT621	WS Scout AH1	Royal Military College of Science, Shrivenham	
XT623	WS Scout AH1	Army Apprentice College, Arborfield	
XT624	WS Scout AH1 [D]	AAC No 660 Sqn, Sek Kong	
XT626	WS Scout AH1 [Q]	AAC No 666 (TA) Sqn, Netheravon	
XT628	WS Scout AH1 [E]	AAC No 660 Sqn, Sek Kong	
XT630	WS Scout AH1 [G]	AAC No 660 Sqn, Sek Kong	
XT631	WS Scout AH1 [D]	MoD(PE) A&AEE Boscombe Down	
XT632	WS Scout AH1	AAC No 658 Sqn, Netheravon	
XT633	WS Scout AH1	Army Apprentice College, Arborfield	
XT634	WS Scout AH1 [T]	AAC No 666(TA) Sqn, Netheravon	
XT636	WS Scout AH1 [F]	AAC No 660 Sqn, Sek Kong	
XT637	WS Scout AH1 (fuselage)	RN, Yeovilton, Fire Section	
XT638	WS Scout AH1 [N]	AAC Fleetlands	
XT639	WS Scout AH1 [Y] (fuselage)	AAC, stored RNAW Almondbank, Perth	
XT640	WS Scout AH1	AAC SAE, Middle Wallop	
XT642	WS Scout AH1 (fuselage)	AAC, stored RNAW Almondbank, Perth	
XT643	WS Scout AH1 [Z]	Army, No 39 Engineer Regt, Waterbeach	
XT644	WS Scout AH1 [Y]	AAC No 666(TA) Sqn, Netheravon	
XT645	WS Scout AH1 (fuselage)	AAC, stored RNAW Almondbank, Perth	
XT646	WS Scout AH1 [Z]	AAC No 666(TA) Sqn, Netheravon	
XT648	WS Scout AH1	AAC, Netheravon, Fire Section	
XT649	WS Scout AH1	AAC, No 658 Sqn, Netheravon	
XT661	Vickers Viscount (9G-AAV)	MoD(PE), stored Bedford	

Notes	Serial	Type	Owner or Operator
	XT667	WS58 Wessex HC2 [F]	RAF No 28 Sqn, Sek Kong
	XT668	WS58 Wessex HC2 [S]	RAF No 72 Sqn, Aldergrove
	XT669	WS58 Wessex HC2 (8894M) [T]	RAF Aldergrove Fire Section
	XT670	WS58 Wessex HC2	RAF No 22 Sqn, E Flt, Coltishall
	XT671	WS58 Wessex HC2 [D]	RAF No 60 Sqn, Benson
	XT672	WS58 Wessex HC2 [WE]	RAF No 2 FTS, Shawbury
	XT673	WS58 Wessex HC2 [G]	RAF No 28 Sqn, Sek Kong
	XT675	WS58 Wessex HC2 [C]	RAF No 28 Sqn, Sek Kong
	XT676	WS58 Wessex HC2 [I]	RAF No 60 Sqn, Benson
	XT677	WS58 Wessex HC2 (8016M)	RAF Brize Norton Fire Section
	XT678	WS58 Wessex HC2 [H]	RAF No 28 Sqn, Sek Kong
	XT680	WS58 Wessex HC2	RAF SARTU, Valley
	XT681	WS58 Wessex HC2 [U]	RAF No 72 Sqn, Aldergrove
	XT752	Fairey Gannet T5 [-/LM] (G-APYO/WN365)	RNAS, stored Lee-on-Solent
	XT755	WS58 Wessex HU5 [V] (9053M)	RAF No 1 SoTT, Halton
	XT756	WS58 Wessex HU5 [ZJ]	RN Lee-on-Solent Fire Section
	XT759	WS58 Wessex HU5 [XY]	RN Fleetlands
	XT760	WS58 Wessex HU5 [418]	RN Engineering Training School, Culdrose
	XT761	WS58 Wessex HU5	RN AES, Lee-on-Solent
	XT762	WS58 Wessex HU5	RNAS Culdrose, SAH
	XT765	WS58 Wessex HU5 [J]	RN AES, Lee-on-Solent
	XT766	WS58 Wessex HU5 [822/CU] (9054M)	RAF No 1 SoTT, Halton
	XT768	WS58 Wessex HU5	Stoney Stanton, ground instruction
	XT769	WS58 Wessex HU5 [823/CU]	RN stored Lee-on-Solent
	XT770	WS58 Wessex HU5 (9055M) [P/VL]	RAF No 1 SoTT, Halton
	XT771	WS58 Wessex HU5 [620/PO]	RN AES, Lee-on-Solent
	XT772	WS58 Wessex HU5 (8805M)	RAF Valley, ground instruction
	XT773	WS58 Wessex HU5 (9123M)	RAF
	XT778	WS Wasp HAS1	RN, AES, Lee-on-Solent
	XT780	WS Wasp HAS1 [636]	RNAY Fleetlands Apprentice School
	XT783	WS Wasp HAS1 [470]	*To Malaysia, Aug 1992*
	XT785	WS Wasp HAS1 [FIR1]	*To Malaysia, Aug 1992*
	XT788	WS Wasp HAS1 [442] (G-BMIR)	Privately owned, Charlwood, Surrey
	XT790	WS Wasp HAS1 (608)	*To Malaysia, Aug 1992*
	XT791	WS Wasp HAS1 [433]	*To Malaysia, Aug 1992*
	XT793	WS Wasp HAS1 [456]	RN, stored Fleetlands
	XT795	WS Wasp HAS1 [476/LE]	RN, stored Fleetlands
	XT803	WS Sioux AH1 [Y]	Privately owned, Panshanger
	XT827	WS Sioux AH1 [D]	AAC HF (spares), Middle Wallop
	XT852	McD Phantom FGR2	MoD(PE) DRA West Freugh Fire Section
	XT853	McD Phantom FGR2 (9071M)	RAF Scampton Fire Section
	XT857	McD Phantom FG1 [MP] (8913M)	RAF Leuchars ground instruction
	XT858	McD Phantom FG1	MoD(PE) Aston Down
	XT859	McD Phantom FG1 [BK] (8999M)	*Scrapped at Leuchars April 1992*
	XT864	McD Phantom FG1 [BJ] (8998M)	RAF Leuchars on display
	XT867	McD Phantom FG1 [BH] (9064M)	RAF Leuchars BDRT
	XT870	McD Phantom FG1 [BS]	*Scrapped at Leuchars April 1992*
	XT873	McD Phantom FG1 [BA]	*Scrapped at Leuchars April 1992*
	XT874	McD Phantom FG1 [BE] (9068M)	RAF Wattisham, Fire Section
	XT891	McD Phantom FGR2 [S]	RAF Coningsby on display
	XT892	McD Phantom FGR2 [J]	RAF Wattisham
	XT894	McD Phantom FGR2 [Y]	RAF Wattisham, Fire Section
	XT895	McD Phantom FGR2 [Q]	RAF Valley for ground instruction
	XT896	McD Phantom FGR2 [V]	RAF, stored Shawbury
	XT897	McD Phantom FGR2 [N]	RAF, stored Shawbury
	XT898	McD Phantom FGR2 [CE]	RAF St Athan Fire Section
	XT899	McD Phantom FGR2	*To Kbely, Czechoslovakia, 16 Jan 1992*
	XT900	McD Phantom FGR2 (9099M) [CO]	RAF Honington, BDRT
	XT902	McD Phantom FGR2 [K]	*Scrapped at Wattisham, March 1992*
	XT903	McD Phantom FGR2 [X]	RAF Wattisham
	XT905	McD Phantom FGR2 [P]	RAF Coningsby, instructional use
	XT906	McD Phantom FGR2 [P]	*Scrapped at Wattisham, 1992*
	XT907	McD Phantom FGR2 [W]	RAF Wattisham
	XT909	McD Phantom FGR2 [A]	*Scrapped at Wattisham, September 1992*
	XT910	McD Phantom FGR2 [O]	RAF, stored Shawbury
	XT911	McD Phantom FGR2 [T]	RAF St Athan Fire Section
	XT914	McD Phantom FGR2 [Z]	RAF Leeming
	XV101	BAC VC10 C1K	MoD(PE) A&AEE, Boscombe Down
	XV102	BAC VC10 C1	RAF No 10 Sqn, Brize Norton

Serial	Type	Owner or Operator	Notes
XV103	BAC VC10 C1K	MoD(PE) FR Aviation, Bournemouth	
XV104	BAC VC10 C1	RAF No 10 Sqn, Brize Norton	
XV105	BAC VC10 C1	RAF No 10 Sqn, Brize Norton	
XV106	BAC VC10 C1	RAF No 10 Sqn, Brize Norton	
XV107	BAC VC10 C1	RAF No 10 Sqn, Brize Norton	
XV108	BAC VC10 C1	RAF No 10 Sqn, Brize Norton	
XV109	BAC VC10 C1	RAF No 10 Sqn, Brize Norton	
XV118	WS Scout AH1	RAF Air Movements School, Brize Norton	
XV119	WS Scout AH1 [T]	AAC, Netheravon, BDRT	
XV121	WS Scout AH1	AAC, Fleetlands	
XV122	WS Scout AH1 [A]	AAC, stored RNAW Almondbank, Perth	
XV123	WS Scout AH1	Army Recruiting Team, Middle Wallop	
XV124	WS Scout AH1	AAC, SAE, Middle Wallop	
XV126	WS Scout AH1 [X]	AAC No 666 (TA) Sqn, Netheravon	
XV127	WS Scout AH1	National Army Museum, Chelsea	
XV128	WS Scout AH1	AAC No 667 Sqn, Middle Wallop	
XV129	WS Scout AH1 [V]	AAC No 666 (TA) Sqn, Netheravon	
XV130	WS Scout AH1 [R]	AAC No 666 (TA) Sqn, Netheravon	
XV131	WS Scout AH1 [Y]	AAC, Middle Wallop	
XV134	WS Scout AH1 [P]	AAC No 666 (TA) Sqn, Netheravon	
XV135	WS Scout AH1	AAC	
XV136	WS Scout AH1 [X]	AAC, stored RNAW Almondbank, Perth	
XV137	WS Scout AH1	AAC No 658 Sqn, Netheravon	
XV138	WS Scout AH1	AAC, stored RNAW Almondbank, Perth	
XV139	WS Scout AH1	Army Apprentice College, Arborfield	
XV140	WS Scout AH1 [K]	AAC No 666 (TA) Sqn, Netheravon	
XV141	WS Scout AH1		
XV147	HS Nimrod MR1 (Mod)	MoD(PE) stored DRA Farnborough	
XV148	HS Nimrod MR1 (Mod)	MoD(PE) BAe Woodford, Fatigue Testing	
XV154	HS Buccaneer S2A (8854M)	RAF Lossiemouth, ground instruction	
XV155	HS Buccaneer S2B (8716M)	Lovaux Ltd, Macclesfield	
XV161	HS Buccaneer S2B (9117M)	RAF Lossiemouth, BDRT	
XV163	HS Buccaneer S2A	RAF, stored Shawbury	
XV165	HS Buccaneer S2B	RAF, stored Shawbury	
XV168	HS Buccaneer S2B	RAF, No 12 Sqn, Lossiemouth	
XV176	Lockheed Hercules C3P	RAF Lyneham Transport Wing	
XV177	Lockheed Hercules C3P	RAF Lyneham Transport Wing	
XV178	Lockheed Hercules C1P	RAF Lyneham Transport Wing	
XV179	Lockheed Hercules C1P	RAF Lyneham Transport Wing	
XV181	Lockheed Hercules C1P	RAF Lyneham Transport Wing	
XV182	Lockheed Hercules C1P	MoD(PE) A&AEE Boscombe Down	
XV183	Lockheed Hercules C3P	RAF Lyneham Transport Wing	
XV184	Lockheed Hercules C3P	RAF Lyneham Transport Wing	
XV185	Lockheed Hercules C1P	RAF Lyneham Transport Wing	
XV186	Lockheed Hercules C1P	RAF Lyneham Transport Wing	
XV187	Lockheed Hercules C1P	RAF Lyneham Transport Wing	
XV188	Lockheed Hercules C3P	RAF Lyneham Transport Wing	
XV189	Lockheed Hercules C3P	RAF Lyneham Transport Wing	
XV190	Lockheed Hercules C3P	RAF Lyneham Transport Wing	
XV191	Lockheed Hercules C1P	RAF Lyneham Transport Wing	
XV192	Lockheed Hercules C1K	RAF Lyneham Transport Wing	
XV193	Lockheed Hercules C3P	RAF Lyneham Transport Wing	
XV195	Lockheed Hercules C1P	RAF Lyneham Transport Wing	
XV196	Lockheed Hercules C1P	RAF Lyneham Transport Wing	
XV197	Lockheed Hercules C3P	RAF Lyneham Transport Wing	
XV199	Lockheed Hercules C3P	RAF Lyneham Transport Wing	
XV200	Lockheed Hercules C1P	RAF Lyneham Transport Wing	
XV201	Lockheed Hercules C1K	RAF Lyneham Transport Wing	
XV202	Lockheed Hercules C3P	RAF Lyneham Transport Wing	
XV203	Lockheed Hercules C1K	RAF No 1312 Flt, Mount Pleasant, FI	
XV204	Lockheed Hercules C1K	RAF Lyneham Transport Wing	
XV205	Lockheed Hercules C1P	RAF Lyneham Transport Wing	
XV206	Lockheed Hercules C1P	RAF Lyneham Transport Wing	
XV207	Lockheed Hercules C3P	RAF Lyneham Transport Wing	
XV208	Lockheed Hercules W2	MoD(PE) MRF Farnborough	
XV209	Lockheed Hercules C3P	RAF Lyneham Transport Wing	
XV210	Lockheed Hercules C1P	RAF Lyneham Transport Wing	
XV211	Lockheed Hercules C1P	RAF Lyneham Transport Wing	
XV212	Lockheed Hercules C3P	RAF Lyneham Transport Wing	
XV213	Lockheed Hercules C1K	RAF No 1312 Flt, Mount Pleasant, FI	
XV214	Lockheed Hercules C3P	RAF Lyneham Transport Wing	
XV215	Lockheed Hercules C1P	RAF Lyneham Transport Wing	
XV217	Lockheed Hercules C3P	RAF Lyneham Transport Wing	

Notes	Serial	Type	Owner or Operator
	XV218	Lockheed Hercules C1P	RAF Lyneham Transport Wing
	XV219	Lockheed Hercules C3P	RAF Lyneham Transport Wing
	XV220	Lockheed Hercules C3P	RAF Lyneham Transport Wing
	XV221	Lockheed Hercules C3P	RAF Lyneham Transport Wing
	XV222	Lockheed Hercules C3P	RAF Lyneham Transport Wing
	XV223	Lockheed Hercules C3P	RAF Lyneham Transport Wing
	XV226	HS Nimrod MR2P	RAF Kinloss MR Wing
	XV227	HS Nimrod MR2P	RAF Kinloss MR Wing
	XV228	HS Nimrod MR2P	RAF Kinloss MR Wing
	XV229	HS Nimrod MR2P	RAF Kinloss MR Wing
	XV230	HS Nimrod MR2P	RAF Kinloss MR Wing
	XV231	HS Nimrod MR2P	RAF Kinloss MR Wing
	XV232	HS Nimrod MR2P	RAF Kinloss MR Wing
	XV233	HS Nimrod MR2P	RAF Kinloss MR Wing
	XV234	HS Nimrod MR2P	RAF Kinloss MR Wing
	XV235	HS Nimrod MR2P	RAF Kinloss MR Wing
	XV236	HS Nimrod MR2P	RAF Kinloss MR Wing
	XV237	HS Nimrod MR2P	*Scrapped, RAF St Mawgan, Jan 1992*
	XV239	HS Nimrod MR2P	RAF Kinloss MR Wing
	XV240	HS Nimrod MR2P	RAF Kinloss MR Wing
	XV241	HS Nimrod MR2	RAF Kinloss MR Wing
	XV242	HS Nimrod MR2P	RAF, stored Kinloss
	XV243	HS Nimrod MR2P	RAF Kinloss MR Wing
	XV244	HS Nimrod MR2P	RAF Kinloss MR Wing
	XV245	HS Nimrod MR2P	RAF Kinloss MR Wing
	XV246	HS Nimrod MR2	RAF Kinloss MR Wing
	XV247	HS Nimrod MR2P	RAF, stored Kinloss
	XV248	HS Nimrod MR2P	RAF Kinloss MR Wing
	XV249	HS Nimrod MR2P	RAF, stored Kinloss
	XV250	HS Nimrod MR2P	RAF Kinloss MR Wing
	XV251	HS Nimrod MR2P	RAF Kinloss MR Wing
	XV252	HS Nimrod MR2P	RAF Kinloss MR Wing
	XV253	HS Nimrod MR2P (9118M)	RAF Kinloss, instructional use
	XV254	HS Nimrod MR2P	RAF Kinloss MR Wing
	XV255	HS Nimrod MR2P	RAF Kinloss MR Wing
	XV257	HS Nimrod MR2	*Scrapped at Woodford, March 1992*
	XV258	HS Nimrod MR2P	RAF Kinloss MR Wing
	XV259	BAe Nimrod AEW3	*Scrapped at Abingdon by June 1992*
	XV260	HS Nimrod MR2P	RAF Kinloss MR Wing
	XV262	BAe Nimrod AEW3	*Scrapped by June 1992*
	XV263	BAe Nimrod AEW3P (8967M)	RAF Air Engineer Sqn, Finningley
	XV269	DHC Beaver AL1 (8011M)	AAC SAE, Middle Wallop
	XV271	DHC Beaver AL1	*To N985P, April 1992*
	XV277	HS Harrier GR3	RN ETS, Yeovilton
	XV279	HS Harrier GR1 [44] (8566M)	RAF Wittering WLT
	XV280	HS Harrier GR1 (nose)	RNAS Yeovilton Fire Section
	XV281	HS Harrier GR3	BAe Kingston
	XV290	Lockheed Hercules C3P	RAF Lyneham Transport Wing
	XV291	Lockheed Hercules C1P	RAF Lyneham Transport Wing
	XV292	Lockheed Hercules C1P	RAF Lyneham Transport Wing
	XV293	Lockheed Hercules C1P	RAF Lyneham Transport Wing
	XV294	Lockheed Hercules C3P	RAF Lyneham Transport Wing
	XV295	Lockheed Hercules C1P	RAF Lyneham Transport Wing
	XV296	Lockheed Hercules C1K	RAF No 1312 Flt, Mount Pleasant, FI
	XV297	Lockheed Hercules C1P	RAF Lyneham Transport Wing
	XV298	Lockheed Hercules C1P	RAF Lyneham Transport Wing
	XV299	Lockheed Hercules C3P	RAF Lyneham Transport Wing
	XV300	Lockheed Hercules C1P	RAF Lyneham Transport Wing
	XV301	Lockheed Hercules C3P	RAF Lyneham Transport Wing
	XV302	Lockheed Hercules C3P	RAF Lyneham Transport Wing
	XV303	Lockheed Hercules C3P	RAF Lyneham Transport Wing
	XV304	Lockheed Hercules C3P	RAF Lyneham Transport Wing
	XV305	Lockheed Hercules C3P	RAF Lyneham Transport Wing
	XV306	Lockheed Hercules C1P	RAF Lyneham Transport Wing
	XV307	Lockheed Hercules C3P	RAF Lyneham Transport Wing
	XV328	BAC Lightning T5 [BZ]	Privately owned, Cranfield
	XV332	HS Buccaneer S2B	RAF No 12 Sqn, Lossiemouth
	XV333	HS Buccaneer S2B	RAF No 208 Sqn, Lossiemouth
	XV337	HS Buccaneer S2C (8852M)	*Scrapped by June 1992*
	XV338	HS Buccaneer S2A (nose only)	*Scrapped during June 1991*
	XV342	HS Buccaneer S2B	*Scrapped, August 1992*
	XV344	HS Buccaneer S2C	MoD(PE) DRA Farnborough
	XV350	HS Buccaneer S2B	BAe Warton

Serial	Type	Owner or Operator	Notes
XV352	HS Buccaneer S2B [U]	RAF No 208 Sqn, Lossiemouth	
XV353	HS Buccaneer S2B	RAF No 12 Sqn, Lossiemouth	
XV355	HS Buccaneer S2B	*Scrapped May 1992*	
XV359	HS Buccaneer S2B	RAF No 208 Sqn, Lossiemouth	
XV361	HS Buccaneer S2B	RAF No 208 Sqn, Lossiemouth	
XV370	Sikorsky SH-3D (G-ATYU)	RN AES, Lee-on-Solent	
XV371	WS61 Sea King HAS1	MoD(PE) DRA Farnborough	
XV372	WS61 Sea King HAS1	Privately owned, Trowbridge, Wilts	
XV393	McD Phantom FGR2 [Q]	RAF Marham, BDRT	
XV398	McD Phantom FGR2 [CI]	RAF Wattisham, Fire Section	
XV399	McD Phantom FGR2 [L]	RAF St Athan, Fire Section	
XV401	McD Phantom FGR2 [I]	RAF Wattisham, stored	
XV402	McD Phantom FGR2 [C]	*Scrapped at St Athan, Oct 1991*	
XV404	McD Phantom FGR2 [I]	RAF Wattisham, stored	
XV406	McD Phantom FGR2 [CK](9098M)	RAF Carlisle, on display	
XV408	McD Phantom FGR2 [Z]	RAF Cranwell	
XV409	McD Phantom FGR2 [H]	*RAF Mount Pleasant, on display*	
XV410	McD Phantom FGR2 [E]	*Scrapped at Wattisham, September 1992*	
XV411	McD Phantom FGR2 (9103M) [L]	FSCTE, RAF Manston	
XV414	McD Phantom FGR2 [R]	*Scrapped at Wattisham*	
XV415	McD Phantom FGR2 [E]	RAF Wattisham, stored	
XV419	McD Phantom FGR2 [AA]	Privately owned, Aston justa Mondrium Cheshire	
XV420	McD Phantom FGR2 [O]	RAF Wattisham, on display	
XV422	McD Phantom FGR2 [T]	Stornaway Airport, on display	
XV423	McD Phantom FGR2 [Y]	RAF Leeming	
XV424	McD Phantom FGR2 [I]	RAF Museum, Hendon	
XV426	McD Phantom FGR2 [P]	RAF Coningsby	
XV430	McD Phantom FGR2	*Scrapped at Wattisham, March 1992*	
XV432	McD Phantom FGR2	*Scrapped at Wattisham, March 1992*	
XV433	McD Phantom FGR2 [E]	RAF, stored Shawbury	
XV435	McD Phantom FGR2 [R]	RAF Wattisham, stored	
XV438	McD Phantom FGR2 [T]	*Scrapped at Wattisham, September 1992*	
XV439	McD Phantom FGR2	*Scrapped at Wattisham, March 1992*	
XV460	McD Phantom FGR2 [R]	RAF Wattisham, on display	
XV461	McD Phantom FGR2 [C]	*Scrapped at Mount Pleasant, August 1992*	
XV464	McD Phantom FGR2 [AN]	*Scrapped at Wattisham, March 1992*	
XV465	McD Phantom FGR2 [S]	RAF Leeming	
XV466	McD Phantom FGR2 [D]	*Scrapped at Mount Pleasant, August 1992*	
XV467	McD Phantom FGR2 [F]	Benbecula Airport, on display	
XV468	McD Phantom FGR2 [H]	RAF Woodvale on display	
XV469	McD Phantom FGR2	RAF, stored Shawbury	
XV470	McD Phantom FGR2 [D]	*RAF Akrotiri, Cyprus on display*	
XV472	McD Phantom FGR2 [F]	*Scrapped at Mount Pleasant, August 1992*	
XV473	McD Phantom FGR2 [N]	RAF Waddington, BDRT	
XV474	McD Phantom FGR2 [T	Imperial War Museum, Duxford	
XV481	McD Phantom FGR2 [H]	*RAF Bruggen for BDRT, Feb 1992*	
XV482	McD Phantom FGR2 (9107M) [T]	RAF Leuchars, Fire Section	
XV486	McD Phantom FGR2 [N]	RAF St Athan, Fire Section	
XV487	McD Phantom FGR2 [G]	RAF, stored Shawbury	
XV488	McD Phantom FGR2 [O]	RAF Wattisham, Fire Section	
XV489	McD Phantom FGR2 [S]	RAF St Athan, Fire Section	
XV490	McD Phantom FGR2 [R]	RAF Wattisham Fire Section	
XV492	McD Phantom FGR2 [W]	*Scrapped Wattisham March 1992*	
XV494	McD Phantom FGR2 [O]	RAF Wattisham, Fire Section	
XV495	McD Phantom FGR2	RAF St Athan Fire Section	
XV496	McD Phantom FGR2 [A]	RAF Wattisham, Fire Section	
XV497	McD Phantom FGR2 [W]	RAF Coningsby	
XV498	McD Phantom FGR2 [U]	*Scrapped at Wattisham, March 1992*	
XV499	McD Phantom FGR2 [I]	RAF Leeming	
XV500	McD Phantom FGR2	RAF St Athan, on display	
XV567	McD Phantom FG1 [AI]	*Scrapped at Leuchars, April 1992*	
XV568	McD Phantom FG1 [AT]	*Scrapped at Leuchars April 1992*	
XV570	McD Phantom FG1 [BN] (9069M)	RAF Wattisham Fire Section	
XV571	McD Phantom FG1 [A]	*Scrapped at Leuchars March 1992*	
XV572	McD Phantom FG1	*Scrapped at Leuchars April 1992*	
XV573	McD Phantom FG1	*Scrapped at Leuchars 1992*	
XV577	McD Phantom FG1 [AM] (9065M)	RAF Leuchars BDRT	
XV579	McD Phantom FG1 [AR]	*Scrapped at Leuchars April 1992*	
XV581	McD Phantom FG1 [AE] (9070M)	RAF Buchan, on display	
XV582	McD Phantom FG1 [M] (9066M)	RAF Leuchars on display	
XV585	McD Phantom FG1 [AP]	RAF	
XV586	McD Phantom FG1 [AJ] (9067M)	RAF Leuchars BDRT	

Notes	Serial	Type	Owner or Operator
	XV587	McD Phantom FG1 (9088M)	*Scrapped at Wattisham 1992*
	XV590	McD Phantom FG1 [AX]	*Scrapped at Leuchars April 1992*
	XV615	BHC SR.N6 Winchester 2	*Reclassified as Ship*
	XV623	WS Wasp HAS1 [601] (A2724)	RN Portland, BDRT
	XV625	WS Wasp HAS1 [471]	RNEC Manadon, for instruction
	XV629	WS Wasp HAS1	AAC Middle Wallop, BDRT
	XV631	WS Wasp HAS1	DRA Farnborough, ground instruction
	XV638	WS Wasp HAS1 (8826M) [430/A]	RAF High Wycombe
	XV639	WS Wasp HAS1 [612]	RN, stored Fleetlands
	XV642	WS61 Sea King HAS2A	RN AES, Lee-on-Solent
	XV643	WS61 Sea King HAS6 [265]	RNAY Fleetlands
	XV644	WS61 Sea King HAS1 (A2664) [664]	RN Lee-on-Solent, Fire Section
	XV647	WS61 Sea King HAR5 [820/CU]	RN No 771 Sqn, Culdrose
	XV648	WS61 Sea King HAS6 [582]	RN No 706 Sqn, Culdrose
	XV649	WS61 Sea King AEW2A [184/R]	RN No 849 Sqn, Culdrose
	XV650	WS61 Sea King AEW2A [186/N]	RN No 849 Sqn, Culdrose
	XV651	WS61 Sea King HAS6 [591]	RN No 706 Sqn, Culdrose
	XV653	WS61 Sea King HAS6	RN No 810 Sqn, Culdrose
	XV654	WS61 Sea King HAS6 [705/PW]	RN No 819 Sqn, Prestwick
	XV655	WS61 Sea King HAS6 [272/N]	RNAY Fleetlands
	XV656	WS61 Sea King AEW2A [187/N]	RN No 849 Sqn, Culdrose
	XV657	WS61 Sea King HAS6 [132]	RNAY Fleetlands
	XV659	WS61 Sea King HAS6 [266]	RNAY Fleetlands
	XV660	WS61 Sea King HAS6 [503]	RN No 810 Sqn, Culdrose
	XV661	WS61 Sea King HAS6 [135]	RN AMG Culdrose
	XV663	WS61 Sea King HAS6 [015]	RNAY Fleetlands
	XV664	WS61 Sea King AEW2A [185/N]	RN No 849 Sqn, Culdrose
	XV665	WS61 Sea King HAS6 [017]	RN No 820 Sqn, Culdrose
	XV666	WS61 Sea King HAR5 [823/CU]	RN No 771 Sqn, Culdrose
	XV669	WS61 Sea King HAS1 [10] (A2659)	RNAS Culdrose, Engineering Training School
	XV670	WS61 Sea King HAS6 [592]	RN No 706 Sqn, Culdrose
	XV671	WS61 Sea King AEW2A [181/CU]	RN No 849 Sqn, Culdrose
	XV672	WS61 Sea King AEW2A [183/R]	RN No 849 Sqn, Culdrose
	XV673	WS61 Sea King HAS6 [588]	RN No 706 Sqn, Culdrose
	XV674	WS61 Sea King HAS6 [135/CL]	RN No 826 Sqn, Culdrose
	XV675	WS61 Sea King HAS6 [594]	RN No 706 Sqn, Culdrose
	XV676	WS61 Sea King HAS6 [707/PW]	RN No 819 Sqn, Prestwick
	XV677	WS61 Sea King HAS6 [501]	RN AMG Culdrose
	XV696	WS61 Sea King HAS6 [132]	RN No 826 Sqn, Culdrose
	XV697	WS61 Sea King AEW2A [157/N]	RN AMG Culdrose
	XV699	WS61 Sea King HAS5 [134]	RN stored, Fleetlands
	XV700	WS61 Sea King HAS6 [508/04]	RN No 810 Sqn, Culdrose
	XV701	WS61 Sea King HAS6 [010]	RN No 820 Sqn, Culdrose
	XV703	WS61 Sea King HAS6	RNAY Fleetlands
	XV704	WS61 Sea King AEW2A [184/R]	RN No 849 Sqn, Culdrose
	XV705	WS61 Sea King HAR5 [821/CU]	RN No 771 Sqn, Culdrose
	XV706	WS61 Sea King HAS6 [505/CU]	RN No 810 Sqn, Culdrose
	XV707	WS61 Sea King AEW2A [182/R]	RN No 849 Sqn, Culdrose
	XV708	WS61 Sea King HAS6 [708/PW]	RN AMG Culdrose
	XV709	WS61 Sea King HAS6 [585]	RN No 706 Sqn, Culdrose
	XV710	WS61 Sea King HAS6 [270/N]	RN No 814 Sqn, Culdrose
	XV711	WS61 Sea King HAS6 [273/N]	RNAY Fleetlands
	XV712	WS61 Sea King HAS6 [502]	RN 810 Sqn, Culdrose
	XV713	WS61 Sea King HAS6 [508]	RNAY Fleetlands
	XV714	WS61 Sea King AEW2A [180/N]	RN No 849 Sqn, Culdrose
	XV720	WS58 Wessex HC2	RAF No 22 Sqn, A Flt, Chivenor
	XV721	WS58 Wessex HC2 [H]	RAF No 72 Sqn, Aldergrove
	XV722	WS58 Wessex HC2 [WH]	RAF No 2 FTS, Shawbury
	XV723	WS58 Wessex HC2 [Q]	RAF No 72 Sqn, Aldergrove
	XV724	WS58 Wessex HC2	RAF No 22 Sqn, E Flt, Coltishall
	XV725	WS58 Wessex HC2 [C]	RAF No 60 Sqn, Benson
	XV726	WS58 Wessex HC2 [J]	RAF No 60 Sqn, Benson
	XV728	WS58 Wessex HC2 [A]	RAF No 72 Sqn, Aldergrove
	XV729	WS58 Wessex HC2	RAF No 22 Sqn, C Flt, Valley
	XV730	WS58 Wessex HC2	RAF No 22 Sqn, C Flt, Valley
	XV731	WS58 Wessex HC2 [Y]	RAF No 72 Sqn, Aldergrove
	XV732	WS58 Wessex HCC4	RAF Queen's Flight, Benson
	XV733	WS58 Wessex HCC4	RAF Queen's Flight, Benson
	XV738	HS Harrier GR3 [B] (9074M)	RAF No 1 SoTT, Halton
	XV740	HS Harrier GR3 (8989M)	*Scrapped at Abingdon, June 1992*
	XV741	HS Harrier GR3	RN SAH, Culdrose

Serial	Type	Owner or Operator	Notes
XV744	HS Harrier GR3 [3K]	RAF, stored St Athan	
XV747	HS Harrier GR3 (8979M)(fuselage)	Privately owned, Bruntingthorpe	
XV748	HS Harrier GR3	MoD(PE) DRA Bedford	
XV751	HS Harrier GR3 [U]	RN AES, Lee-on-Solent	
XV752	HS Harrier GR3 [S] (9078M)	RAF No 2 SoTT, Cosford	
XV753	HS Harrier GR3 [3F] (9075M)	RAF No 1 SoTT, Halton	
XV755	HS Harrier GR3 [M]	RN ETS Yeovilton	
XV759	HS Harrier GR3 [O]	RAF St Athan Fire Section	
XV760	HS Harrier GR3 [K]	RNAS, Yeovilton, on display	
XV778	HS Harrier GR3 (9001M)	RAF Valley Fire Section	
XV779	HS Harrier GR3 (8931M) [01/A]	RAF Wittering on display	
XV783	HS Harrier GR3 [N]	RN SAH, Culdrose	
XV784	HS Harrier GR3 (8909M) (nose only)	A&AEE, Boscombe Down	
XV786	HS Harrier GR3 [S]	RN SAH, Culdrose	
XV806	HS Harrier GR3 [E]	RN SAH, Culdrose	
XV808	HS Harrier GR3 [3J] (9076M)	RAF No 1 SoTT, Halton	
XV810	HS Harrier GR3 [K] (9038M)	RAF St Athan	
XV814	DH Comet 4 (G-APDF)	MoD(PE) DRA Farnborough	
XV859	BHC SR.N6 Winchester 6	*Reclassified as Ship*	
XV863	HS Buccaneer S2B [S] (9115M)	RAF Lossiemouth, on display	
XV864	HS Buccaneer S2B	RAF No 12 Sqn, Lossiemouth	
XV865	HS Buccaneer S2B	RAF No 208 Sqn, Lossiemouth	
XV867	HS Buccaneer S2B	RAF No 208 Sqn, Lossiemouth	
XV869	HS Buccaneer S2B	RAF, stored Shawbury	
XW175	HS Harrier T4A	MoD(PE) DRA Bedford	
XW198	WS Puma HC1 [DL]	RAF No 230 Sqn, Aldergrove	
XW199	WS Puma HC1 [DU]	RAF No 33 Sqn, Odiham	
XW200	WS Puma HC1 [FA]	RAF No 240 OCU, Odiham	
XW201	WS Puma HC1 [FB]	RAF No 240 OCU, Odiham	
XW202	WS Puma HC1 [CE]	RAF No 33 Sqn, Odiham	
XW204	WS Puma HC1 [CA]	RAF No 33 Sqn, Odiham	
XW206	WS Puma HC1 [CC]	RAF No 33 Sqn, Odiham	
XW207	WS Puma HC1 [CD]	RAF No 33 Sqn, Odiham	
XW208	WS Puma HC1 [C]	RAF No 230 Sqn, Aldergrove	
XW209	WS Puma HC1 [CF]	RAF No 33 Sqn, Odiham	
XW210	WS Puma HC1 [CG]	RAF No 33 Sqn, Odiham	
XW211	WS Puma HC1 [CH]	Westlands, Yeovil	
XW212	WS Puma HC1 [CI]	RAF No 1563 Flt, Belize	
XW213	WS Puma HC1 [CJ]	RAF No 33 Sqn, Odiham	
XW214	WS Puma HC1 [CK]	RAF No 230 Sqn, Aldergrove	
XW215	WS Puma HC1 [R]	Westland, Yeovil (on rebuild)	
XW216	WS Puma HC1	RAF No 230 Sqn, Aldergrove	
XW217	WS Puma HC1 [CS]	RAF No 33 Sqn, Odiham	
XW218	WS Puma HC1 [BW]	RAF No 18 Sqn, Laarbruch	
XW219	WS Puma HC1 [DC]	RAF No 18 Sqn, Laarbruch	
XW220	WS Puma HC1 [CZ]	RAF No 33 Sqn, Odiham	
XW221	WS Puma HC1 [CM]	RAF No 33 Sqn, Odiham	
XW222	WS Puma HC1 [BX]	RAF No 18 Sqn, Laarbruch	
XW223	WS Puma HC1 [CB]	RAF No 33 Sqn, Odiham	
XW224	WS Puma HC1 [DH]	RAF No 33 Sqn, Odiham	
XW225	WS Puma HC1 [FE]	RAF No 240 OCU, Odiham	
XW226	WS Puma HC1 [DK]	RAF No 18 Sqn, Laarbruch	
XW227	WS Puma HC1 [DN]	RAF No 33 Sqn, Odiham	
XW229	WS Puma HC1 [DB]	RAF No 33 Sqn, Odiham	
XW231	WS Puma HC1 [FD]	RAF No 240 OCU Odiham	
XW232	WS Puma HC1 [DJ]	RAF No 18 Sqn, Laarbruch	
XW233	WS Puma HC1 [CN]	RAF No 230 Sqn, Aldergrove	
XW234	WS Puma HC1 [CO]	RAF No 33 Sqn, Odiham	
XW235	WS Puma HC1 [CP]	RAF No 230 Sqn, Aldergrove	
XW236	WS Puma HC1 [BZ]	RAF No 18 Sqn, Laarbruch	
XW237	WS Puma HC1	RAF No 230 Sqn, Aldergrove	
XW241	Sud SA330E Puma (F-ZJUX)	MoD(PE), stored DRA Farnborough	
XW249	Cushioncraft CC7	Flambards Village Theme Park, Helston	
XW255	BHC BH-7 Wellington	*Reclassified as Ship*	
XW264	HS Harrier T2 (forward fuselage)	CARG store, RAF Innsworth	
XW265	HS Harrier T4A [W]	RAF HOCU/No 20(R) Sqn, Wittering	
XW266	HS Harrier T4N [719]	RN No 899 Sqn, Yeovilton	
XW267	HS Harrier T4 [SA]	RAF SAOEU, Boscombe Down	
XW268	HS Harrier T4N [720]	RN No 899 Sqn, Yeovilton	
XW269	HS Harrier T4 [BD]	RAF SAOEU, Boscombe Down	

Notes	Serial	Type	Owner or Operator
	XW270	HS Harrier T4 [T]	RAF HOCU/No 20(R) Sqn, Wittering
	XW271	HS Harrier T4 [X]	RAF HOCU/No 20(R) Sqn, Wittering
	XW272	HS Harrier T4 (8783M) (nose only)	BAe, Kingston-upon-Thames
	XW276	Aerospatiale SA341 (F-ZWRI)	North-East Aircraft Museum, Usworth
	XW280	WS Scout AH1	AAC No 660 Sqn, Brunei
	XW281	WS Scout AH1 [U]	AAC No 666(TA) Sqn, Netheravon
	XW282	WS Scout AH1 [W]	AAC No 666(TA) Sqn, Netheravon
	XW283	WS Scout AH1	AAC No 658 Sqn, Netheravon
	XW284	WS Scout AH1 [A] (fuselage only)	AAC, stored Perth
	XW287	BAC Jet Provost T5 [P]	RAF No 6 FTS, Finningley
	XW289	BAC Jet Provost T5A [73]	RAF No 1 FTS, Linton-on-Ouse
	XW290	BAC Jet Provost T5A [41]	RAF, stored Shawbury
	XW291	BAC Jet Provost T5 [N]	RAF No 6 FTS, Finningley
	XW292	BAC Jet Provost T5A (9128M) [32]	RAF No 1 SoTT, Halton
	XW293	BAC Jet Provost T5 [Z]	RAF, stored Finningley
	XW294	BAC Jet Provost T5A (9129M) [45]	RAF No 1 SoTT, Halton
	XW295	BAC Jet Provost T5A [29]	RAF, stored Shawbury
	XW296	BAC Jet Provost T5 [Q]	RAF No 6 FTS, Finningley
	XW298	BAC Jet Provost T5 (9013M)	*Scrapped at Abingdon by June 1992*
	XW299	BAC Jet Provost T5A [60]	RAF No 1 SoTT, Halton
	XW301	BAC Jet Provost T5A [63]	RAF No 1 SoTT, Halton
	XW302	BAC Jet Provost T5 [T]	RAF No 6 FTS, Finningley
	XW303	BAC Jet Provost T5A (9119M)[127]	RAF No 1 SoTT, Halton
	XW304	BAC Jet Provost T5 [X]	RAF No 6 FTS, Finningley
	XW305	BAC Jet Provost T5A [42]	RAF, stored Shawbury
	XW306	BAC Jet Provost T5 [O]	RAF No 6 FTS, Finningley
	XW307	BAC Jet Provost T5 [S]	RAF No 6 FTS, Finningley
	XW309	BAC Jet Provost T5 [V]	RAF No 6 FTS, Finningley
	XW310	BAC Jet Provost T5A [37]	RAF, stored Shawbury
	XW311	BAC Jet Provost T5 [W]	RAF No 6 FTS, Finningley
	XW312	BAC Jet Provost T5A (9109M) [64]	RAF No 1 SoTT, Halton
	XW313	BAC Jet Provost T5A [85]	RAF No 1 FTS, Linton-on-Ouse
	XW315	BAC Jet Provost T5A	Stratford Aircraft Collection, Long Marston
	XW316	BAC Jet Provost T5A [28]	RAF, stored Shawbury
	XW317	BAC Jet Provost T5A [79]	RAF, stored Shawbury
	XW318	BAC Jet Provost T5A [78]	RAF No 1 FTS, Linton-on-Ouse
	XW319	BAC Jet Provost T5A [76]	RAF No 1 FTS, Scampton
	XW320	BAC Jet Provost T5A [71] (9018M)	RAF No 1 SoTT, Halton
	XW321	BAC Jet Provost T5A [62]	RAF No 1 FTS, Linton-on-Ouse
	XW322	BAC Jet Provost T5B [D]	RAF No 6 FTS, Finningley
	XW323	BAC Jet Provost T5A [86]	RAF Museum, Hendon
	XW324	BAC Jet Provost T5 [U]	RAF No 6 FTS, Finningley
	XW325	BAC Jet Provost T5B [E]	RAF No 6 FTS, Finningley
	XW326	BAC Jet Provost T5A [62]	RAF, stored Shawbury
	XW327	BAC Jet Provost T5A (9130M) [62]	RAF No 1 SoTT, Halton
	XW328	BAC Jet Provost T5A [75]	RAF No 1 FTS, Linton-on-Ouse
	XW330	BAC Jet Provost T5A [82]	RAF No 1 FTS, Linton-on-Ouse
	XW332	BAC Jet Provost T5A [34]	RAF, stored Shawbury
	XW333	BAC Jet Provost T5A [79]	RAF No 1 FTS, Linton-on-Ouse
	XW334	BAC Jet Provost T5A [18]	RAF, stored Shawbury
	XW335	BAC Jet Provost T5A [74] (9061M)	RAF No 1 SoTT, Halton
	XW336	BAC Jet Provost T5A [67]	RAF No 1 FTS, Linton-on-Ouse
	XW351	BAC Jet Provost T5A (9062M) [31]	RAF No 1 SoTT, Halton
	XW352	BAC Jet Provost T5 [R]	RAF No 6 FTS, Finningley
	XW353	BAC Jet Provost T5A (9090M) [3]	RAF Cranwell, on display
	XW354	BAC Jet Provost T5A [70]	RAF No 1 FTS, Linton-on-Ouse
	XW355	BAC Jet Provost T5A [20]	RAF, stored Shawbury
	XW357	BAC Jet Provost T5A [5]	RAF, stored Shawbury
	XW358	BAC Jet Provost T5A [59]	RAF CFS, Scampton
	XW359	BAC Jet Provost T5B [65]	RAF No 1 FTS, Linton-on-Ouse
	XW360	BAC Jet Provost T5A [61]	RAF No 1 FTS, Linton-on-Ouse
	XW361	BAC Jet Provost T5A [81]	RAF, No 1 FTS, Linton-on-Ouse
	XW362	BAC Jet Provost T5A [17]	RAF, stored Shawbury
	XW363	BAC Jet Provost T5A [36]	BAe, Training School, Warton
	XW364	BAC Jet Provost T5A [35]	RAF, stored Shawbury
	XW365	BAC Jet Provost T5A (9015M) [73]	RAF No 1 SoTT, Halton
	XW366	BAC Jet Provost T5A [75](9097M)	RAF No 1 SoTT, Halton
	XW367	BAC Jet Provost T5A [64]	RAF No 1 FTS, Linton-on-Ouse
	XW368	BAC Jet Provost T5A [66]	RAF No 1 FTS, Linton-on-Ouse
	XW369	BAC Jet Provost T5A [69]	RAF No 1 FTS, Linton-on-Ouse

Serial	Type	Owner or Operator	Notes
XW370	BAC Jet Provost T5A [72]	RAF No 1 FTS, Linton-on-Ouse	
XW372	BAC Jet Provost T5A [M]	RAF, stored Shawbury	
XW373	BAC Jet Provost T5A [11]	RAF, stored Shawbury	
XW374	BAC Jet Provost T5A [38]	RAF, stored Shawbury	
XW375	BAC Jet Provost T5A [52]	RAF No 1 SoTT, Halton	
XW404	BAC Jet Provost T5A (9049M) [77]	RAF CTTS, St Athan	
XW405	BAC Jet Provost T5A [J]	RAF, stored Shawbury	
XW406	BAC Jet Provost T5A [23]	Privately owned	
XW408	BAC Jet Provost T5A [24]	RAF, stored Shawbury	
XW409	BAC Jet Provost T5A (9047M) [123]	RAF CTTS, St Athan	
XW410	BAC Jet Provost T5A (9125M) [80]	RAF No 1 SoTT, Halton	
XW412	BAC Jet Provost T5A [74]	RAF No 1 FTS, Linton-on-Ouse	
XW413	BAC Jet Provost T5A (9126M) [69]	RAF No 1 SoTT, Halton	
XW415	BAC Jet Provost T5A [80]	RAF No 1 FTS, Linton-on-Ouse	
XW416	BAC Jet Provost T5A [84]	RAF No 1 FTS, Linton-on-Ouse	
XW418	BAC Jet Provost T5A [60]	RAF No 1 FTS, Linton-on-Ouse	
XW419	BAC Jet Provost T5A (9120M)[125]	RAF No 1 SoTT, Halton	
XW420	BAC Jet Provost T5A [83]	RAF No 1 FTS, Linton-on-Ouse	
XW421	BAC Jet Provost T5A (9111M) [60]	RAF No 1 SoTT, Halton	
XW422	BAC Jet Provost T5A [3]	RAF, stored Shawbury	
XW423	BAC Jet Provost T5A [14]	RAF, stored Shawbury	
XW425	BAC Jet Provost T5A [H]	RAF, stored Shawbury	
XW427	BAC Jet Provost T5A (9124M) [67]	RAF No 1 SoTT, Halton	
XW428	BAC Jet Provost T5A	MoD(PE) DRA Farnborough	
XW429	BAC Jet Provost T5B [C]	RAF No 6 FTS, Finningley	
XW430	BAC Jet Provost T5A [77]	RAF No 1 FTS, Linton-on-Ouse	
XW431	BAC Jet Provost T5B [A]	RAF No 6 FTS, Finningley	
XW432	BAC Jet Provost T5A (9127M) [76]	RAF No 1 SoTT, Halton	
XW433	BAC Jet Provost T5A [63]	RAF, stored Shawbury	
XW434	BAC Jet Provost T5A (9091M) [78]	RAF No 1 SoTT, Halton	
XW435	BAC Jet Provost T5A [4]	RAF, stored Shawbury	
XW436	BAC Jet Provost T5A [68]	RAF No 1 SoTT, Halton	
XW437	BAC Jet Provost T5A [71]	RAF No 1 FTS, Linton-on-Ouse	
XW438	BAC Jet Provost T5B [B]	RAF No 6 FTS, Finningley	
XW527	HS Buccaneer S2B	RAF No 12 Sqn Lossiemouth	
XW528	HS Buccaneer S2B (8861M) [C]	RAF Coningsby Fire Section	
XW529	HS Buccaneer S2B	RAF Lossiemouth, Fire Section	
XW530	HS Buccaneer S2B	RAF No 12 Sqn, Lossiemouth	
XW533	HS Buccaneer S2B	*Scrapped 1992*	
XW534	HS Buccaneer S2B	RAF, stored Shawbury	
XW542	HS Buccaneer S2B	RAF No 208 Sqn, Lossiemouth	
XW543	HS Buccaneer S2B	*Scrapped at St Mawgan, Aug 1992*	
XW544	HS Buccaneer S2B (8857M) [Y]	RAF No 2 SoTT, Cosford	
XW546	HS Buccaneer S2B	*Scrapped at Lossiemouth, Aug 1992*	
XW547	HS Buccaneer S2B [R]	RAF, stored Shawbury	
XW549	HS Buccaneer S2B (8860M)	RAF Kinloss, BDRT	
XW550	HS Buccaneer S2B [X]	*Scrapped at St Athan, Oct 1992*	
XW566	SEPECAT Jaguar T2	MoD(PE), stored DRA Farnborough	
XW612	WS Scout AH1 [A]	AAC No 660 Sqn, Sek Kong	
XW613	WS Scout AH1 [B]	AAC No 660 Sqn, Sek Kong	
XW614	WS Scout AH1	AAC Historic Flight, Middle Wallop	
XW616	WS Scout AH1	AAC, No 70 MU Middle Wallop	
XW626	DH Comet 4AEW (G-APDS)	MoD(PE), DRA Bedford derelict	
XW630	HS Harrier GR3	RN AES, Lee-on-Solent	
XW635	Beagle D5/180 (G-AWSW)	Privately owned, Cranwell North	
XW664	HS Nimrod R1P	RAF No 51 Sqn, Wyton	
XW665	HS Nimrod R1P	RAF No 51 Sqn, Wyton	
XW666	HS Nimrod R1P	RAF No 51 Sqn, Wyton	
XW750	HS748 Series 107 (G-ASJT)	MoD(PE) DRA Bedford	
XW764	HS Harrier GR3 (8981M)	RAF Leeming, BDRT	
XW768	HS Harrier GR3 [N] (9072M)	RAF No 1 SoTT, Halton	
XW784	Mitchell-Procter Kittiwake I (G-BBRN)	Privately owned, Haverfordwest	
XW788	HS125 CC1	RAF No 32 Sqn, Northolt	
XW789	HS125 CC1	RAF No 32 Sqn, Northolt	
XW790	HS125 CC1	RAF No 32 Sqn, Northolt	
XW791	HS125 CC1	RAF No 32 Sqn, Northolt	
XW795	WS Scout AH1	AAC, stored Perth	
XW796	WS Scout AH1 [X]	AAC	
XW797	WS Scout AH1 [G]	AAC No 660 Sqn, Sek Kong	
XW798	WS Scout AH1 [H]	AAC No 660 Sqn, Brunei	
XW799	WS Scout AH1	AAC No 658 Sqn, Netheravon	
XW836	WS Lynx	RNAS Lee-on-Solent, Fire Section	

Notes	Serial	Type	Owner or Operator
	XW837	WS Lynx (fuselage)	IHM, Weston-super-Mare
	XW838	WS Lynx [TAD 009]	AAC SAE, Middle Wallop
	XW839	WS Lynx	RNEC Manadon
	XW843	WS Gazelle AH1	AAC SAE, Middle Wallop
	XW844	WS Gazelle AH1	AAC Fleetlands
	XW845	WS Gazelle HT2 [47/CU]	RN No 705 Sqn, Culdrose
	XW846	WS Gazelle AH1 [M]	AAC No 670 Sqn, Middle Wallop
	XW847	WS Gazelle AH1	AAC No 667 Sqn, Middle Wallop
	XW848	WS Gazelle AH1 [D]	AAC No 670 Sqn, Middle Wallop
	XW849	WS Gazelle AH1 [G]	RM 3 CBAS, Yeovilton
	XW851	WS Gazelle AH1	RM Fleetlands
	XW852	WS Gazelle HT3	RAF No 32 Sqn, Northolt
	XW853	WS Gazelle HT2 [53/CU]	RN No 705 Sqn, Culdrose
	XW854	WS Gazelle HT2 [46/CU]	RN No 705 Sqn, Culdrose
	XW855	WS Gazelle HCC4	RAF No 32 Sqn, Northolt
	XW856	WS Gazelle HT2 [49/CU]	RN No 705 Sqn, Culdrose
	XW857	WS Gazelle HT2 [55/CU]	RN No 705 Sqn, Culdrose
	XW858	WS Gazelle HT3 [C]	RAF No 2 FTS, Shawbury
	XW860	WS Gazelle HT2	RNAY Fleetlands
	XW861	WS Gazelle HT2 [52/CU]	RN No 705 Sqn, Culdrose
	XW862	WS Gazelle HT3 [D]	RAF No 2 FTS, Shawbury
	XW863	WS Gazelle HT2 [42/CU]	AAC, SAE, Middle Wallop
	XW864	WS Gazelle HT2 [54/CU]	RN No 705 Sqn, Culdrose
	XW865	WS Gazelle AH1 [C]	AAC No 670 Sqn, Middle Wallop
	XW866	WS Gazelle HT3 [E]	RAF No 2 FTS, Shawbury
	XW868	WS Gazelle HT2 [50/CU]	RN No 705 Sqn, Culdrose
	XW870	WS Gazelle HT3 [F]	RAF No 2 FTS, Shawbury
	XW871	WS Gazelle HT2 [44/CU]	RN No 705 Sqn, Culdrose
	XW884	WS Gazelle HT2 [41/CU]	RN No 705 Sqn, Culdrose
	XW885	WS Gazelle AH1 [B]	AAC No 670 Sqn, Middle Wallop
	XW887	WS Gazelle HT2 [FL]	RNAS, Station Flight Fleetlands
	XW888	WS Gazelle AH1	AAC SAE, Middle Wallop
	XW889	WS Gazelle AH1	AAC SAE, Middle Wallop
	XW890	WS Gazelle HT2 [53/CU]	RNAY Fleetlands
	XW891	WS Gazelle HT2 [49] (fuselage)	*Scrapped at Culdrose*
	XW892	WS Gazelle AH1 [L]	AAC No 662 Sqn, Soest
	XW893	WS Gazelle AH1	RNAY Fleetlands
	XW894	WS Gazelle HT2 [52/CU]	MoD(PE) DRA Farnborough
	XW895	WS Gazelle HT2 [51/CU]	RN No 705 Sqn, Culdrose
	XW897	WS Gazelle AH1 [Z]	AAC No 670 Sqn, Middle Wallop
	XW898	WS Gazelle HT3 [G]	RAF No 2 FTS, Shawbury
	XW899	WS Gazelle AH1 [K]	AAC No 662 Sqn, Soest
	XW900	WS Gazelle AH1 (TAD-900)	AAC SAE, Middle Wallop
	XW902	WS Gazelle HT3 [H]	RAF No 2 FTS, Shawbury
	XW903	WS Gazelle AH1 [E]	AAC, stored Fleetlands
	XW904	WS Gazelle AH1	AAC No 7 Flight, Gatow
	XW906	WS Gazelle HT3 [J]	RAF No 2 FTS, Shawbury
	XW907	WS Gazelle HT2 [48/CU]	RN No 705 Sqn, Culdrose
	XW908	WS Gazelle AH1 [E]	AAC No 670 Sqn, Middle Wallop
	XW909	WS Gazelle AH1 [A]	AAC No 664 Sqn, Dishforth
	XW910	WS Gazelle HT3 [K]	RAF No 2 FTS, Shawbury
	XW911	WS Gazelle AH1 [I]	AAC No 670 Sqn, Middle Wallop
	XW912	WS Gazelle AH1	AAC SAE, Middle Wallop
	XW913	WS Gazelle AH1 [U]	AAC No 657 Sqn, Dishforth
	XW916	HS Harrier GR3 [W]	RAF Wittering Fire Section
	XW919	HS Harrier GR3 [W]	RN SAH, Culdrose
	XW923	HS Harrier GR3 (cockpit) (8724M)	RAF Wittering for rescue training
	XW924	HS Harrier GR3 [G] (9073M)	RAF No 1 SoTT, Halton
	XW927	HS Harrier T4 [02]	RAF Laarbruch
	XW930	HS125-1B (G-ATPC)	MoD(PE) DRA Farnborough, stored
	XW934	HS Harrier T4 [Y]	RAF HOCU/No 20(R) Sqn Wittering
	XW986	HS Buccaneer S2B	MoD(PE) DRA Farnborough
	XW987	HS Buccaneer S2B	MoD(PE) A&AEE, Boscombe Down
	XW988	HS Buccaneer S2B	MoD(PE) A&AEE, Boscombe Down
	XX101	Cushioncraft CC7	IHM, Weston-super-Mare
	XX102	Cushioncraft CC7	Museum of Army Transport, Beverley
	XX105	BAC 1-11/201 (G-ASJD)	MoD(PE) DRA Bedford
	XX108	SEPECAT Jaguar GR1 (G27-313)	MoD(PE) BAe Warton/A&AEE Boscombe Down
	XX109	SEPECAT Jaguar GR1 (8918M)	RAF Coltishall, ground instruction
	XX110	SEPECAT Jaguar GR1 Replica (BAPC 169)	RAF No 1 SoTT, Halton

Serial	Type	Owner or Operator	Notes
XX110	SEPECAT Jaguar GR1 [EP] (8955M)	RAF No 2 SoTT, Cosford	
XX112	SEPECAT Jaguar GR1A [EC]	RAF stored Shawbury	
XX115	SEPECAT Jaguar GR1 (JI005) (8821M) (fuselage only)	*Scrapped at Abingdon by June 1992*	
XX116	SEPECAT Jaguar GR1A [02] (J1008)	RAF No 16(R) Sqn, Lossiemouth	
XX117	SEPECAT Jaguar GR1A [06] (JI004)	MoD(PE) DRA, Farnborough	
XX118	SEPECAT Jaguar GR1 (JI018) (8815M) (fuselage only)	*Scrapped at Abingdon by June 1992*	
XX119	SEPECAT Jaguar GR1 [01] (8898M)	RAF No 16(R) Sqn, Lossiemouth	
XX139	SEPECAT Jaguar T2A [C]	RAF No 16(R) Sqn, Lossiemouth	
XX140	SEPECAT Jaguar T2 [D] (9008M)	RAF No 2 SoTT, Cosford	
XX141	SEPECAT Jaguar T2A [Z]	RAF No 16(R) Sqn, Lossiemouth	
XX143	SEPECAT Jaguar T2A [GS] (JI002)	RAF No 54 Sqn, Coltishall	
XX144	SEPECAT Jaguar T2A [I]	RAF No 16(R) Sqn, Lossiemouth	
XX145	SEPECAT Jaguar T2	MoD(PE) ETPS, Boscombe Down	
XX146	SEPECAT Jaguar T2A [Y]	RAF No 41 Sqn, Coltishall	
XX150	SEPECAT Jaguar T2A [W]	RAF No 16(R) Sqn, Lossiemouth	
XX154	HS Hawk T1 [1]	RAF, stored St Athan	
XX156	HS Hawk T1	MoD(PE) A&AEE, Boscombe Down	
XX157	HS Hawk T1A [B]	RAF No 7 FTS/92(R) Sqn, Chivenor	
XX158	HS Hawk T1A	RAF No 7 FTS/19(R) Sqn, Chivenor	
XX159	HS Hawk T1A [TO]	RAF No 4 FTS/74(R) Sqn, Valley	
XX160	HS Hawk T1	MoD(PE) T&EE Llanbedr	
XX161	HS Hawk T1	RAF No 4 FTS/234(R) Sqn, Valley	
XX162	HS Hawk T1	RAF No 4 FTS/74(R) Sqn, Valley	
XX163	HS Hawk T1	RAF CFS, Valley	
XX164	HS Hawk T1 [CN]	RAF No 100 Sqn, Wyton	
XX165	HS Hawk T1 [TM]	RAF No 4 FTS/74(R) Sqn, Valley	
XX167	HS Hawk T1 [Q]	RAF No 7 FTS/92(R) Sqn, Chivenor	
XX168	HS Hawk T1 [TL]	RAF No 4 FTS/74(R) Sqn, Valley	
XX169	HS Hawk T1	RAF No 4 FTS/234(R) Sqn, Valley	
XX170	HS Hawk T1	RAF No 4 FTS/234(R) Sqn, Valley	
XX171	HS Hawk T1 [TK]	RAF No 4 FTS/74(R) Sqn, Valley	
XX172	HS Hawk T1	RAF St Athan Station Flight	
XX173	HS Hawk T1	RAF No 6 FTS, Finningley	
XX174	HS Hawk T1	RAF No 4 FTS/74(R) Sqn, St Athan (repair)	
XX175	HS Hawk T1	RAF No 7 FTS/19(R) Sqn, Chivenor	
XX176	HS Hawk T1	RAF No 7 FTS/19(R) Sqn, Chivenor	
XX177	HS Hawk T1	RAF No 100 Sqn, Wyton	
XX178	HS Hawk T1 [M]	RAF No 7 FTS/92(R) Sqn, Chivenor	
XX179	HS Hawk T1 [E]	RAF No 7 FTS/92(R) Sqn, Chivenor	
XX181	HS Hawk T1 [CB]	RAF No 100 Sqn, Wyton	
XX183	HS Hawk T1	RAF No 4 FTS/234(R) Sqn, Valley	
XX184	HS Hawk T1	RAF, stored St Athan	
XX185	HS Hawk T1	RAF No 7 FTS/19(R) Sqn, Chivenor	
XX186	HS Hawk T1A	RAF No 7 FTS/19(R) Sqn, Chivenor	
XX187	HS Hawk T1A	RAF CFS, Valley	
XX188	HS Hawk T1A [CG]	RAF No 100 Sqn, Wyton	
XX189	HS Hawk T1A [TB]	RAF No 4 FTS/74(R) Sqn, Valley	
XX190	HS Hawk T1A [TA]	RAF No 4 FTS/74(R) Sqn, Valley	
XX191	HS Hawk T1A	RAF No 4 FTS/234(R) Sqn, Valley	
XX193	HS Hawk T1A	RAF	
XX194	HS Hawk T1A [TI]	RAF No 4FTS/74(R) Sqn, Valley	
XX195	HS Hawk T1A [CA]	RAF No 100 Sqn, Wyton	
XX196	HS Hawk T1A	RAF No 4 FTS/234(R) Sqn, Valley	
XX198	HS Hawk T1A	RAF No 4 FTS/234(R) Sqn, Valley	
XX199	HS Hawk T1A [TG]	RAF No 4 FTS/74(R) Sqn, Valley	
XX200	HS Hawk T1A	RAF No 4 FTS/74(R) Sqn, Valley	
XX201	HS Hawk T1A [N]	RAF No 7 FTS/92(R) Sqn, Chivenor	
XX202	HS Hawk T1A [P]	RAF No 7 FTS/92(R) Sqn, Chivenor	
XX203	HS Hawk T1A	RAF No 4 FTS/234(R) Sqn, Valley	
XX204	HS Hawk T1A [H]	RAF No 7 FTS/92(R) Sqn, Chivenor	
XX205	HS Hawk T1A [V]	RAF No 7 FTS/92(R) Sqn, Chivenor	
XX217	HS Hawk T1A	RAF No 7 FTS/19(R) Sqn, Chivenor	
XX218	HS Hawk T1A	RAF No 4 FTS/234(R) Sqn, Valley	
XX219	HS Hawk T1A	RAF No 7 FTS/19(R) Sqn, Chivenor	
XX220	HS Hawk T1A	RAF CFS, Valley	

Serial	Type	Owner or Operator	Notes
XX221	HS Hawk T1A	RAF No 4 FTS/234(R) Sqn, Chivenor	
XX222	HS Hawk T1A [TJ]	RAF No 4 FTS/74(R) Sqn, Valley	
XX223	HS Hawk T1 (fuselage)	Privately owned, Charlwood	
XX224	HS Hawk T1	RAF CFS, Valley	
XX225	HS Hawk T1	RAF No 7 FTS/19(R) Sqn, Chivenor	
XX226	HS Hawk T1	RAF No 4 FTS/74(R) Sqn, Valley	
XX227	HS Hawk T1A	RAF *Red Arrows*, Scampton	
XX228	HS Hawk T1 [CC]	RAF No 100 Sqn, Wyton	
XX230	HS Hawk T1A	RAF No 7 FTS/19(R) Sqn, Chivenor	
XX231	HS Hawk T1 [X]	RAF No 7 FTS/92(R) Sqn, Chivenor	
XX232	HS Hawk T1	RAF No 4 FTS/74(R) Sqn, Valley	
XX233	HS Hawk T1	RAF CFS/*Red Arrows*, Scampton	
XX234	HS Hawk T1	RAF No 4 FTS/234(R) Sqn, Valley	
XX235	HS Hawk T1	RAF CFS, Valley	
XX236	HS Hawk T1	RAF CFS, Valley	
XX237	HS Hawk T1	RAF *Red Arrows*, Scampton	
XX238	HS Hawk T1	RAF CFS, Valley	
XX239	HS Hawk T1	RAF CFS, Valley	
XX240	HS Hawk T1	RAF No 6 FTS, Finningley	
XX242	HS Hawk T1 [Y]	RAF No 7 FTS/92(R) Sqn, Chivenor	
XX244	HS Hawk T1	RAF No 4 FTS/234(R) Sqn, Valley	
XX245	HS Hawk T1	RAF No 7 FTS/19(R) Sqn, Chivenor	
XX246	HS Hawk T1A	RAF No 7 FTS/19(R) Sqn, Chivenor	
XX247	HS Hawk T1A [CM]	RAF No 100 Sqn, Wyton	
XX248	HS Hawk T1A [CJ]	RAF No 100 Sqn, Wyton	
XX249	HS Hawk T1	RAF No 4 FTS/234(R) Sqn, Valley	
XX250	HS Hawk T1	RAF No 6 FTS, Finningley	
XX252	HS Hawk T1A	RAF *Red Arrows*, Scampton	
XX253	HS Hawk T1A	RAF *Red Arrows*, Scampton	
XX254	HS Hawk T1A	RAF No 7 FTS/19(R) Sqn, Chivenor	
XX255	HS Hawk T1A [TE]	RAF No 4 FTS/74(R) Sqn, Valley	
XX256	HS Hawk T1A	RAF No 7 FTS/19(R) Sqn, Chivenor	
XX257	HS Hawk T1	RAF Chivenor, BDRT	
XX258	HS Hawk T1A	RAF CFS, Valley	
XX260	HS Hawk T1A	RAF *Red Arrows*, Scampton	
XX261	HS Hawk T1A	RAF No 4 FTS/234(R) Sqn, Valley	
XX263	HS Hawk T1A	RAF No 7 FTS/19(R) Sqn, Chivenor	
XX263	HS Hawk T1 Replica (BAPC 152)	RAF Exhibition Flight, Abingdon	
XX264	HS Hawk T1A	RAF *Red Arrows*, Scampton	
XX265	HS Hawk T1A [U]	RAF No 7 FTS/92(R) Sqn, Chivenor	
XX266	HS Hawk T1A	RAF *Red Arrows*, Scampton	
XX278	HS Hawk T1A	RAF No 7 FTS/19(R) Sqn, Chivenor	
XX280	HS Hawk T1A	RAF No 4 FTS/234(R) Sqn, Valley	
XX281	HS Hawk T1A [O]	RAF No 7 FTS/92(R) Sqn, Chivenor	
XX282	HS Hawk T1A	RAF No 7 FTS/19(R) Sqn, Chivenor	
XX283	HS Hawk T1A [CD]	RAF No 100 Sqn, Wyton	
XX284	HS Hawk T1A [CL]	RAF No 100 Sqn, Wyton	
XX285	HS Hawk T1A [CH]	RAF No 100 Sqn, Wyton	
XX286	HS Hawk T1A	RAF No 4 FTS/234(R) Sqn, Valley	
XX287	HS Hawk T1A [S]	RAF No 7 FTS/92(R) Sqn, Chivenor	
XX288	HS Hawk T1	RAF No 100 Sqn, Wyton	
XX289	HS Hawk T1A	RAF No 7 FTS/19(R) Sqn, Chivenor	
XX290	HS Hawk T1	RAF No 4 FTS/234(R) Sqn, Valley	
XX292	HS Hawk T1 [R]	RAF No 7 FTS/92(R) Sqn, Chivenor	
XX294	HS Hawk T1	RAF *Red Arrows*, Scampton	
XX295	HS Hawk T1	RAF No 6 FTS, Finningley	
XX296	HS Hawk T1 [TR]	RAF No 4 FTS/74(R) Sqn, Valley	
XX297	HS Hawk T1A (8933M)	RAF Finningley Fire Section	
XX297	HS Hawk T1 Replica (BAPC171)	RAF Exhibition Flight, Abingdon	
XX299	HS Hawk T1 [J]	RAF No 7 FTS/92(R) Sqn, Chivenor	
XX301	HS Hawk T1A [L]	RAF No 7 FTS/92(R) Sqn, Chivenor	
XX302	HS Hawk T1A	RAF St Athan	
XX303	HS Hawk T1A [TH]	RAF No 4 FTS/74(R) Sqn, Valley	
XX304	HS Hawk T1A (fuselage)	RAF stored, Shawbury	
XX306	HS Hawk T1A	RAF *Red Arrows*, Scampton	
XX307	HS Hawk T1	RAF *Red Arrows*, Scampton	
XX308	HS Hawk T1	RAF *Red Arrows*, Scampton	
XX309	HS Hawk T1	RAF No 4 FTS/234(R) Sqn, Valley	
XX310	HS Hawk T1W [TN]	RAF No 4 FTS/74(R) Sqn, Valley	
XX311	HS Hawk T1 [T]	RAF No 7 FTS/92(R) Sqn, Chivenor	
XX312	HS Hawk T1 [CF]	RAF No 100 Sqn, Wyton	
XX313	HS Hawk T1	RAF No 7 FTS/19(R) Sqn, Chivenor	
XX314	HS Hawk T1 [TP]	RAF No 4 FTS/74(R) Sqn, Valley	

Serial	Type	Owner or Operator	Notes
XX315	HS Hawk T1A	RAF No 4 FTS/234(R) Sqn, Valley	
XX316	HS Hawk T1A [TQ]	RAF No 4 FTS/74(R) Sqn, Valley	
XX317	HS Hawk T1A	RAF No 4 FTS/234(R) Sqn, Valley	
XX318	HS Hawk T1A	RAF St Athan	
XX319	HS Hawk T1A [TF]	RAF No 4 FTS/74(R) Sqn, Valley	
XX320	HS Hawk T1A	RAF No 7 FTS/19(R) Sqn, Chivenor	
XX321	HS Hawk T1A	RAF No 7 FTS/19(R) Sqn, Chivenor	
XX322	HS Hawk T1A [W]	RAF No 7 FTS/92(R) Sqn, Chivenor	
XX323	HS Hawk T1A [TD]	RAF No 4 FTS/74(R) Sqn, Valley	
XX324	HS Hawk T1A	RAF No 4 FTS/234(R) Sqn, Valley	
XX325	HS Hawk T1A [CE]	RAF No 100 Sqn, Wyton	
XX326	HS Hawk T1A	RAF No 7 FTS/19(R) Sqn, Chivenor	
XX327	HS Hawk T1	MoD(PE) DRA Farnborough	
XX329	HS Hawk T1A [C]	RAF No 7 FTS/92(R) Sqn, Chivenor	
XX330	HS Hawk T1A [D]	RAF No 7 FTS/92(R) Sqn, Chivenor	
XX331	HS Hawk T1A [CK]	RAF No 100 Sqn, Wyton	
XX332	HS Hawk T1A [F]	RAF No 7 FTS/92(R) Sqn, Chivenor	
XX334	HS Hawk T1A	*Crashed at Chivenor 30 Sep 1992*	
XX335	HS Hawk T1A [I]	RAF No 7 FTS/92(R) Sqn, Chivenor	
XX337	HS Hawk T1A [K]	RAF No 7 FTS/92(R) Sqn, Chivenor	
XX338	HS Hawk T1	RAF No 7 FTS/19(R) Sqn, Chivenor	
XX339	HS Hawk T1A [TS]	RAF No 4 FTS/74(R) Sqn, Valley	
XX341	HS Hawk T1 ASTRA [1]	MoD(PE) ETPS, Boscombe Down	
XX342	HS Hawk T1 [2]	MoD(PE) ETPS, Boscombe Down	
XX343	HS Hawk T1 [3]	MoD(PE) ETPS, Boscombe Down	
XX344	HS Hawk T1 (8847M) fuselage	MoD(PE) DRA Farnborough, Fire Section	
XX345	HS Hawk T1A	RAF No 7 FTS/19(R) Sqn, Chivenor	
XX346	HS Hawk T1A	RAF No 7 FTS/19(R) Sqn, Chivenor	
XX348	HS Hawk T1A	RAF No 4 FTS/234(R) Sqn, Valley	
XX349	HS Hawk T1	RAF No 100 Sqn, Wyton	
XX350	HS Hawk T1A [TC]	RAF No 4 FTS/74(R) Sqn, Valley	
XX351	HS Hawk T1A	RAF No 4 FTS/234(R) Sqn, Valley	
XX352	HS Hawk T1A	RAF No 7 FTS/19(R) Sqn, Chivenor	
XX370	WS Gazelle AH1 [A]	AAC No 658 Sqn, Netheravon	
XX371	WS Gazelle AH1	AAC No 12 Flt, Wildenrath	
XX372	WS Gazelle AH1 [B]	AAC No 658 Sqn, Netheravon	
XX375	WS Gazelle AH1 [C]	AAC No 658 Sqn, Netheravon	
XX378	WS Gazelle AH1 [Q]	AAC No 670 Sqn, Middle Wallop	
XX379	WS Gazelle AH1 [D]	AAC No 658 Sqn, Netheravon	
XX380	WS Gazelle AH1 [A]	RM 3 CBAS, Yeovilton	
XX381	WS Gazelle AH1	AAC No 2 Flt, Netheravon	
XX382	WS Gazelle HT3 [M]	RAF No 2 FTS, Shawbury	
XX383	WS Gazelle AH1 [E]	AAC No 658 Sqn, Netheravon	
XX384	WS Gazelle AH1	AAC No 664 Sqn, Minden	
XX385	WS Gazelle AH1 [X]	AAC No 670 Sqn, Middle Wallop	
XX386	WS Gazelle AH1	AAC No 12 Flt, Wildenrath	
XX387	WS Gazelle AH1	AAC No 651 Sqn, Hildesheim	
XX388	WS Gazelle AH1	AAC No 652 Sqn, Hildesheim	
XX389	WS Gazelle AH1 [C]	AAC No 653 Sqn, Soest	
XX391	WS Gazelle HT2 [56/CU]	RN No 705 Sqn, Culdrose	
XX392	WS Gazelle AH1 [A1]	AAC No 670 Sqn, Middle Wallop	
XX393	WS Gazelle AH1	AAC No 2 Flt, Netheravon	
XX394	WS Gazelle AH1	AAC Fleetlands	
XX395	WS Gazelle AH1 [J]	AAC No 662 Sqn, Soest	
XX396	WS Gazelle HT3 (8718M) [N]	RAF Exhibition Flight, Henlow	
XX398	WS Gazelle AH1	AAC 4 Regiment, Detmold	
XX399	WS Gazelle AH1	AAC No 2 Flt, Netheravon	
XX403	WS Gazelle AH1 [Y]	Westland, Weston-super-Mare (on rebuild)	
XX405	WS Gazelle AH1 [C1]	AAC No 670 Sqn, Middle Wallop	
XX406	WS Gazelle HT3 [P]	RAF No 2 FTS, Shawbury	
XX407	WS Gazelle AH1 [D1]	AAC No 670 Sqn, Middle Wallop	
XX408	WS Gazelle AH1 [Y]	AAC No 670 Sqn, Middle Wallop	
XX409	WS Gazelle AH1	AAC No 656 Sqn, Netheravon	
XX410	WS Gazelle HT2 [58/CU]	RN AES, Lee-on-Solent	
XX411	WS Gazelle AH1 [X]	AAC Middle Wallop, BDRT	
XX411	WS Gazelle AH1 (tail only)	FAA Museum, RNAS Yeovilton	
XX412	WS Gazelle AH1 [B]	RM 3 CBAS, Yeovilton	
XX413	WS Gazelle AH1 [C]	RM 3 CBAS, Yeovilton	
XX414	WS Gazelle AH1 [N]	AAC No 662 Sqn, Soest	
XX416	WS Gazelle AH1 [I]	AAC No 656 Sqn, Netheravon	
XX417	WS Gazelle AH1	AAC No 665 Sqn, Aldergrove	
XX418	WS Gazelle AH1	AAC No 651 Sqn, Hildesheim	
XX419	WS Gazelle AH1 [A]	AAC No 664 Sqn, Dishforth	

Notes	Serial	Type	Owner or Operator
	XX431	WS Gazelle HT2 [43/CU]	RN No 705 Sqn, Culdrose
	XX432	WS Gazelle AH1	AAC No 665 Sqn, Aldergrove
	XX433	WS Gazelle AH1	AAC No 665 Sqn, Aldergrove
	XX434	WS Gazelle AH1	*Scrapped by June 1992*
	XX435	WS Gazelle AH1 [B]	AAC No 653 Sqn, Soest
	XX436	WS Gazelle HT2 [39/CU]	RN No 705 Sqn, Culdrose
	XX437	WS Gazelle AH1 [G]	AAC No 653 Sqn, Soest
	XX438	WS Gazelle AH1 [B]	AAC No 664 Sqn, Dishforth
	XX439	WS Gazelle AH1	AAC No 651 Sqn, Hildesheim
	XX440	WS Gazelle AH1 (G-BCHN)	AAC No 665 Sqn, Aldergrove
	XX441	WS Gazelle HT2 [38/CU]	RN No 705 Sqn, Culdrose
	XX442	WS Gazelle AH1	AAC Fleetlands
	XX443	WS Gazelle AH1 [T]	AAC No 663 Sqn, Soest
	XX444	WS Gazelle AH1 [E]	AAC No 656 Sqn, Netheravon
	XX445	WS Gazelle AH1	AAC No 657 Sqn, Dishforth
	XX446	WS Gazelle HT2 [57/CU]	RN No 705 Sqn, Culdrose
	XX447	WS Gazelle AH1 [U]	AAC No 663 Sqn, Soest
	XX448	WS Gazelle AH1 [A]	AAC No 669 Sqn, Detmold
	XX449	WS Gazelle AH1 [T]	AAC No 669 Sqn, Detmold
	XX450	WS Gazelle AH1	RM 3 CBAS, Yeovilton
	XX451	WS Gazelle HT2 [58/CU]	RN No 705 Sqn, Culdrose
	XX452	WS Gazelle AH1	AAC Middle Wallop Fire Section
	XX453	WS Gazelle AH1 [K]	AAC 4 Regiment, Detmold
	XX454	WS Gazelle AH1	AAC No 659 Sqn, Detmold
	XX455	WS Gazelle AH1	AAC No 651 Sqn, Hildesheim
	XX456	WS Gazelle AH1	AAC No 7 Flt, Gatow
	XX457	WS Gazelle AH1	AAC No 2 Flt, Netheravon
	XX460	WS Gazelle AH1	AAC 4 Regiment, Detmold
	XX462	WS Gazelle AH1	AAC No 661 Sqn, Hildesheim
	XX466	HS Hunter T66B/T7 [830/DD]	RNAS Culdrose, SAH
	XX467	HS Hunter T66B/T7	Air Service Training, Perth
	XX469	WS Lynx HAS2 (G-BNCL) (A2657)	Helicopter Museum of GB, Blackpool
	XX475	SA Jetstream T2 [572/CU] (G-AWVJ/N1036S)	RN No 750 Sqn, Culdrose
	XX476	SA Jetstream T2 [561/CU] (G-AXGL/N1037S)	RN No 750 Sqn, Culdrose
	XX477	SA Jetstream T1 (8462M) (G-AXXS) (fuselage)	RAF Finningley for ground instruction
	XX478	SA Jetstream T2 [564/CU] (G-AXXT)	RN No 750 Sqn, Culdrose
	XX479	SA Jetstream T2 [563/CU] (G-AXUR)	RN No 750 Sqn, Culdrose
	XX480	SA Jetstream T2 [565/CU] (G-AXXU)	RN No 750 Sqn, Culdrose
	XX481	SA Jetstream T2 [560/CU] (G-AXUP)	RN No 750 Sqn, Culdrose
	XX482	SA Jetstream T1 [J]	RAF No 6 FTS/45(R) Sqn, Finningley
	XX483	SA Jetstream T2 [562/CU]	RN No 750 Sqn, Culdrose
	XX484	SA Jetstream T2 [566/CU]	RN No 750 Sqn, Culdrose
	XX485	SA Jetstream T2 [567/CU]	RN No 750 Sqn, Culdrose
	XX486	SA Jetstream T2 [569/CU]	RN No 750 Sqn, Culdrose
	XX487	SA Jetstream T2 [568/CU]	RN No 750 Sqn, Culdrose
	XX488	SA Jetstream T2 [571/CU]	RN No 750 Sqn, Culdrose
	XX490	SA Jetstream T2 [570/CU]	RN No 750 Sqn, Culdrose
	XX491	SA Jetstream T1 [K]	RAF No 6 FTS/45(R) Sqn, Finningley
	XX492	SA Jetstream T1 [A]	RAF No 6 FTS/45(R) Sqn, Finningley
	XX493	SA Jetstream T1 [L]	RAF No 6 FTS/45(R) Sqn, Finningley
	XX494	SA Jetstream T1 [B]	RAF No 6 FTS/45(R) Sqn, Finningley
	XX495	SA Jetstream T1 [C]	RAF No 6 FTS/45(R) Sqn, Finningley
	XX496	SA Jetstream T1 [D]	RAF No 6 FTS/45(R) Sqn, Finningley
	XX497	SA Jetstream T1 [E]	RAF No 6 FTS/45(R) Sqn, Finningley
	XX498	SA Jetstream T1 [F]	RAF No 6 FTS/45(R) Sqn, Finningley
	XX499	SA Jetstream T1 [G]	RAF No 6 FTS/45(R) Sqn, Finningley
	XX500	SA Jetstream T1 [H]	RAF No 6 FTS/45(R) Sqn, Finningley
	XX507	HS125 CC2	RAF No 32 Sqn, Northolt
	XX508	HS125 CC2	RAF No 32 Sqn, Northolt
	XX510	WS Lynx HAS2 [69/LS]	RN AES, Lee-on-Solent
	XX513	SA Bulldog T1 [A]	RAF No 1 FTS/RNEFTS, Topcliffe
	XX515	SA Bulldog T1 [U]	RAF, Northumbria UAS, Leeming
	XX516	SA Bulldog T1 [C]	RAF No 1 FTS/RNEFTS, Topcliffe
	XX518	SA Bulldog T1 [Z]	RAF, Cambridge UAS, Cambridge
	XX519	SA Bulldog T1 [I]	RAF No 1 FTS/RNEFTS, Topcliffe

Serial	Type	Owner or Operator	Notes
XX520	SA Bulldog T1 [2]	RAF CFS, Scampton	
XX521	SA Bulldog T1 [01]	RAF, East Lowlands UAS, Turnhouse	
XX522	SA Bulldog T1 [E]	RAF No 1 FTS/RNEFTS, Topcliffe	
XX523	SA Bulldog T1 [F]	RAF No 1 FTS/RNEFTS, Topcliffe	
XX524	SA Bulldog T1 [04]	RAF, London UAS, Benson	
XX525	SA Bulldog T1 [03]	RAF, East Lowlands UAS, Turnhouse	
XX526	SA Bulldog T1 [C]	RAF, Oxford UAS, Benson	
XX527	SA Bulldog T1 [G]	RAF No 1 FTS/RNEFTS, Topcliffe	
XX528	SA Bulldog T1 [D]	RAF, Oxford UAS, Benson	
XX529	SA Bulldog T1 [W]	RAF No 6 FTS, Finningley	
XX530	SA Bulldog T1 [12]	*Scrapped at RAF Manston*	
XX531	SA Bulldog T1 [B]	RAF No 1 FTS/RNEFTS, Topcliffe	
XX532	SA Bulldog T1 [D]	RAF, Yorkshire UAS, Finningley	
XX533	SA Bulldog T1 [J]	RAF No 1 FTS/RNEFTS, Topcliffe	
XX534	SA Bulldog T1 [B]	RAF, Birmingham UAS, Cosford	
XX535	SA Bulldog T1 [10]	RAF, London UAS, Benson	
XX536	SA Bulldog T1 [D]	RAF No 1 FTS/RNEFTS, Topcliffe	
XX537	SA Bulldog T1 [02]	RAF, East Lowlands UAS, Turnhouse	
XX538	SA Bulldog T1 [P]	RAF No 1 FTS/RNEFTS, Topcliffe	
XX539	SA Bulldog T1 [1]	RAF CFS, Scampton	
XX540	SA Bulldog T1 [K]	RAF No 1 FTS/RNEFTS, Topcliffe	
XX541	SA Bulldog T1 [L]	RAF No 1 FTS/RNEFTS, Topcliffe	
XX543	SA Bulldog T1 [F]	RAF, Yorkshire UAS, Finningley	
XX544	SA Bulldog T1 [01]	RAF, London UAS, Benson	
XX545	SA Bulldog T1 PAX [02]	RAF, East Lowlands UAS, Turnhouse	
XX546	SA Bulldog T1 [03]	RAF, London UAS, Benson	
XX547	SA Bulldog T1 [05]	RAF, London UAS, Benson	
XX548	SA Bulldog T1 [06]	RAF, London UAS, Benson	
XX549	SA Bulldog T1 [T]	RAF No 1 FTS/RNEFTS, Topcliffe	
XX550	SA Bulldog T1 [Z]	RAF, Yorkshire UAS, Finningley	
XX551	SA Bulldog T1 [M]	RAF No 1 FTS/RNEFTS, Topcliffe	
XX552	SA Bulldog T1 [08]	RAF, London UAS, Benson	
XX553	SA Bulldog T1 [07]	RAF, London UAS, Benson	
XX554	SA Bulldog T1 [09]	RAF, London UAS, Benson	
XX555	SA Bulldog T1 [10]	RAF CFS, Scampton	
XX556	SA Bulldog T1 [S]	RAF, East Midlands UAS, Newton	
XX557	SA Bulldog T1 PAX	RAF Topcliffe, ground instruction	
XX558	SA Bulldog T1 [A]	RAF, Birmingham UAS, Cosford	
XX559	SA Bulldog T1	RAF, Glasgow & Strathclyde UAS, Glasgow	
XX560	SA Bulldog T1	RAF, Glasgow & Strathclyde UAS, Glasgow	
XX561	SA Bulldog T1 [A]	RAF, Aberdeen, Dundee & St Andrews UAS, Leuchars	
XX562	SA Bulldog T1 [E]	RAF No 13 AEF, Sydenham	
XX611	SA Bulldog T1	RAF, Glasgow & Strathclyde UAS, Glasgow	
XX612	SA Bulldog T1 [05]	RAF, Wales UAS, St Athan	
XX613	SA Bulldog T1 [A]	*Written off Cumber, NI 16 Oct 1992*	
XX614	SA Bulldog T1 [11]	RAF CFS, Scampton	
XX615	SA Bulldog T1 [2]	RAF, Manchester UAS, Woodvale	
XX616	SA Bulldog T1 [3]	RAF, Manchester UAS, Woodvale	
XX617	SA Bulldog T1 [4]	RAF, Manchester UAS, Woodvale	
XX619	SA Bulldog T1 [B]	RAF, Yorkshire UAS, Finningley	
XX620	SA Bulldog T1 [C]	RAF, Yorkshire UAS, Finningley	
XX621	SA Bulldog T1 [X]	RAF No 6 FTS, Finningley	
XX622	SA Bulldog T1 [E]	RAF, Yorkshire UAS, Finningley	
XX623	SA Bulldog T1 [M]	RAF, East Midlands UAS, Newton	
XX624	SA Bulldog T1 [Y]	RAF No 6 FTS, Finningley	
XX625	SA Bulldog T1 [01]	RAF, Wales UAS, St Athan	
XX626	SA Bulldog T1 [02]	RAF, Wales UAS, St Athan	
XX627	SA Bulldog T1 [03]	RAF, Wales UAS, St Athan	
XX628	SA Bulldog T1 [04]	RAF, Wales UAS, St Athan	
XX629	SA Bulldog T1 [V]	RAF, Northumbria UAS, Leeming	
XX630	SA Bulldog T1 [A]	RAF, Liverpool UAS, Woodvale	
XX631	SA Bulldog T1 [W]	RAF, Northumbria UAS, Leeming	
XX632	SA Bulldog T1 [D]	RAF, Bristol UAS, Colerne	
XX633	SA Bulldog T1 [X]	RAF, Northumbria UAS, Leeming	
XX634	SA Bulldog T1 [C]	RAF, Cambridge UAS, Cambridge	
XX635	SA Bulldog T1 (8767M) [S]	RAF St Athan, CTTS	
XX636	SA Bulldog T1 [Y]	RAF, Northumbria UAS, Leeming	
XX637	SA Bulldog T1 [U]	RAF, Northumbria UAS, Leeming	
XX638	SA Bulldog T1 [N]	RAF No 1 FTS/RNEFTS, Topcliffe	

Notes	Serial	Type	Owner or Operator
	XX639	SA Bulldog T1 [02]	RAF, London UAS, Benson
	XX640	SA Bulldog T1 [B]	RAF, Queen's UAS, Sydenham
	XX653	SA Bulldog T1 [E]	RAF, Bristol UAS, Colerne
	XX654	SA Bulldog T1 [A]	RAF, Bristol UAS, Colerne
	XX655	SA Bulldog T1 [B]	RAF, Bristol UAS, Colerne
	XX656	SA Bulldog T1 [C]	RAF, Bristol UAS, Colerne
	XX657	SA Bulldog T1 [U]	RAF, Cambridge UAS, Cambridge
	XX658	SA Bulldog T1 [A]	RAF, Cambridge UAS, Cambridge
	XX659	SA Bulldog T1 [S]	RAF, Cambridge UAS, Cambridge
	XX660	SA Bulldog T1 [A]	BAe Prestwick, spares recovery
	XX661	SA Bulldog T1 [B]	RAF, Oxford UAS, Benson
	XX663	SA Bulldog T1 [B]	RAF, Aberdeen, Dundee & St Andrews UAS, Leuchars
	XX664	SA Bulldog T1 [04]	RAF, East Lowlands UAS, Turnhouse
	XX665	SA Bulldog T1 [E]	RAF, Aberdeen, Dundee & St Andrews UAS, Leuchars
	XX666	SA Bulldog T1 [V]	RAF No 1 FTS/RNEFTS, Topcliffe
	XX667	SA Bulldog T1 [D]	RAF, Aberdeen, Dundee & St Andrews UAS, Leuchars
	XX668	SA Bulldog T1 [1]	RAF, Manchester UAS, Woodvale
	XX669	SA Bulldog T1 [B] (8997M)	Privately owned, Bruntingthorpe
	XX670	SA Bulldog T1 [C]	RAF, Birmingham UAS, Cosford
	XX671	SA Bulldog T1 [D]	RAF, Birmingham UAS, Cosford
	XX672	SA Bulldog T1 [E]	RAF, Birmingham UAS, Cosford
	XX685	SA Bulldog T1 [L]	RAF, Liverpool UAS, Woodvale
	XX686	SA Bulldog T1 [U]	RAF, Liverpool UAS, Woodvale
	XX687	SA Bulldog T1 [A]	RAF, East Midlands UAS, Newton
	XX688	SA Bulldog T1 [S]	RAF, Liverpool UAS, Woodvale
	XX689	SA Bulldog T1 [3]	RAF CFS, Scampton
	XX690	SA Bulldog T1 [A]	RAF, Yorkshire UAS, Finningley
	XX691	SA Bulldog T1 [G]	RAF, Yorkshire UAS, Finningley
	XX692	SA Bulldog T1 [5]	RAF CFS, Scampton
	XX693	SA Bulldog T1 [4]	RAF CFS, Scampton
	XX694	SA Bulldog T1 [E]	RAF, East Midlands UAS, Newton
	XX695	SA Bulldog T1 [A]	RAF, Oxford UAS, Benson
	XX696	SA Bulldog T1 [8]	RAF CFS, Scampton
	XX697	SA Bulldog T1 [C]	RAF, Queen's UAS, Sydenham
	XX698	SA Bulldog T1 [9]	RAF CFS, Scampton
	XX699	SA Bulldog T1 [Q]	RAF No 1 FTS/RNEFTS, Topcliffe
	XX700	SA Bulldog T1 [R]	RAF No 1 FTS/RNEFTS, Topcliffe
	XX701	SA Bulldog T1 [02]	RAF, Southampton UAS, Lee-on-Solent
	XX702	SA Bulldog T1 [bl]	RAF, Glasgow & Strathclyde UAS, Glasgow
	XX704	SA Bulldog T1 [U]	RAF, East Midlands UAS, Newton
	XX705	SA Bulldog T1 [05]	RAF, Southampton UAS, Lee-on-Solent
	XX706	SA Bulldog T1 [01]	RAF, Southampton UAS, Lee-on-Solent
	XX707	SA Bulldog T1 [04]	RAF, Southampton UAS, Lee-on-Solent
	XX708	SA Bulldog T1 [03]	RAF, Southampton UAS, Lee-on-Solent
	XX709	SA Bulldog T1 [C]	RAF, Aberdeen, Dundee St Andrews UAS, Leuchars
	XX710	SA Bulldog T1 [5]	RAF, Manchester UAS, Woodvale
	XX711	SA Bulldog T1 [D]	RAF, Queen's UAS, Sydenham
	XX713	SA Bulldog T1 [6]	RAF CFS, Scampton
	XX713	SA Bulldog T1 [Z]	RAF No 6 FTS, Finningley
	XX714	SA Bulldog T1 [12]	RAF CFS, Scampton
	XX718	SEPECAT Jaguar GR1 Replica (BAPC150) [GA]	RAF Exhibition Flight, St Athan
	XX719	SEPECAT Jaguar GR1A [EE]	RAF No 6 Sqn, Coltishall
	XX720	SEPECAT Jaguar GR1A (JI003) [EN]	RAF No 6 Sqn, Coltishall
	XX723	SEPECAT Jaguar GR1A [GQ]	RAF No 54 Sqn, Coltishall
	XX724	SEPECAT Jaguar GR1A [GA]	RAF, stored Shawbury
	XX725	SEPECAT Jaguar GR1A [GU] (JI010)	RAF No 54 Sqn, Coltishall
	XX726	SEPECAT Jaguar GR1 [EB] (8947M)	RAF No 1 SoTT, Halton
	XX727	SEPECAT Jaguar GR1 [ER] (8951M)	RAF No 2 SoTT, Cosford
	XX729	SEPECAT Jaguar GR1A (JI012) [GC]	RAF No 54 Sqn, Coltishall
	XX730	SEPECAT Jaguar GR1 [EC] (8952M)	RAF No 2 SoTT, Cosford
	XX733	SEPECAT Jaguar GR1A [ER]	RAF No 6 Sqn, Coltishall

Serial	Type	Owner or Operator	Notes
XX734	SEPECAT Jaguar GR1 (JI014) (8816M)	Privately owned, Charlwood	
XX736	SEPECAT Jaguar GR1 (JI013/9110M)	RAF Coltishall, BDRT	
XX737	SEPECAT Jaguar GR1A [EG] (JI015)	RAF, stored Shawbury	
XX738	SEPECAT Jaguar GR1A [GJ] (JI016)	RAF Coltishall Fire Section	
XX739	SEPECAT Jaguar GR1 (8902M) [I]	RAF No 1 SoTT, Halton	
XX741	SEPECAT Jaguar GR1A [EJ]	RAF No 6 Sqn, Coltishall	
XX743	SEPECAT Jaguar GR1 [EG] (8949M)	RAF No 1 SoTT, Halton	
XX745	SEPECAT Jaguar GR1A [04]	RAF No 16(R) Sqn, Lossiemouth	
XX746	SEPECAT Jaguar GR1A [09] (8895M)	RAF No 1 SoTT, Halton	
XX747	SEPECAT Jaguar GR1 (8903M)	RAF Cranwell, Engineering Wing	
XX748	SEPECAT Jaguar GR1A [GK]	RAF No 54 Sqn, Coltishall	
XX751	SEPECAT Jaguar GR1 [10] (8937M)	RAF No 2 SoTT, Cosford	
XX752	SEPECAT Jaguar GR1A [EQ]	RAF No 6 Sqn, Coltishall	
XX753	SEPECAT Jaguar GR1 [05] (9087M) (cockpit only)	RAF Exhibition Flt, St Athan	
XX756	SEPECAT Jaguar GR1 [AM] (8899M)	RAF No 2 SoTT, Cosford	
XX757	SEPECAT Jaguar GR1 [CU] (8948M)	RAF No 1 SoTT, Halton	
XX763	SEPECAT Jaguar GR1 [24] (9009M)	RAF CTTS, St Athan	
XX764	SEPECAT Jaguar GR1 [13] (9010M)	RAF CTTS, St Athan	
XX765	SEPECAT Jaguar ACT	Loughborough University	
XX766	SEPECAT Jaguar GR1A [EA]	RAF No 6 Sqn, Coltishall	
XX767	SEPECAT Jaguar GR1A [GE]	RAF No 54 Sqn, Coltishall	
XX818	SEPECAT Jaguar GR1 [DE] (8945M)	RAF No 1 SoTT, Halton	
XX819	SEPECAT Jaguar GR1 [CE] (8923M)	RAF No 2 SoTT, Cosford	
XX821	SEPECAT Jaguar GR1 [P] (8896M)	RAF SIF, Cranwell	
XX824	SEPECAT Jaguar GR1 [AD] (9019M)	RAF No 1 SoTT, Halton	
XX825	SEPECAT Jaguar GR1 [BN] (9020M)	RAF No 1 SoTT, Halton	
XX826	SEPECAT Jaguar GR1 [34] (9021M)	RAF No 2 SoTT, Cosford	
XX829	SEPECAT Jaguar T2A [ET]	RAF No 6 Sqn, Coltishall	
XX830	SEPECAT Jaguar T2	MoD(PE) ETPS, Boscombe Down	
XX832	SEPECAT Jaguar T2A [S]	RAF, stored Shawbury	
XX833	SEPECAT Jaguar T2A [N]	RAF SAOEU, Boscombe Down	
XX835	SEPECAT Jaguar T2	MoD(PE) DRA Farnborough	
XX836	SEPECAT Jaguar T2A	RAF, stored Shawbury	
XX837	SEPECAT Jaguar T2 [Z] (8978M)	RAF No 1 SoTT, Halton	
XX838	SEPECAT Jaguar T2A [X]	RAF No 16(R) Sqn, Lossiemouth	
XX839	SEPECAT Jaguar T2A [Y]	RAF No 16(R) Sqn, Lossiemouth	
XX840	SEPECAT Jaguar T2A [X]	RAF, stored Shawbury	
XX841	SEPECAT Jaguar T2 [ES]	RAF No 6 Sqn, Coltishall	
XX842	SEPECAT Jaguar T2A [EW]	RAF No 6 Sqn, Coltishall	
XX844	SEPECAT Jaguar T2 [F] (9023M)	RAF No 2 SoTT, Cosford	
XX845	SEPECAT Jaguar T2A [V]	RAF No 41 Sqn, Coltishall	
XX846	SEPECAT Jaguar T2A [A]	RAF No 16(R) Sqn, Lossiemouth	
XX847	SEPECAT Jaguar T2A [X]	RAF No 41 Sqn, Coltishall	
XX885	HS Buccaneer S2B [L]	RAF No 12 Sqn, Lossiemouth	
XX886	HS Buccaneer S2B	RAF Honington, WLT use	
XX888	HS Buccaneer S2B (nose)	Privately owned, Ottershaw	
XX889	HS Buccaneer S2B	RAF No 12 Sqn, Lossiemouth	
XX892	HS Buccaneer S2B	RAF No 208 Sqn, Lossiemouth	
XX893	HS Buccaneer S2B	RAF No 208 Sqn, Lossiemouth	
XX894	HS Buccaneer S2B	RAF No 12 Sqn, Lossiemouth	
XX895	HS Buccaneer S2B	RAF No 208 Sqn, Lossiemouth	
XX897	HS Buccaneer S2B	MoD(PE) DRA Bedford	
XX899	HS Buccaneer S2B	RAF No 12 Sqn, Lossiemouth	

Notes	Serial	Type	Owner or Operator
	XX900	HS Buccaneer S2B	RAF No 208 Sqn, Lossiemouth
	XX901	HS Buccaneer S2B	RAF No 208 Sqn, Lossiemouth
	XX910	WS Lynx HAS2	DRA, stored Farnborough
	XX914	BAC VC10 srs 1103 (G-ATDJ/ 8777M) (rear fuselage)	RAF AMS, Brize Norton
	XX919	BAC 1-11/402 (PI-C 1121)	MoD(PE) DRA Farnborough
	XX946	Panavia Tornado (P02) [WT] (8883M)	RAF Honington, WLT
	XX947	Panavia Tornado (P03) (8797M)	*Scrapped at Marham 1991*
	XX948	Panavia Tornado (P06) (8879M)	RAF No 2 SoTT, Cosford
	XX955	SEPECAT Jaguar GR1A [GK]	RAF, stored Shawbury
	XX956	SEPECAT Jaguar GR1 [BE] (8950M)	RAF No 1 SoTT, Halton
	XX958	SEPECAT Jaguar GR1 [BK] (9022M)	RAF No 2 SoTT, Cosford
	XX959	SEPECAT Jaguar GR1 [CJ] (8953M)	RAF No 2 SoTT, Cosford
	XX962	SEPECAT Jaguar GR1A [EK]	RAF No 6 Sqn, Coltishall
	XX965	SEPECAT Jaguar GR1A [07]	RAF No 16(R) Sqn, Lossiemouth
	XX966	SEPECAT Jaguar GR1A (8904M) [EL]	RAF No 1 SoTT, Halton
	XX967	SEPECAT Jaguar GR1 [AC] (9006M)	RAF No 2 SoTT, Cosford
	XX968	SEPECAT Jaguar GR1 [AJ] (9007M)	RAF No 2 SoTT, Cosford
	XX969	SEPECAT Jaguar GR1A (8897M) [01]	RAF No 2 SoTT, Cosford
	XX970	SEPECAT Jaguar GR1A [EH]	RAF No 6 Sqn, Coltishall
	XX974	SEPECAT Jaguar GR1A [GH]	RAF No 54 Sqn, Coltishall
	XX975	SEPECAT Jaguar GR1 [07] (8905M)	RAF No 1 SoTT, Halton
	XX976	SEPECAT Jaguar GR1 (8906M) [BD]	RAF No 1 SoTT, Halton
	XX977	SEPECAT Jaguar GR1 [DL]	*Scrapped at Abingdon by June 1992*
	XX979	SEPECAT Jaguar GR1A	MoD(PE) A&AEE Boscombe Down
	XZ101	SEPECAT Jaguar GR1A [Q]	RAF No 41 Sqn, Coltishall
	XZ103	SEPECAT Jaguar GR1A [P]	RAF No 41 Sqn, Coltishall
	XZ104	SEPECAT Jaguar GR1A [M]	RAF No 41 Sqn, Coltishall
	XZ106	SEPECAT Jaguar GR1A [FR]	RAF No 41 Sqn, Coltishall
	XZ107	SEPECAT Jaguar GR1A [H]	RAF No 41 Sqn, Coltishall
	XZ108	SEPECAT Jaguar GR1A [GD]	RAF No 54 Sqn, Coltishall
	XZ109	SEPECAT Jaguar GR1A [GL]	RAF No 54 Sqn, Coltishall
	XZ111	SEPECAT Jaguar GR1A [EL]	RAF No 6 Sqn, Coltishall
	XZ112	SEPECAT Jaguar GR1A [GA]	RAF No 54 Sqn, Coltishall
	XZ113	SEPECAT Jaguar GR1A [FD]	RAF No 41 Sqn, Coltishall
	XZ114	SEPECAT Jaguar GR1A [FB]	RAF No 41 Sqn, Coltishall
	XZ115	SEPECAT Jaguar GR1A [FC]	RAF No 41 Sqn, Coltishall
	XZ117	SEPECAT Jaguar GR1A [GG]	RAF No 54 Sqn, Coltishall
	XZ118	SEPECAT Jaguar GR1A [F]	RAF No 41 Sqn, Coltishall
	XZ119	SEPECAT Jaguar GR1A [FG]	RAF No 41 Sqn, Coltishall
	XZ129	HS Harrier GR3 [ETS]	RN ETS, Yeovilton
	XZ130	HS Harrier GR3 [A] (9079M)	RAF No 2 SoTT, Cosford
	XZ131	HS Harrier GR3	RAF, stored St Athan
	XZ132	HS Harrier GR3 [C]	RAF, stored St Athan
	XZ133	HS Harrier GR3	RAF
	XZ133	HS Harrier GR3	Imperial War Museum, Lambeth
	XZ135	HS Harrier GR3 (8848M) (nose only)	RAF Exhibition Flight, St Athan
	XZ138	HS Harrier GR3 [14] (9040M)	RAF Cranwell Instructional use
	XZ145	HS Harrier T4 [14]	RAF No 1 Sqn, Wittering
	XZ146	HS Harrier T4 [04] [S]	RAF HOCU/No 20(R) Sqn, Wittering
	XZ170	WS Lynx AH7 (mod)	MoD(PE)/Westland, Yeovil
	XZ171	WS Lynx AH7	MoD(PE), DRA Farnborough
	XZ172	WS Lynx AH7	MoD(PE)/Westland, Yeovil
	XZ173	WS Lynx AH1 [X]	AAC No 656 Sqn, Netheravon
	XZ174	WS Lynx AH7	AAC No 665 Sqn, Aldergrove
	XZ175	WS Lynx AH1 [A]	AAC No 671 Sqn, Middle Wallop
	XZ176	WS Lynx AH7	AAC, Fleetlands
	XZ177	WS Lynx AH1 [B]	AAC
	XZ178	WS Lynx AH7	AAC, Fleetlands
	XZ179	WS Lynx AH7	AAC No 667 Sqn, Middle Wallop
	XZ180	WS Lynx AH7 [R]	RM 3 CBAS, Yeovilton

Serial	Type	Owner or Operator	Notes
XZ181	WS Lynx AH1 [W]	AAC No 663 Sqn, Soest	
XZ182	WS Lynx AH7 [M]	RM 3 CBAS, Yeovilton	
XZ183	WS Lynx AH1 [M]	AAC No 657 Sqn, Dishforth	
XZ184	WS Lynx AH1 [K]	AAC No 662 Sqn, Soest	
XZ185	WS Lynx AH1 [L]	AAC No 662 Sqn, Soest	
XZ186	WS Lynx AH7	*Crashed Co Tyrone 14 Nov 1991*	
XZ187	WS Lynx AH7	AAC No 655 Sqn, Aldergrove	
XZ188	WS Lynx AH7	AAC No 655 Sqn, Aldergrove	
XZ190	WS Lynx AH1 [N]	AAC No 671 Sqn, Middle Wallop	
XZ191	WS Lynx AH1 [X]	AAC No 663 Sqn, Soest	
XZ192	WS Lynx AH7 [N]	AAC No 657 Sqn, Dishforth	
XZ193	WS Lynx AH7 [I]	AAC No 663 Sqn, Soest	
XZ194	WS Lynx AH1 [T]	AAC No 663 Sqn, Soest	
XZ195	WS Lynx AH7	AAC No 655 Sqn, Aldergrove	
XZ196	WS Lynx AH1 [A]	AAC No 653 Sqn, Soest	
XZ197	WS Lynx AH7	AAC No 671 Sqn, Middle Wallop	
XZ198	WS Lynx AH7	AAC No 655 Sqn, Ballykelly	
XZ199	WS Lynx AH7	AAC No 654 Sqn, Detmold	
XZ203	WS Lynx AH1 [C]	AAC No 653 Sqn, Soest	
XZ205	WS Lynx AH7	AAC SAE, Middle Wallop	
XZ206	WS Lynx AH1 [B]	AAC No 671 Sqn, Middle Wallop	
XZ207	WS Lynx AH9	MoD(PE)/Westland, Yeovil	
XZ208	WS Lynx AH7	AAC No 659 Sqn, Detmold	
XZ209	WS Lynx AH7	AAC No 656 Sqn, Netheravon	
XZ210	WS Lynx AH7	AAC, Fleetlands	
XZ211	WS Lynx AH1 [L]	AAC No 657 Sqn, Dishforth	
XZ212	WS Lynx AH1 [O]	AAC No 657 Sqn, Dishforth	
XZ213	WS Lynx AH1 [TAD213]	AAC, SAE, Middle Wallop	
XZ214	WS Lynx AH7	AAC No 669 Sqn, Detmold	
XZ215	WS Lynx AH7	AAC No 669 Sqn, Detmold	
XZ216	WS Lynx AH7 [V]	AAC No 656 Sqn, Netheravon	
XZ217	WS Lynx AH7	AAC No 657 Sqn, Dishforth	
XZ218	WS Lynx AH7	AAC No 655 Sqn, Aldergrove	
XZ219	WS Lynx AH7	AAC No 654 Sqn, Detmold	
XZ220	WS Lynx AH1 [P]	AAC No 657 Sqn, Dishforth	
XZ221	WS Lynx AH7 [J]	AAC No 669 Sqn, Detmold	
XZ222	WS Lynx AH7 [K]	AAC No 654 Sqn, Detmold	
XZ227	WS Lynx HAS3 [345/NC]	RN No 815 Sqn, Portland	
XZ228	WS Lynx HAS3 [303/JP]	RN No 815 Sqn, Portland	
XZ229	WS Lynx HAS3 [336]	RNAY Fleetlands	
XZ230	WS Lynx HAS3 [336]	RN No 829 Sqn, Portland	
XZ231	WS Lynx HAS3 [604]	RN No 829 Sqn, Portland	
XZ232	WS Lynx HAS3S	RN AMG Portland	
XZ233	WS Lynx HAS3 [435/ED]	RN No 829 Sqn, Portland	
XZ234	WS Lynx HAS3S [336/CV]	RN No 829 Sqn, Portland	
XZ235	WS Lynx HAS3	RNAY Fleetlands	
XZ236	WS Lynx HAS8	MoD(PE)/Westland, Yeovil	
XZ237	WS Lynx HAS3 [326]	RN AMG Portland	
XZ238	WS Lynx HAS3 [645]	RN No 702 Sqn, Portland	
XZ239	WS Lynx HAS3 [374/VB]	RN No 829 Sqn, Portland	
XZ240	WS Lynx HAS3 [363/MA]	RN No 829 Sqn, Portland	
XZ241	WS Lynx HAS3S [365/AY]	RN No 829 Sqn, Portland	
XZ243	WS Lynx HAS3 [635] (wreck)	RNAS Portland, BDRT	
XZ245	WS Lynx HAS3 [405/LO]	RN No 815 Sqn, Portland	
XZ246	WS Lynx HAS3 [434/ED]	RN No 829 Sqn, Portland	
XZ248	WS Lynx HAS3S	RNAY Fleetlands	
XZ249	WS Lynx HAS2	RN Predannack Fire School	
XZ250	WS Lynx HAS3S [603]	RN AMG, Portland	
XZ252	WS Lynx HAS3 [644]	RN No 702 Sqn, Portland	
XZ254	WS Lynx HAS3 [410/GC]	RN No 815 Sqn, Portland	
XZ255	WS Lynx HAS3 [335/CF]	RN No 815 Sqn, Portland	
XZ256	WS Lynx HAS3 [450/SS]	RN No 815 Sqn, Portland	
XZ257	WS Lynx HAS3 [634]	RNAY Fleetlands	
XZ280	BAe Nimrod AEW3	*Scrapped at Abingdon by June 1992*	
XZ281	BAe Nimrod AEW3	*Scrapped at Abingdon by June 1992*	
XZ282	BAe Nimrod AEW3 (9000M)	RAF NMSU, Kinloss	
XZ283	BAe Nimrod AEW3	*Scrapped at Abingdon by June 1992*	
XZ284	HS Nimrod MR2P	RAF Kinloss MR Wing	
XZ285	BAe Nimrod AEW3	*Scrapped at Abingdon by June 1992*	
XZ287	BAe Nimrod AEW3 (fuselage)	RAF Stafford, BDRT	
XZ290	WS Gazelle AH1 [F]	AAC No 670 Sqn, Middle Wallop	
XZ291	WS Gazelle AH1	AAC No 12 Flt, Wildenrath	
XZ292	WS Gazelle AH1	AAC No 654 Sqn, Detmold	

Notes	Serial	Type	Owner or Operator
	XZ294	WS Gazelle AH1	AAC No 658 Sqn, Netheravon
	XZ295	WS Gazelle AH1	AAC No 12 Flt, Wildenrath
	XZ296	WS Gazelle AH1	AAC 4 Regiment, Detmold
	XZ298	WS Gazelle AH1	AAC No 659 Sqn, Detmold
	XZ299	WS Gazelle AH1 [G1]	AAC No 670 Sqn, Middle Wallop
	XZ300	WS Gazelle AH1 [L]	AAC No 670 Sqn, Middle Wallop
	XZ301	WS Gazelle AH1 [U]	AAC No 670 Sqn, Middle Wallop
	XZ302	WS Gazelle AH1	AAC, stored Fleetlands
	XZ303	WS Gazelle AH1 [S]	AAC No 663 Sqn, Soest
	XZ304	WS Gazelle AH1	AAC, Fleetlands
	XZ305	WS Gazelle AH1	AAC No 665 Sqn, Aldergrove
	XZ307	WS Gazelle AH1	AAC No 665 Sqn, Aldergrove
	XZ308	WS Gazelle AH1	AAC 4 Regiment, Detmold
	XZ309	WS Gazelle AH1	AAC, Fleetlands
	XZ310	WS Gazelle AH1	AAC No 651 Sqn, Hildesheim
	XZ311	WS Gazelle AH1	AAC, Fleetlands
	XZ312	WS Gazelle AH1	AAC No 2 Flight, Netheravon
	XZ313	WS Gazelle AH1 [S]	AAC No 670 Sqn, Middle Wallop
	XZ314	WS Gazelle AH1 [A]	AAC No 656 Sqn, Netheravon
	XZ315	WS Gazelle AH1	AAC No 665 Sqn, Aldergrove
	XZ316	WS Gazelle AH1 [R]	AAC No 670 Sqn, Middle Wallop
	XZ317	WS Gazelle AH1 [Q]	Westland, Weston-super-Mare (on rebuild)
	XZ318	WS Gazelle AH1 [V]	AAC No 656 Sqn, Netheravon
	XZ320	WS Gazelle AH1	RM, Fleetlands
	XZ321	WS Gazelle AH1	AAC No 665 Sqn, Aldergrove
	XZ322	WS Gazelle AH1 [N]	AAC No 670 Sqn, Middle Wallop
	XZ323	WS Gazelle AH1	AAC, Fleetlands
	XZ324	WS Gazelle AH1	AAC, Fleetlands
	XZ325	WS Gazelle AH1 [T]	AAC No 670 Sqn, Middle Wallop
	XZ326	WS Gazelle AH1 [C]	RM, Fleetlands
	XZ327	WS Gazelle AH1 [B1]	AAC No 670 Sqn, Middle Wallop
	XZ328	WS Gazelle AH1	AAC 4 Regiment, Detmold
	XZ329	WS Gazelle AH1 [J]	AAC No 670 Sqn, Middle Wallop
	XZ330	WS Gazelle AH1	AAC No 7 Flt, Gatow
	XZ331	WS Gazelle AH1 [C]	AAC No 664 Sqn, Dishforth
	XZ332	WS Gazelle AH1 [O]	AAC No 670 Sqn, Middle Wallop
	XZ333	WS Gazelle AH1 [A]	AAC No 670 Sqn, Middle Wallop
	XZ334	WS Gazelle AH1 [M]	AAC No 662 Sqn, Soest
	XZ335	WS Gazelle AH1	AAC No 654 Sqn, Detmold
	XZ337	WS Gazelle AH1	AAC 4 Regiment, Detmold
	XZ338	WS Gazelle AH1 [X]	AAC No 670 Sqn, Middle Wallop
	XZ339	WS Gazelle AH1 [F1]	AAC No 670 Sqn, Middle Wallop
	XZ340	WS Gazelle AH1	AAC No 29 Flt, BATUS, Canada
	XZ341	WS Gazelle AH1	AAC No 667 Sqn, Middle Wallop
	XZ342	WS Gazelle AH1 [E]	AAC No 653 Sqn, Soest
	XZ343	WS Gazelle AH1 [U]	AAC 4 Regiment, Detmold
	XZ344	WS Gazelle AH1 [F1]	AAC No 670 Sqn, Middle Wallop
	XZ345	WS Gazelle AH1	AAC Fleetlands
	XZ346	WS Gazelle AH1	AAC No 665 Sqn, Aldergrove
	XZ347	WS Gazelle AH1 [T]	AAC No 657 Sqn, Dishforth
	XZ348	WS Gazelle AH1 (wreckage)	AAC, stored Fleetlands
	XZ349	WS Gazelle AH1 [M]	AAC, stored Fleetlands
	XZ355	SEPECAT Jaguar GR1A [FJ]	RAF No 41 Sqn, Coltishall
	XZ356	SEPECAT Jaguar GR1A [EP]	RAF No 6 Sqn, Coltishall
	XZ357	SEPECAT Jaguar GR1A [K]	RAF No 41 Sqn, Coltishall
	XZ358	SEPECAT Jaguar GR1A [L]	RAF No 41 Sqn, Coltishall
	XZ360	SEPECAT Jaguar GR1A [N]	RAF No 41 Sqn, Coltishall
	XZ361	SEPECAT Jaguar GR1A	RAF No 41 Sqn, Coltishall
	XZ362	SEPECAT Jaguar GR1A [E]	RAF No 41 Sqn, Coltishall
	XZ363	SEPECAT Jaguar GR1A [O]	RAF No 41 Sqn, Coltishall
	XZ363	SEPECAT Jaguar GR1A Replica [A] (BAPC 151)	RAF Exhibition Flight, St Athan
	XZ364	SEPECAT Jaguar GR1A [GJ]	RAF No 54 Sqn, Coltishall
	XZ366	SEPECAT Jaguar GR1A [S]	RAF No 41 Sqn, Coltishall
	XZ367	SEPECAT Jaguar GR1A [GP]	RAF No 54 Sqn, Coltishall
	XZ368	SEPECAT Jaguar GR1 [8900M] [AG]	RAF No 2 SoTT, Cosford
	XZ369	SEPECAT Jaguar GR1A [EF]	RAF No 6 Sqn, Coltishall
	XZ370	SEPECAT Jaguar GR1 [BN] (9004M)	RAF No 2 SoTT, Cosford
	XZ371	SEPECAT Jaguar GR1 (8907M) [AP]	RAF No 2 SoTT, Cosford
	XZ372	SEPECAT Jaguar GR1A [ED]	RAF No 6 Sqn, Coltishall

Serial	Type	Owner or Operator	Notes
XZ373	SEPECAT Jaguar GR1A [GF]	RAF No 54 Sqn, Coltishall	
XZ374	SEPECAT Jaguar GR1 (9005M) [JC]	RAF No 2 SoTT, Cosford	
XZ375	SEPECAT Jaguar GR1A [GR]	RAF No 54 Sqn, Coltishall	
XZ377	SEPECAT Jaguar GR1A [EB]	RAF No 6 Sqn, Coltishall	
XZ378	SEPECAT Jaguar GR1A [EP]	RAF, stored Shawbury	
XZ381	SEPECAT Jaguar GR1A [GB]	RAF No 54 Sqn, Coltishall	
XZ382	SEPECAT Jaguar GR1 (8908M) [AE]	RAF No 1 SoTT, Halton	
XZ383	SEPECAT Jaguar GR1 (8901M) [AF]	RAF No 2 SoTT, Cosford	
XZ384	SEPECAT Jaguar GR1 (8954M) [BC]	RAF No 2 SoTT, Cosford	
XZ385	SEPECAT Jaguar GR1A [GM]	RAF No 54 Sqn, Coltishall	
XZ389	SEPECAT Jaguar GR1 (8946M) [BL]	RAF No 1 SoTT, Halton	
XZ390	SEPECAT Jaguar GR1A (9003M) [35]	RAF No 2 SoTT, Cosford	
XZ391	SEPECAT Jaguar GR1A [05]	RAF No 16(R) Sqn, Lossiemouth	
XZ392	SEPECAT Jaguar GR1A [GQ]	RAF, stored Shawbury	
XZ394	SEPECAT Jaguar GR1A [GN]	RAF No 54 Sqn, Coltishall	
XZ396	SEPECAT Jaguar GR1A [EM]	RAF No 6 Sqn, Coltishall	
XZ398	SEPECAT Jaguar GR1A [FA] (JI007)	RAF No 41 Sqn, Coltishall	
XZ399	SEPECAT Jaguar GR1A [03]	RAF No 16(R) Sqn, Lossiemouth	
XZ400	SEPECAT Jaguar GR1A [EG]	RAF, stored Shawbury	
XZ431	HS Buccaneer S2B	RAF No 12 Sqn, Lossiemouth	
XZ432	HS Buccaneer S2B	RAF Lossiemouth, Fire Section	
XZ439	BAe Sea Harrier FRS2 [2]	MoD(PE)/BAe Dunsfold	
XZ440	BAe Sea Harrier FRS1 [126/N]	RN, BAe, Brough on rebuild	
XZ445	BAe Harrier T4A [721]	RN No 899 Sqn, Yeovilton	
XZ455	BAe Sea Harrier FRS1 [715]	RN, stored St Athan	
XZ457	BAe Sea Harrier FRS1 [710]	RN No 899 Sqn, Yeovilton	
XZ459	BAe Sea Harrier FRS1 [125/N]	RN No 800 Sqn, Yeovilton	
XZ492	BAe Sea Harrier FRS1 [002/R]	RN No 801 Sqn, Yeovilton	
XZ493	BAe Sea Harrier FRS1 [714]	RN No 899 Sqn, Yeovilton	
XZ494	BAe Sea Harrier FRS1 [004]	RN No 801 Sqn, Yeovilton	
XZ495	BAe Sea Harrier FRS1 [129/N]	RN, stored St Athan	
XZ497	BAe Sea Harrier FRS2	MoD(PE)/BAe Dunsfold	
XZ498	BAe Sea Harrier FRS1 [002]	RN AMG Yeovilton	
XZ499	BAe Sea Harrier FRS1 [003/R]	RN No 801 Sqn, Yeovilton	
XZ551	Slingsby Venture T2	*To G-BUGT, April 1992*	
XZ552	Slingsby Venture T2 [A]	*To PH-940, April 1992*	
XZ553	Slingsby Venture T2 [1]	*To G-BUJX, July 1992*	
XZ554	Slingsby Venture T2	*To G-BUHR, April 1992*	
XZ556	Slingsby Venture T2	*To G-BUIH, April 1992*	
XZ557	Slingsby Venture T2 [7]	RAF, stored Little Rissington	
XZ558	Slingsby Venture T2 [8]	Privately owned, Rufforth	
XZ559	Slingsby Venture T2	*To G-BUEK, March 1992*	
XZ560	Slingsby Venture T2	*To G-BUFR, April 1992*	
XZ562	Slingsby Venture T2	*To G-BUJI, May 1992*	
XZ563	Slingsby Venture T2 [3]	*To G-BUDT, February 1992*	
XZ564	Slingsby Venture T2	*To G-BUGV, April 1992*	
XZ570	WS61 Sea King HAS5 (mod)	MoD(PE)/Westland, Yeovil	
XZ571	WS61 Sea King HAS5 [136]	RN No 826 Sqn, Culdrose	
XZ574	WS61 Sea King HAS6 [506]	RNAY Fleetlands	
XZ575	WS61 Sea King HAS6 [599]	RN No 706 Sqn, Culdrose	
XZ575	WS61 Sea King HAS5 [599]	RN No 706 Sqn, Culdrose	
XZ576	WS61 Sea King HAS6	MoD(PE), A&AEE Boscombe Down	
XZ577	WS61 Sea King HAS5 [138]	RN No 826 Sqn, Culdrose	
XZ578	WS61 Sea King HAS5 [581]	RN Culdrose, AMG	
XZ579	WS61 Sea King HAS6 [011/R]	RN No 820 Sqn, Culdrose	
XZ580	WS61 Sea King HAS6 [507]	RN No 810 Sqn, Culdrose	
XZ581	WS61 Sea King HAS6 [133/BD]	RN No 826 Sqn, Boscombe Down	
XZ585	WS61 Sea King HAR3	RAF No 202 Sqn, St Mawgan	
XZ586	WS61 Sea King HAR3	RAF No 202 Sqn, C Flt, Manston	
XZ587	WS61 Sea King HAR3	RAF No 202 Sqn, B Flt, Brawdy	
XZ588	WS61 Sea King HAR3	RAF No 202 Sqn, C Flt, Manston	
XZ589	WS61 Sea King HAR3	RAF No 202 Sqn, A Flt, Boulmer	
XZ590	WS61 Sea King HAR3	RAF No 202 Sqn, A Flt, Boulmer	
XZ591	WS61 Sea King HAR3 [S]	RAF No 78 Sqn, Mount Pleasant, FI	
XZ592	WS61 Sea King HAR3	RAF No 202 Sqn, St Mawgan	
XZ593	WS61 Sea King HAR3	RAF, Fleetlands for repair	

Notes	Serial	Type	Owner or Operator
	XZ594	WS61 Sea King HAR3	RAF No 202 Sqn, D Flt, Lossiemouth
	XZ595	WS61 Sea King HAR3	RAF No 202 Sqn, E Flt, Leconfield
	XZ596	WS61 Sea King HAR3	RAF No 202 Sqn, B Flt, Brawdy
	XZ597	WS61 Sea King HAR3 [S]	RAF SKTU, Culdrose
	XZ598	WS61 Sea King HAR3	RAF SKTU, Culdrose
	XZ599	WS61 Sea King HAR3 [S]	RAF No 202 Sqn, St Mawgan
	XZ605	WS Lynx AH7 [Y]	RM 3 CBAS, Yeovilton
	XZ606	WS Lynx AH7	AAC No 654 Sqn, Detmold
	XZ607	WS Lynx AH7 [M]	AAC No 671 Sqn, Middle Wallop
	XZ608	WS Lynx AH7	AAC No 671 Sqn, Middle Wallop
	XZ609	WS Lynx AH1 [R]	AAC No 657 Sqn, Dishforth
	XZ610	WS Lynx AH7	AAC No 669 Sqn, Detmold
	XZ611	WS Lynx AH1 [H]	AAC No 671 Sqn, Middle Wallop
	XZ612	WS Lynx AH7 [N]	RM 3 CBAS, Yeovilton
	XZ613	WS Lynx AH7	AAC No 655 Sqn, Aldergrove
	XZ614	WS Lynx AH7 [X]	RM 3 CBAS, Yeovilton
	XZ615	WS Lynx AH7	AAC 1 Regiment, Hildesheim
	XZ616	WS Lynx AH7 [P]	AAC No 654 Sqn, Detmold
	XZ617	WS Lynx AH7 [Q]	AAC 4 Regiment, Detmold
	XZ631	Panavia Tornado GR1T	MoD(PE)/BAe Warton
	XZ641	WS Lynx AH7	AAC SAE, Middle Wallop
	XZ642	WS Lynx AH7	AAC No 654 Sqn, Detmold
	XZ643	WS Lynx AH1 [N]	AAC No 662 Sqn, Soest
	XZ644	WS Lynx AH7	AAC 3 Regiment, Soest
	XZ645	WS Lynx AH7	AAC, Fleetlands
	XZ646	WS Lynx AH7 [S]	AAC No 669 Sqn, Detmold
	XZ647	WS Lynx AH7	AAC No 665 Sqn, Aldergrove
	XZ648	WS Lynx AH7 [D]	AAC No 671 Sqn, Middle Wallop
	XZ649	WS Lynx AH7	AAC No 665 Sqn, Aldergrove
	XZ650	WS Lynx AH7	AAC No 659 Sqn, Detmold
	XZ651	WS Lynx AH7	AAC No 651 Sqn, Hildesheim
	XZ652	WS Lynx AH1 [Z]	AAC No 656 Sqn, Netheravon
	XZ653	WS Lynx AH7	AAC No 659 Sqn, Detmold
	XZ654	WS Lynx AH7	AAC No 654 Sqn, Detmold
	XZ655	WS Lynx AH7 [K]	AAC No 671 Sqn, Middle Wallop
	XZ655	WS Lynx AH1	AAC No 652 Sqn, Hildesheim
	XZ661	WS Lynx AH1 [L]	AAC No 671 Sqn, Middle Wallop
	XZ662	WS Lynx AH7	AAC, Fleetlands
	XZ663	WS Lynx AH7	AAC No 665 Sqn, Aldergrove
	XZ664	WS Lynx AH7	AAC No 671 Sqn, Middle Wallop
	XZ665	WS Lynx AH7	AAC No 655 Sqn, Aldergrove
	XZ666	WS Lynx AH7	AAC No 655 Sqn, Aldergrove
	XZ667	WS Lynx AH7	AAC No 665 Sqn, Aldergrove
	XZ668	WS Lynx AH7 [E]	MoD(PE)/Westland, Yeovil
	XZ669	WS Lynx AH7	AAC, Fleetlands
	XZ670	WS Lynx AH7 [X]	AAC 659 Sqn, Detmold
	XZ671	WS Lynx AH9	MoD(PE)/Westland, Yeovil
	XZ672	WS Lynx AH7	AAC 4 Regiment, Detmold
	XZ673	WS Lynx AH7	AAC 3 Regiment, Soest
	XZ674	WS Lynx AH1 [Y]	AAC No 663 Sqn, Soest
	XZ675	WS Lynx AH7 [E]	AAC No 671 Sqn, Middle Wallop
	XZ676	WS Lynx AH7	AAC, Fleetlands
	XZ677	WS Lynx AH7	AAC, Fleetlands
	XZ678	WS Lynx AH7	AAC, Fleetlands
	XZ679	WS Lynx AH7 [Y]	AAC No 659 Sqn, Detmold
	XZ680	WS Lynx AH7 [Z]	AAC No 659 Sqn Detmold
	XZ681	WS Lynx AH1	AAC Middle Wallop, BDRT
	XZ689	WS Lynx HAS3S [353/SD]	RN No 829 Sqn, Portland
	XZ690	WS Lynx HAS3 [635]	RN No 702 Sqn, Portland
	XZ691	WS Lynx HAS3S [432/SC]	RN No 815 Sqn, Portland
	XZ692	WS Lynx HAS3S [643]	RN No 702 Sqn, Portland
	XZ693	WS Lynx HAS3 [640]	RNAY Fleetlands
	XZ694	WS Lynx HAS3 [407/YK]	RN No 815 Sqn, Portland
	XZ695	WS Lynx HAS3S [472/AM]	RN No 829 Sqn, Portland
	XZ696	WS Lynx HAS3S [332/LP]	RN No 815 Sqn, Portland
	XZ697	WS Lynx HAS3CTS [323/AB]	RN No 815 Sqn, Portland
	XZ698	WS Lynx HAS3 [479]	RN No 829 Sqn, Portland
	XZ699	WS Lynx HAS3 [301]	RN No 815 Sqn, Portland
	XZ719	WS Lynx HAS3S [638]	RN No 702 Sqn, Portland
	XZ720	WS Lynx HAS3 [411/EB]	RN No 815 Sqn, Portland
	XZ721	WS Lynx HAS3S [41/NM]	RN No 815 Sqn, Portland
	XZ722	WS Lynx HAS3S	RN No 815 Sqn, Portland
	XZ723	WS Lynx HAS3S [600]	RN No 829 Sqn, Portland

Serial	Type	Owner or Operator	Notes
XZ724	WS Lynx HAS3 [475/HM]	RN No 815 Sqn, Portland	
XZ725	WS Lynx HAS3S [327/AL]	RN No 815 Sqn, Portland	
XZ726	WS Lynx HAS3 [344/GW]	RN No 815 Sqn, Portland	
XZ727	WS Lynx HAS3 [334]	RN No 815 Sqn, Portland	
XZ728	WS Lynx HAS3 [300]	RN No 815 Sqn, Portland	
XZ729	WS Lynx HAS3 [466/AT]	RN No 829 Sqn, Portland	
XZ730	WS Lynx HAS3	RN No 815 Sqn, Portland	
XZ731	WS Lynx HAS3S [641/PO]	RN No 702 Sqn, Portland	
XZ732	WS Lynx HAS3S [417/NM]	RN No 815 Sqn, Portland	
XZ733	WS Lynx HAS3S [328/BA]	RN No 829 Sqn, Portland	
XZ735	WS Lynx HAS3 [602/PO]	RN No 829 Sqn, Portland	
XZ736	WS Lynx HAS3S [424]	RN No 829 Sqn, Portland	
XZ918	WS61 Sea King HAS6 [589]	RN No 706 Sqn, Culdrose	
XZ920	WS61 Sea King HAR5 [822]	RN No 771 Sqn, Culdrose	
XZ921	WS61 Sea King HAS6 [593/CU]	RN No 706 Sqn, Culdrose	
XZ922	WS61 Sea King HAS6 [134/BD]	RN No 826 Sqn, Boscombe Down	
XZ930	WS Gazelle HT3 [Q]	RAF No 2 FTS, Shawbury	
XZ931	WS Gazelle HT3 [R]	RAF No 2 FTS, Shawbury	
XZ932	WS Gazelle HT3 [S]	RAF No 2 FTS, Shawbury	
XZ933	WS Gazelle HT3 [T]	RAF No 2 FTS, Shawbury	
XZ934	WS Gazelle HT3 [U]	RAF No 2 FTS, Shawbury	
XZ935	WS Gazelle HT3	RAF No 32 Sqn, Northolt	
XZ936	WS Gazelle HT2	MoD(PE), ETPS, Boscombe Down	
XZ937	WS Gazelle HT2 [Y]	RAF No 2 FTS, Shawbury	
XZ938	WS Gazelle HT2 [45/CU]	RN No 705 Sqn, Culdrose	
XZ939	WS Gazelle HT2 [Z]	MoD(PE), ETPS, Boscombe Down	
XZ940	WS Gazelle HT2	RAF No 7 Sqn, Odiham	
XZ941	WS Gazelle HT2 [B]	RAF No 2 FTS, Shawbury	
XZ942	WS Gazelle HT2 [42/CU]	RN No 705 Sqn, Culdrose	
XZ964	BAe Harrier GR3	RAF, stored St Athan	
XZ965	BAe Harrier GR3 [L]	RAF, stored St Athan	
XZ966	BAe Harrier GR3 [F]	RAF No 1417 Flt, Belize	
XZ967	BAe Harrier GR3 [F] (9077M)	RAF No 1 SoTT, Halton	
XZ968	BAe Harrier GR3	RAF, stored St Athan	
XZ969	BAe Harrier GR3 [D]	RNEC Manadon	
XZ970	BAe Harrier GR3 [H]	RAF, stored St Athan	
XZ971	BAe Harrier GR3 [U]	RAF HOCU/No 20(R) Sqn, Wittering	
XZ987	BAe Harrier GR3 [C]	RAF Wittering	
XZ990	BAe Harrier GR3 [3D]	*Written off, 14 May 1992*	
XZ991	BAe Harrier GR3 [3A]	RAF Wittering	
XZ993	BAe Harrier GR3 [M]	RAF, stored St Athan	
XZ994	BAe Harrier GR3	RAF, stored Wittering	
XZ995	BAe Harrier GR3 [3G]	RAF HOCU/No 20(R) Sqn, Wittering	
XZ996	BAe Harrier GR3 [G]	RAF No 1417 Flt, Belize	
XZ997	BAe Harrier GR3 [V]	RAF Museum, Hendon	
XZ998	BAe Harrier GR3	RAF, stored St Athan	
ZA101	BAe Hawk 100 (G-HAWK/XX155)	BAe Warton	
ZA105	WS61 Sea King HAR3 [S]	RAF No 78 Sqn, Mount Pleasant, FI	
ZA110	BAe Jetstream T2 [573/CU] (G-AXUO)	RN No 750 Sqn, Culdrose	
ZA111	BAe Jetstream T2 [574/CU] (G-AXFV)	RN No 750 Sqn, Culdrose	
ZA126	WS61 Sea King HAS6 [509]	RN No 810 Sqn, Culdrose	
ZA127	WS61 Sea King HAS6 [504]	RNAY Fleetlands	
ZA128	WS61 Sea King HAS5 [598]	RN No 706 Sqn, Culdrose	
ZA129	WS61 Sea King HAS6 [012/R]	RN No 820 Sqn, Culdrose	
ZA130	WS61 Sea King HAS6 [701/PW]	RN, AMG Culdrose	
ZA131	WS61 Sea King HAS6 [133]	RNAY Fleetlands	
ZA133	WS61 Sea King HAS5 [139]	RN No 826 Sqn, Culdrose	
ZA134	WS61 Sea King HAS6 [702/PW]	RN No 819 Sqn, Prestwick	
ZA135	WS61 Sea King HAS6 [015]	RN No 820 Sqn, Culdrose	
ZA136	WS61 Sea King HAS6 [706/PW]	RN No 819 Sqn, Prestwick	
ZA137	WS61 Sea King HAS6 [137]	RN No 826 Sqn, Culdrose	
ZA140	BAe VC10 K2 (G-ARVL) [A]	RAF No 101 Sqn, Brize Norton	
ZA141	BAe VC10 K2 (G-ARVG) [B]	RAF No 101 Sqn, Brize Norton	
ZA142	BAe VC10 K2 (G-ARVI) [C]	RAF No 101 Sqn, Brize Norton	
ZA143	BAe VC10 K2 (G-ARVK) [D]	RAF No 101 Sqn, Brize Norton	
ZA144	BAe VC10 K2 (G-ARVC) [E]	RAF No 101 Sqn, Brize Norton	
ZA147	BAe VC10 K3 (5H-MMT) [F]	RAF No 101 Sqn, Brize Norton	
ZA148	BAe VC10 K3 (5Y-ADA) [G]	RAF No 101 Sqn, Brize Norton	
ZA149	BAe VC10 K3 (5X-UVJ) [H]	RAF No 101 Sqn, Brize Norton	
ZA150	BAe VC10 K3 (5H-MOG) [J]	RAF No 101 Sqn, Brize Norton	

Notes	Serial	Type	Owner or Operator
	ZA166	WS61 Sea King HAS6 [590]	RN No 706 Sqn, Culdrose
	ZA167	WS61 Sea King HAS5 [131/CM]	RN No 826 Sqn, Culdrose
	ZA168	WS61 Sea King HAS6 [703/PW]	RN No 819 Sqn, Prestwick
	ZA169	WS61 Sea King HAS6 [266/N]	RN No 814 Sqn, Culdrose
	ZA170	WS61 Sea King HAS6 [584]	RN No 706 Sqn, Culdrose
	ZA175	BAe Sea Harrier FRS1 [717]	RN No 899 Sqn, Yeovilton
	ZA176	BAe Sea Harrier FRS2	MoD(PE)/BAe Dunsfold
	ZA193	BAe Sea Harrier FRS1 [126/N]	*Crashed in Sea off Cyprus 28 May 1992*
	ZA195	BAe Sea Harrier FRS2	MoD(PE)/BAe, Dunsfold
	ZA250	BAe Harrier T52 (G-VTOL)	Brooklands Aviation Museum
	ZA254	Panavia Tornado F2	MoD(PE), BAe Warton
	ZA267	Panavia Tornado F2T	MoD(PE), A&AEE Boscombe Down
	ZA283	Panavia Tornado F2	MoD(PE), BAe Warton
	ZA291	WS61 Sea King HC4 [VN]	RN No 846 Sqn, Yeovilton
	ZA292	WS61 Sea King HC4 [ZW]	RN No 707 Sqn, Yeovilton
	ZA293	WS61 Sea King HC4 [VO]	RN No 846 Sqn, Yeovilton
	ZA295	WS61 Sea King HC4 [ZU]	RN No 707 Sqn, Yeovilton
	ZA296	WS61 Sea King HC4 [VK]	RNAY Fleetlands
	ZA297	WS61 Sea King HC4 [25]	RN No 772 Sqn, Portland
	ZA298	WS61 Sea King HC4 (G-BJNM)	RN No 846 Sqn, Yeovilton
	ZA299	WS61 Sea King HC4 [26/PO]	RN No 772 Sqn, Portland
	ZA310	WS61 Sea King HC4 [ZV]	RN No 707 Sqn, Yeovilton
	ZA312	WS61 Sea King HC4 [B]	RN No 845 Sqn, Yeovilton
	ZA313	WS61 Sea King HC4 [E]	RN No 845 Sqn, Yeovilton
	ZA314	WS61 Sea King HC4 [D]	RN No 845 Sqn, Yeovilton
	ZA319	Panavia Tornado GR1T [B-11]	RAF TTTE, Cottesmore
	ZA320	Panavia Tornado GR1T [B-01]	RAF TTTE, Cottesmore
	ZA321	Panavia Tornado GR1 [B-58]	RAF TTTE, Cottesmore
	ZA322	Panavia Tornado GR1 [B-50]	RAF TTTE, Cottesmore
	ZA323	Panavia Tornado GR1T [B-14]	RAF TTTE, Cottesmore
	ZA324	Panavia Tornado GR1T [B-02]	RAF TTTE, Cottesmore
	ZA325	Panavia Tornado GR1T [B-03]	RAF TTTE, Cottesmore
	ZA326	Panavia Tornado GR1T	MoD(PE), DRA Bedford
	ZA327	Panavia Tornado GR1 [B-51]	RAF TTTE, Cottesmore
	ZA328	Panavia Tornado GR1	MoD(PE), BAe Warton
	ZA330	Panavia Tornado GR1T [B-08]	RAF TTTE, Cottesmore
	ZA352	Panavia Tornado GR1T [B-04]	RAF TTTE, Cottesmore
	ZA353	Panavia Tornado GR1 [B-53]	RAF TTTE, Cottesmore
	ZA354	Panavia Tornado GR1	MoD(PE), A&AEE Boscombe Down
	ZA355	Panavia Tornado GR1 [B-54]	RAF TTTE, Cottesmore
	ZA356	Panavia Tornado GR1T [B-07]	RAF TTTE, Cottesmore
	ZA357	Panavia Tornado GR1T [B-05]	RAF TTTE, Cottesmore
	ZA358	Panavia Tornado GR1T [B-06]	RAF TTTE, Cottesmore
	ZA359	Panavia Tornado GR1 [B-55]	RAF TTTE, Cottesmore
	ZA360	Panavia Tornado GR1	RAF, stored St Athan
	ZA361	Panavia Tornado GR1 [B-57]	RAF TTTE, Cottesmore
	ZA362	Panavia Tornado GR1T [B-09]	RAF TTTE, Cottesmore
	ZA365	Panavia Tornado GR1T [GZ]	RAF No 27 Sqn, Marham
	ZA367	Panavia Tornado GR1 [GX]	RAF No 617 Sqn, Marham
	ZA368	Panavia Tornado GR1 [AJ-P]	RAF No 617 Sqn, Marham
	ZA369	Panavia Tornado GR1A [II]	RAF No 2 Sqn, Marham
	ZA370	Panavia Tornado GR1A [A]	RAF No 2 Sqn, Marham
	ZA371	Panavia Tornado GR1A [C]	RAF No 2 Sqn, Marham
	ZA372	Panavia Tornado GR1A [E]	RAF No 2 Sqn, Marham
	ZA373	Panavia Tornado GR1A [H]	RAF No 2 Sqn, Marham
	ZA374	Panavia Tornado GR1 [CN]	RAF No 17 Sqn, Bruggen
	ZA375	Panavia Tornado GR1 [BN]	RAF No 25 Sqn, Marham
	ZA393	Panavia Tornado GR1 [AJ-T]	RAF No 617 Sqn, Marham
	ZA395	Panavia Tornado GR1A [N]	RAF No 2 Sqn, Marham
	ZA397	Panavia Tornado GR1A [O]	RAF No 2 Sqn, Marham
	ZA398	Panavia Tornado GR1A [S]	RAF No 2 Sqn, Marham
	ZA399	Panavia Tornado GR1 [GA]	RAF No 27 Sqn, Marham
	ZA400	Panavia Tornado GR1A [T]	RAF No 2 Sqn, Marham
	ZA401	Panavia Tornado GR1A [R]	RAF No 2 Sqn, Marham
	ZA402	Panavia Tornado GR1	MoD(PE), A&AEE Boscombe Down
	ZA404	Panavia Tornado GR1A [W]	RAF No 2 Sqn, Marham
	ZA405	Panavia Tornado GR1A [Y]	RAF No 2 Sqn, Marham
	ZA406	Panavia Tornado GR1 [AJ-Z]	RAF No 617 Sqn, Marham
	ZA407	Panavia Tornado GR1 [AJ-G]	RAF No 617 Sqn, Marham
	ZA409	Panavia Tornado GR1T [JQ]	RAF No 27 Sqn, Marham
	ZA410	Panavia Tornado GR1T [BZ]	RAF No 14 Sqn, Bruggen
	ZA411	Panavia Tornado GR1T [Z]	RAF No 2 Sqn, Marham

Serial	Type	Owner or Operator	Notes
ZA412	Panavia Tornado GR1T [FX]	RAF, stored St Athan	
ZA446	Panavia Tornado GR1	RAF No 17 Sqn, Bruggen	
ZA446	Panavia Tornado GR1 (Replica) (BAPC155) [F]	RAF Exhibition Flight, St Athan	
ZA447	Panavia Tornado GR1 [JH]	RAF No 27 Sqn, Marham	
ZA449	Panavia Tornado GR1	MoD(PE), BAe Warton	
ZA450	Panavia Tornado GR1 [JC]	RAF No 27 Sqn, Marham	
ZA452	Panavia Tornado GR1 [BP]	RAF No 14 Sqn, Bruggen	
ZA453	Panavia Tornado GR1 [CG]	RAF No 17 Sqn, Bruggen	
ZA455	Panavia Tornado GR1 [DP]	RAF No 31 Sqn, Bruggen	
ZA456	Panavia Tornado GR1 [GB]	RAF No 27 Sqn, Marham	
ZA457	Panavia Tornado GR1 [CE]	RAF No 17 Sqn, Bruggen	
ZA458	Panavia Tornado GR1 [AJ-A]	RAF No 617 Sqn, Marham	
ZA459	Panavia Tornado GR1 [JF]	RAF No 27 Sqn, Marham	
ZA460	Panavia Tornado GR1 [JG]	RAF No 27 Sqn, Marham	
ZA461	Panavia Tornado GR1 [DK]	RAF No 31 Sqn, Bruggen	
ZA462	Panavia Tornado GR1 [AJ-M]	RAF No 617 Sqn, Marham	
ZA463	Panavia Tornado GR1 [AJ-J]	RAF No 617 Sqn, Marham	
ZA465	Panavia Tornado GR1 [CD]	RAF No 17 Sqn, Bruggen	
ZA469	Panavia Tornado GR1 [AJ-O]	RAF No 617 Sqn, Marham	
ZA470	Panavia Tornado GR1 [AJ-L]	RAF No 617 Sqn, Marham	
ZA471	Panavia Tornado GR1 [ER]	RAF No 15(R) Sqn, Honington	
ZA472	Panavia Tornado GR1 [AJ-E]	RAF No 617 Sqn, Marham	
ZA473	Panavia Tronado GR1 [JJ]	RAF No 27 Sqn, Marham	
ZA474	Panavia Tornado GR1 [JK]	RAF No 27 Sqn, Marham	
ZA475	Panavia Tornado GR1 [JL]	RAF No 27 Sqn, Marham	
ZA490	Panavia Tornado GR1 [JM]	RAF No 27 Sqn, Marham	
ZA491	Panavia Tornado GR1 [JN]	RAF No 27 Sqn, Marham	
ZA492	Panavia Tornado GR1 [DJ]	RAF No 31 Sqn, Bruggen	
ZA541	Panavia Tornado GR1T [TO]	RAF No 15(R) Sqn, Honington	
ZA542	Panavia Tornado GR1 [JA]	RAF, stored St Athan	
ZA543	Panavia Tornado GR1	RAF, stored St Athan	
ZA544	Panavia Tornado GR1T [TP]	RAF No 15(R) Sqn, Honington	
ZA546	Panavia Tornado GR1 [JB]	RAF, stored St Athan	
ZA547	Panavia Tornado GR1 [JC]	RAF, stored St Athan	
ZA548	Panavia Tornado GR1T [TQ]	RAF No 15(R) Sqn, Honington	
ZA549	Panavia Tornado GR1T [TR]	RAF No 15(R) Sqn, Honington	
ZA550	Panavia Tornado GR1 [JD]	RAF, stored St Athan	
ZA551	Panavia Tornado GR1T [TS]	RAF No 15(R) Sqn, Honington	
ZA552	Panavia Tornado GR1T [X]	RAF, stored St Athan	
ZA553	Panavia Tornado GR1 [JE]	RAF, stored St Athan	
ZA554	Panavia Tornado GR1 [DM]	RAF No 31 Sqn, Bruggen	
ZA556	Panavia Tornado GR1 [TA]	RAF No 15(R) Sqn, Honington	
ZA557	Panavia Tornado GR1 [TB]	RAF No 15(R) Sqn, Honington	
ZA559	Panavia Tornado GR1	RAF No 15(R) Sqn, Honington	
ZA560	Panavia Tornado GR1 [MC]	RAF, stored St Athan	
ZA562	Panavia Tornado GR1T [TT]	RAF No 15(R) Sqn, Honington	
ZA563	Panavia Tornado GR1 [TC]	RAF No 15(R) Sqn, Honington	
ZA564	Panavia Tornado GR1 [JK]	RAF, stored St Athan	
ZA585	Panavia Tornado GR1 [TD]	RAF No 15(R) Sqn, Honington	
ZA587	Panavia Tornado GR1 [TE]	RAF No 15(R) Sqn, Honington	
ZA588	Panavia Tornado GR1 [B-52]	RAF, stored St Athan	
ZA589	Panavia Tornado GR1 [DC]	RAF No 31 Sqn, Bruggen	
ZA590	Panavia Tornado GR1 [TF]	RAF No 15(R) Sqn, Honington	
ZA591	Panavia Tornado GR1 [JH]	RAF, stored St Athan	
ZA592	Panavia Tornado GR1 [B]	RAF No 617 Sqn, Marham	
ZA594	Panavia Tornado GR1T [TU]	RAF No 15(R) Sqn, Honington	
ZA595	Panavia Tornado GR1T [TV]	RAF No 15(R) Sqn, Honington	
ZA596	Panavia Tornado GR1 [TG]	RAF No 15(R) Sqn, Honington	
ZA597	Panavia Tornado GR1 [M]	MoD(PE)/BAe, on rebuild	
ZA598	Panavia Tornado GR1T [S]	RAF, stored St Athan	
ZA599	Panavia Tornado GR1T [TW]	RAF No 15(R) Sqn, Honington	
ZA600	Panavia Tornado GR1 [TH]	RAF No 15(R) Sqn, Honington	
ZA601	Panavia Tornado GR1 [TI]	RAF No 15(R) Sqn, Honington	
ZA602	Panavia Tornado GR1T [TX]	RAF No 15(R) Sqn, Honington	
ZA604	Panavia Tornado GR1T [TY]	RAF No 15(R) Sqn, Honington	
ZA606	Panavia Tornado GR1	RAF, stored St Athan	
ZA607	Panavia Tornado GR1 [TJ]	RAF No 15(R) Sqn, Honington	
ZA608	Panavia Tornado GR1 [TK]	RAF No 15(R) Sqn, Honington	
ZA609	Panavia Tornado GR1 [J]	RAF, stored St Athan	
ZA611	Panavia Tornado GR1 [A]	MoD(PE), BAe Warton	
ZA612	Panavia Tornado GR1T [TZ]	RAF No 15(R) Sqn, Honington	
ZA613	Panavia Tornado GR1 [N]	RAF, stored St Athan	

Notes	Serial	Type	Owner or Operator
	ZA614	Panavia Tornado GR1 [F]	RAF No 15(R) Sqn, Honington
	ZA626	Slingsby Venture T2	*To G-BUGW, April 1992*
	ZA627	Slingsby Venture T2	*To G-BUDA, February 1992*
	ZA628	Slingsby Venture T2	*To G-BUDB, February 1992*
	ZA630	Slingsby Venture T2	*To G-BUGL, April 1992*
	ZA631	Slingsby Venture T2	*To G-BUFN, April 1992*
	ZA633	Slingsby Venture T2 [3]	*To G-BUNB, August 1992*
	ZA634	Slingsby Venture T2	*To G-BUHA, April 1992*
	ZA652	Slingsby Venture T2	*To G-BUDC, February 1992*
	ZA653	Slingsby Venture T2	*To G-BUJA, May 1992*
	ZA654	Slingsby Venture T2 [4]	RAF, stored Little Rissington
	ZA658	Slingsby Venture T2	*To G-BUFG, April 1992*
	ZA659	Slingsby Venture T2	*To G-BUJB, May 1992*
	ZA660	Slingsby Venture T2	*To G-BUED, March 1992*
	ZA662	Slingsby Venture T2	*To G-BUGZ, April 1992*
	ZA663	Slingsby Venture T2	*To G-BUFP, April 1992*
	ZA664	Slingsby Venture T2 [X]	*To SE-UCF, July 1992*
	ZA665	Slingsby Venture T2	RAF, stored Little Rissington
	ZA670	B-V Chinook HC1 [BG]	RAF No 18 Sqn, Laarbruch
	ZA671	B-V Chinook HC1 [EO]	RAF No 7 Sqn, Odiham
	ZA673	B-V Chinook HC1 [BF]	RAF No 18 Sqn, Laarbruch
	ZA674	B-V Chinook HC2 [BA]	Boeing, Philadelphia (conversion)
	ZA675	B-V Chinook HC1 [BB]	RAF Fleetlands
	ZA676	B-V Chinook HC1 [FG]	RAF, stored Fleetlands
	ZA677	B-V Chinook HC1 [EU]	MoD(PE), A&AEE Boscombe Down
	ZA678	B-V Chinook HC1 [EZ] (wreck)	RAF, stored Fleetlands
	ZA679	B-V Chinook HC1 [EZ]	RAF No 7 Sqn, Odiham
	ZA680	B-V Chinook HC1 [C]	RAF No 78 Sqn, Mount Pleasant, FI
	ZA681	B-V Chinook HC2 [ES]	Boeing, Philadelphia (conversion)
	ZA682	B-V Chinook HC1 [EM]	RAF No 7 Sqn, Odiham
	ZA683	B-V Chinook HC1 [EW]	RAF No 7 Sqn, Odiham
	ZA684	B-V Chinook HC1 [BE]	RAF No 18 Sqn, Laarbruch
	ZA704	B-V Chinook HC1 [EJ]	RAF No 7 Sqn, Odiham
	ZA705	B-V Chinook HC1 [A]	RAF No 78 Sqn, Mount Pleasant, FI
	ZA707	B-V Chinook HC1 [EV]	RAF No 7 Sqn, Odiham
	ZA708	B-V Chinook HC1 [EK]	RAF No 7 Sqn, Odiham
	ZA709	B-V Chinook HC1	RAF No 78 Sqn, Mount Pleasant, FI
	ZA710	B-V Chinook HC2	Boeing, Philadelphia (conversion)
	ZA711	B-V Chinook HC1 [ET]	RAF No 7 Sqn, Odiham
	ZA712	B-V Chinook HC1 [ER]	RAF No 7 Sqn, Odiham
	ZA713	B-V Chinook HC1 [EN]	RAF No 7 Sqn, Odiham
	ZA714	B-V Chinook HC2	Boeing, Philadelphia (conversion)
	ZA717	B-V Chinook HC1 [C] (wreck)	RAF, stored Fleetlands
	ZA718	B-V Chinook HC2	Boeing, Philadelphia (conversion)
	ZA720	B-V Chinook HC1 [EP]	RAF No 7 Sqn, Odiham
	ZA726	WS Gazelle AH1 [V]	AAC No 663 Sqn, Soest
	ZA728	WS Gazelle AH1 [E]	RM 3 CBAS, Yeovilton
	ZA729	WS Gazelle AH1	AAC 1 Regiment, Hildesheim
	ZA730	WS Gazelle AH1	AAC No 665 Sqn, Aldergrove
	ZA731	WS Gazelle AH1 [A]	AAC No 29 Flt, Suffield, Canada
	ZA733	WS Gazelle AH1	AAC No 665 Sqn, Aldergrove
	ZA734	WS Gazelle AH1	AAC, stored Fleetlands
	ZA735	WS Gazelle AH1	AAC No 25 Flt, Belize
	ZA736	WS Gazelle AH1 [S]	AAC No 29 Flt, Suffield, Canada
	ZA737	WS Gazelle AH1 [V]	AAC No 670 Sqn, Middle Wallop
	ZA765	WS Gazelle AH1	AAC
	ZA766	WS Gazelle AH1 [W]	AAC No 663 Sqn, Soest
	ZA767	WS Gazelle AH1	AAC No 25 Flt, Belize
	ZA768	WS Gazelle AH1 [F] (wreck)	*Scrapped at Wroughton*
	ZA769	WS Gazelle AH1 [K]	AAC No 670 Sqn, Middle Wallop
	ZA771	WS Gazelle AH1 [D]	AAC No 664 Sqn, Dishforth
	ZA772	WS Gazelle AH1 [F]	AAC No 656 Sqn, Netheravon
	ZA773	WS Gazelle AH1	AAC No 665 Sqn, Aldergrove
	ZA774	WS Gazelle AH1	AAC No 665 Sqn, Aldergrove
	ZA775	WS Gazelle AH1 [H]	AAC No 656 Sqn, Netheravon
	ZA776	WS Gazelle AH1 [F]	RM 3 CBAS, Yeovilton
	ZA777	WS Gazelle AH1	AAC No 661 Sqn, Hildesheim
	ZA802	WS Gazelle HT3 [W]	RAF No 2 FTS, Shawbury
	ZA803	WS Gazelle HT3 [X]	RAF No 2 FTS, Shawbury
	ZA804	WS Gazelle HT3 [I]	RAF No 2 FTS, Shawbury
	ZA934	WS Puma HC1 [FC]	RAF No 240 OCU, Odiham
	ZA935	WS Puma HC1	RAF No 230 Sqn, Aldergrove
	ZA936	WS Puma HC1 [CU]	RAF No 33 Sqn, Odiham

Serial	Type	Owner or Operator	Notes
ZA937	WS Puma HC1 [CV]	RAF No 33 Sqn, Odiham	
ZA938	WS Puma HC1 [CW]	RAF No 33 Sqn, Odiham	
ZA939	WS Puma HC1	RAF No 230 Sqn, Aldergrove	
ZA940	WS Puma HC1	RAF No 230 Sqn, Algergrove	
ZA947	Douglas Dakota C3	MoD(PE), DRA Farnborough	
ZB506	WS61 Sea King Mk 4X	MoD(PE), DRA Bedford	
ZB507	WS61 Sea King Mk 4X	MoD(PE), DRA Farnborough	
ZB600	BAe Harrier T4 [Z]	RAF HOCU/No 20(R) Sqn, Wittering	
ZB601	BAe Harrier T4	RN AMG, Yeovilton	
ZB602	BAe Harrier T4 [R]	RAF Wittering	
ZB603	BAe Harrier T4 [720]	RN No 899 Sqn, Yeovilton	
ZB604	BAe Harrier T4N [722]	RN No 899 Sqn, Yeovilton	
ZB605	BAe Harrier T8 [721]	MoD(PE), BAe Dunsfold (conversion)	
ZB615	SEPECAT Jaguar T2A	MoD(PE), IAM, Farnborough	
ZB625	WS Gazelle HT3 [N]	RAF No 2 FTS, Shawbury	
ZB626	WS Gazelle HT3 [L]	RAF No 2 FTS, Shawbury	
ZB627	WS Gazelle HT3 [A]	RAF No 7 Sqn, Odiham	
ZB628	WS Gazelle HT3 [V]	RAF No 2 FTS, Shawbury	
ZB629	WS Gazelle HT3	RAF No 32 Sqn, Northolt	
ZB646	WS Gazelle HT2	MoD(PE), DRA Farnborough	
ZB647	WS Gazelle HT2 [59/CU]	RN No 705 Sqn, Culdrose	
ZB648	WS Gazelle HT2 [40/CU]	RN No 705 Sqn, Culdrose	
ZB649	WS Gazelle HT2 [VL]	RN FONA, Yeovilton	
ZB665	WS Gazelle AH1 [AH]	AAC, UNFICYP, Nicosia	
ZB666	WS Gazelle AH1 [G]	AAC No 670 Sqn, Middle Wallop	
ZB667	WS Gazelle AH1	AAC, UNFICYP, Nicosia	
ZB668	WS Gazelle AH1	AAC, UNFICYP, Nicosia	
ZB669	WS Gazelle AH1 [A1]	AAC No 670 Sqn, Middle Wallop	
ZB670	WS Gazelle AH1	AAC No 665 Sqn, Aldergrove	
ZB671	WS Gazelle AH1	AAC No 29 Flt, Suffield, Canada	
ZB672	WS Gazelle AH1 [X]	AAC No 657 Sqn, Dishforth	
ZB673	WS Gazelle AH1 [P]	AAC No 670 Sqn, Middle Wallop	
ZB674	WS Gazelle AH1	AAC No 665 Sqn, Aldergrove	
ZB675	WS Gazelle AH1 [C1]	*Written off, 10 Jan 1991*	
ZB676	WS Gazelle AH1 [E1]	AAC No 670 Sqn, Middle Wallop	
ZB677	WS Gazelle AH1	AAC No 29 Flt, Suffield, Canada	
ZB678	WS Gazelle AH1	*Written off, 11 Feb 1990*	
ZB679	WS Gazelle AH1	AAC No 16 Flt, Dhekelia, Cyprus	
ZB680	WS Gazelle AH1 [B]	AAC No 29 Flt, Suffield, Canada	
ZB681	WS Gazelle AH1	AAC No 665 Sqn, Aldergrove	
ZB682	WS Gazelle AH1	AAC No 665 Sqn, Aldergrove	
ZB683	WS Gazelle AH1	Westland, Weston-super-Mare (rebuild)	
ZB684	WS Gazelle AH1	AAC No 655 Sqn, Ballykelly	
ZB685	WS Gazelle AH1	AAC No 665 Sqn, Aldergrove	
ZB686	WS Gazelle AH1	AAC No 655 Sqn, Ballykelly	
ZB687	WS Gazelle AH1 (wreck)	*Scrapped at Thatcham*	
ZB688	WS Gazelle AH1 [H]	AAC No 670 Sqn, Middle Wallop	
ZB689	WS Gazelle AH1 [W]	AAC No 670 Sqn, Middle Wallop	
ZB690	WS Gazelle AH1	AAC No 16 Flt, Dhekelia, Cyprus	
ZB691	WS Gazelle AH1 [Y]	AAC No 657 Sqn Dishforth	
ZB692	WS Gazelle AH1 [E]	AAC No 664 Sqn, Dishforth	
ZB693	WS Gazelle AH1	AAC No 670 Sqn, Middle Wallop	
ZD230	BAC Super VC10 K4 (G-ASGA)	MoD(PE), BAe Filton	
ZD232	BAC Super VC10 (G-ASGD) (8699M)	RAF Brize Norton Fire Section	
ZD233	BAC Super VC10 (G-ASGE)	*Burnt by June 1992 at Manston*	
ZD235	BAC Super VC10 K4 (G-ASGG)	MoD(PE), BAe Filton	
ZD239	BAC Super VC10 (G-ASGK)	FSCTE, RAF Manston	
ZD240	BAC Super VC10 K4 (G-ASGL)	MoD(PE), BAe Filton	
ZD241	BAC Super VC10 K4 (G-ASGM)	MoD(PE), BAe Filton	
ZD242	BAC Super VC10 K4 (G-ASGP)	MoD(PE), BAe Filton	
ZD243	BAC Super VC10 (G-ASGR)	BAe, Filton (spares use)	
ZD249	WS Lynx HAS3	MoD(PE), A&AEE Boscombe Down	
ZD250	WS Lynx HAS3 [601]	RN No 829 Sqn, Portland	
ZD251	WS Lynx HAS3 [403/BX]	RN No 829 Sqn, Portland	
ZD252	WS Lynx HAS3S [302]	RN No 815 Sqn, Portland	
ZD253	WS Lynx HAS3S [376/XB]	RN No 829 Sqn, Portland	
ZD254	WS Lynx HAS3 [631]	RN, AMG Portland	
ZD255	WS Lynx HAS3S	RN, AMG Portland	
ZD256	WS Lynx HAS3 [333/BM]	RN No 815 Sqn, Portland	
ZD257	WS Lynx HAS3S [420/EX]	RN No 815 Sqn, Portland	

Notes	Serial	Type	Owner or Operator
	ZD258	WS Lynx HAS3S [330/BZ]	RN No 829 Sqn, Portland
	ZD259	WS Lynx HAS3S [342/BT]	RN No 829 Sqn, Portland
	ZD260	WS Lynx HAS3 [348/CM]	RN No 829 Sqn, Portland
	ZD261	WS Lynx HAS3S [636]	RN No 702 Sqn, Portland
	ZD262	WS Lynx HAS3 [632/PO]	RN No 702 Sqn, Portland
	ZD263	WS Lynx HAS3S [630/PO]	RN No 702 Sqn, Portland
	ZD264	WS Lynx HAS3 [361/NF]	RN No 829 Sqn, Portland
	ZD265	WS Lynx HAS3 [338/CT]	RN No 829 Sqn, Portland
	ZD266	WS Lynx HAS3CTS	MoD(PE), Westland, Yeovil
	ZD267	WS Lynx HAS8	MoD(PE), Westland, Yeovil
	ZD268	WS Lynx HAS3S [407]	RN No 815 Sqn, Portland
	ZD272	WS Lynx AH7	AAC, Fleetlands
	ZD273	WS Lynx AH7	AAC, Fleetlands
	ZD274	WS Lynx AH7 [M]	AAC No 662 Sqn, Soest
	ZD275	WS Lynx AH7 [F]	AAC No 653 Sqn, Soest
	ZD276	WS Lynx AH1 [W]	AAC No 656 Sqn, Netheravon
	ZD277	WS Lynx AH7	AAC, Fleetlands
	ZD278	WS Lynx AH1 [P]	AAC No 662 Sqn, Soest
	ZD279	WS Lynx AH7 [C]	AAC No 671 Sqn, Middle Wallop
	ZD280	WS Lynx AH1 [Y]	AAC No 656 Sqn, Netheravon
	ZD281	WS Lynx AH1 [K]	AAC No 671 Sqn, Middle Wallop
	ZD282	WS Lynx AH7 [L]	RM 3 CBAS, Yeovilton
	ZD283	WS Lynx AH1 [Z]	AAC No 663 Sqn, Soest
	ZD284	WS Lynx AH7 [H]	AAC No 653 Sqn, Soest
	ZD285	WS Lynx AH7	MoD(PE), DRA Farnborough
	ZD318	BAe Harrier GR7	MoD(PE), A&AEE Boscombe Down
	ZD319	BAe Harrier GR5	MoD(PE), BAe Dunsfold
	ZD320	BAe Harrier GR5	MoD(PE), BAe Dunsfold
	ZD321	BAe Harrier GR5	MoD(PE), A&AEE Boscombe Down
	ZD322	BAe Harrier GR5 [A]	RAF HOCU/No 20(R) Sqn, Wittering
	ZD323	BAe Harrier GR5 [C]	RAF HOCU/No 20(R) Sqn, Wittering
	ZD324	BAe Harrier GR5 [B]	RAF HOCU/No 20(R) Sqn, Wittering
	ZD326	BAe Harrier GR5 [I]	RAF HOCU/No 20(R) Sqn, Wittering
	ZD327	BAe Harrier GR5 [L]	RAF HOCU/No 20(R) Sqn, Wittering
	ZD328	BAe Harrier GR5 [D]	RAF HOCU/No 20(R) Sqn, Wittering
	ZD329	BAe Harrier GR5 [M]	RAF HOCU/No 20(R) Sqn, Wittering
	ZD330	BAe Harrier GR5 [J]	RAF HOCU/No 20(R) Sqn, Wittering
	ZD345	BAe Harrier GR5	RAF Wittering
	ZD346	BAe Harrier GR5 [E]	RAF HOCU/No 20(R) Sqn, Wittering
	ZD347	BAe Harrier GR5 [K]	RAF HOCU/No 20(R) Sqn, Wittering
	ZD348	BAe Harrier GR5 [G]	RAF HOCU/No 20(R) Sqn, Wittering
	ZD349	BAe Harrier GR5 [L]	RAF HOCU/No 20(R) Sqn, Wittering
	ZD350	BAe Harrier GR5 [05]	RAF Wittering
	ZD351	BAe Harrier GR5	RAF, ASF Wittering
	ZD352	BAe Harrier GR5 [15]	RAF No 1 Sqn, Wittering
	ZD353	BAe Harrier GR5 [H]	RAF, ASF Wittering
	ZD354	BAe Harrier GR5 [C]	RAF HOCU/No 20(R) Sqn, Wittering
	ZD375	BAe Harrier GR5 [AO]	*Crashed 8 August 1992*
	ZD376	BAe Harrier GR5 [F]	RAF Wittering (repair)
	ZD377	BAe Harrier GR5	RAF HOCU/No 20(R) Sqn, Wittering
	ZD378	BAe Harrier GR5	RAF HOCU/No 20(R) Sqn, Wittering
	ZD379	BAe Harrier GR5 [H]	RAF HOCU/No 20(R) Sqn, Wittering
	ZD380	BAe Harrier GR7 [06]	RAF No 1 Sqn, Wittering
	ZD400	BAe Harrier GR5 [02]	RAF
	ZD401	BAe Harrier GR5 [AA]	RAF HOCU/No 20(R) Sqn, Wittering
	ZD402	BAe Harrier GR5	RAF
	ZD403	BAe Harrier GR5	RAF, ASF Wittering
	ZD404	BAe Harrier GR5	RAF
	ZD405	BAe Harrier GR5	RAF
	ZD406	BAe Harrier GR5 [AB]	RAF HOCU/No 20(R) Sqn, Wittering
	ZD407	BAe Harrier GR5 [AC]	RAF HOCU/No 20(R) Sqn, Wittering
	ZD408	BAe Harrier GR5	RAF
	ZD409	BAe Harrier GR5 [06]	RAF
	ZD410	BAe Harrier GR5	RAF
	ZD411	BAe Harrier GR5 [AG]	RAF No RAF HOCU/No 20(R) Sqn, Wittering
	ZD430	BAe Harrier GR7 [AO]	RAF No 3 Sqn, Laarbruch
	ZD431	BAe Harrier GR7 [02]	RAF No 1 Sqn, Wittering
	ZD432	BAe Harrier GR7 [N]	RAF HOCU/No 20(R) Sqn, Wittering
	ZD433	BAe Harrier GR7 [AD]	RAF No 3 Sqn, Laarbruch
	ZD434	BAe Harrier GR7	RAF
	ZD435	BAe Harrier GR7 [04]	RAF No 1 Sqn, Wittering
	ZD436	BAe Harrier GR7	MoD(PE), BAe Dunsfold

Serial	Type	Owner or Operator	Notes
ZD437	BAe Harrier GR7 [05]	RAF No 1 Sqn, Wittering	
ZD438	BAe Harrier GR7 [03]	RAF No 1 Sqn, Wittering	
ZD461	BAe Harrier GR7 [01]	RAF No 1 Sqn, Wittering	
ZD462	BAe Harrier GR7 [07]	RAF No 1 Sqn, Wittering	
ZD463	BAe Harrier GR7 [09]	RAF No 1 Sqn, Wittering	
ZD464	BAe Harrier GR7 [10]	RAF No 1 Sqn, Wittering	
ZD465	BAe Harrier GR7 [11]	RAF No 1 Sqn, Wittering	
ZD466	BAe Harrier GR7	MoD(PE), BAe Dunsfold	
ZD467	BAe Harrier GR7 [AV]	RAF No 3 Sqn, Laarbruch	
ZD468	BAe Harrier GR7 [12]	RAF No 1 Sqn, Wittering	
ZD469	BAe Harrier GR7 [08]	RAF No 1 Sqn, Wittering	
ZD470	BAe Harrier GR7 [01]	RAF No 1 Sqn, Wittering	
ZD472	Harrier GR5 Replica [01] (BAPC191)	RAF Exhibition Flight, St Athan	
ZD476	WS61 Sea King HC4 [ZX]	RN No 707 Sqn, Yeovilton	
ZD477	WS61 Sea King HC4 [A]	RN No 845 Sqn, Yeovilton	
ZD478	WS61 Sea King HC4 [VM]	RNAY Fleetlands	
ZD479	WS61 Sea King HC4 [ZS]	RN No 707 Sqn, Yeovilton	
ZD480	WS61 Sea King HC4 [C]	RN No 845 Sqn, Yeovilton	
ZD559	WS Lynx AH5	MoD(PE), DRA Bedford	
ZD560	WS Lynx AH7	MoD(PE), ETPS, Boscombe Down	
ZD565	WS Lynx HAS3	RN No 702 Sqn, Portland	
ZD566	WS Lynx HAS3S [637]	RN, AMG Portland	
ZD567	WS Lynx HAS3S [322/AV]	RN No 815 Sqn, Portland	
ZD574	B-V Chinook HC1 [EH]	RAF No 7 Sqn, Odiham	
ZD575	B-V Chinook HC1 [BL]	RAF No 18 Sqn, Laarbruch	
ZD576	B-V Chinook HC1 [BC]	RAF No 18 Sqn, Laarbruch	
ZD578	BAe Sea Harrier FRS1 [715]	RN No 899 Sqn, Yeovilton	
ZD579	BAe Sea Harrier FRS1 [124/N]	RN No 800 Sqn, Yeovilton	
ZD580	BAe Sea Harrier FRS1 [122/N]	RN No 800 Sqn, Yeovilton	
ZD581	BAe Sea Harrier FRS1 [718]	RN No 899 Sqn, Yeovilton	
ZD582	BAe Sea Harrier FRS1 [002/R]	RN No 801 Sqn, Yeovilton	
ZD607	BAe Sea Harrier FRS1 [001]	RN No 801 Sqn, Yeovilton	
ZD608	BAe Sea Harrier FRS1 [000/R]	RN No 801 Sqn, Yeovilton	
ZD610	BAe Sea Harrier FRS1 [713]	RN, St Athan	
ZD611	BAe Sea Harrier FRS1 [123/N]	RN No 800 Sqn, Yeovilton	
ZD612	BAe Sea Harrier FRS1 [128/N]	RN, St Athan	
ZD613	BAe Sea Harrier FRS1 [712]	RN No 899 Sqn, Yeovilton	
ZD614	BAe Sea Harrier FRS1 [004/R]	RN No 801 Sqn, Yeovilton	
ZD615	BAe Sea Harrier FRS1 [124/N]	RN, St Athan	
ZD620	BAe 125 CC3	RAF No 32 Sqn, Northolt	
ZD621	BAe 125 CC3	RAF No 32 Sqn, Northolt	
ZD625	WS61 Sea King HC4 [ZZ]	RN No 707 Sqn, Yeovilton	
ZD626	WS61 Sea King HC4 [ZY]	RN No 707 Sqn, Yeovilton	
ZD627	WS61 Sea King HC4	RN No 707 Sqn, Yeovilton	
ZD630	WS61 Sea King HAS6 [265]	RN No 814 Sqn, Culdrose	
ZD631	WS61 Sea King HAS6 [266/N] (wreck)	RN, Fleetlands	
ZD633	WS61 Sea King HAS6 [267]	RN No 814 Sqn, Culdrose	
ZD634	WS61 Sea King HAS6 [506/CU]	RN No 810 Sqn, Culdrose	
ZD636	WS61 Sea King HAS5 [704/PW]	RN No 819 Sqn, Prestwick	
ZD637	WS61 Sea King HAS6 [268]	RN No 814 Sqn, Culdrose	
ZD657	Schleicher Valiant TX1 (BGA2893)	RAF No 622 VGS, Upavon	
ZD658	Schleicher Valiant TX1 (BGA2894)	RAF stored, Syerston	
ZD659	Schleicher Valiant TX1 (BGA2895)	RAF No 645 VGS, Catterick	
ZD660	Schleicher Valiant TX1 (BGA2896)	RAF stored, Syerston	
ZD667	BAe Harrier GR3	RAF Wittering	
ZD668	BAe Harrier GR3 [3E]	RAF HOCU/No 20(R) Sqn, Wittering	
ZD669	BAe Harrier GR3 [3B]	RAF Wittering	
ZD670	BAe Harrier GR3 [3C]	RAF HOCU/No 20(R) Sqn, Wittering	
ZD703	BAe 125 CC3	RAF No 32 Sqn, Northolt	
ZD704	BAe 125 CC3	RAF No 32 Sqn, Northolt	
ZD707	Panavia Tornado GR1 [BK]	RAF No 14 Sqn, Bruggen	
ZD708	Panavia Tornado GR1	MoD(PE), BAe Warton/A&AEE Boscombe Down	
ZD709	Panavia Tornado GR1 [BR]	RAF No 14 Sqn, Bruggen	
ZD711	Panavia Tornado GR1T [DY]	RAF No 31 Sqn, Bruggen	
ZD712	Panavia Tornado GR1T [BY]	RAF No 14 Sqn, Bruggen	
ZD713	Panavia Tornado GR1T [BX]	RAF No 14 Sqn, Bruggen	

Notes	Serial	Type	Owner or Operator
	ZD714	Panavia Tornado GR1 [AP]	RAF No 9 Sqn, Bruggen
	ZD715	Panavia Tornado GR1 [DB]	RAF No 31 Sqn, Bruggen
	ZD716	Panavia Tornado GR1 [O]	RAF SAOEU, Boscombe Down
	ZD719	Panavia Tornado GR1 [AD]	RAF No 9 Sqn, Bruggen
	ZD720	Panavia Tornado GR1 [AG]	RAF No 9 Sqn, Bruggen
	ZD739	Panavia Tornado GR1 [AC]	RAF No 9 Sqn, Bruggen
	ZD740	Panavia Tornado GR1 [DA]	RAF No 31 Sqn, Bruggen
	ZD741	Panavia Tornado GR1T [DZ]	RAF No 31 Sqn, Bruggen
	ZD742	Panavia Tornado GR1T [CZ]	RAF No 17 Sqn, Bruggen
	ZD743	Panavia Tornado GR1 [CX]	RAF No 17 Sqn, Bruggen
	ZD744	Panavia Tornado GR1 [BD]	RAF No 14 Sqn, Bruggen
	ZD745	Panavia Tornado GR1 [BM]	RAF No 14 Sqn, Bruggen
	ZD746	Panavia Tornado GR1 [AB]	RAF No 9 Sqn, Bruggen
	ZD747	Panavia Tornado GR1 [AL]	RAF No 9 Sqn, Bruggen
	ZD748	Panavia Tornado GR1 [AK]	RAF No 9 Sqn, Bruggen
	ZD749	Panavia Tornado GR1 [U]	RAF SAOEU, Boscombe Down
	ZD788	Panavia Tornado GR1 [CB]	RAF No 17 Sqn, Bruggen
	ZD789	Panavia Tornado GR1 [AM]	RAF No 9 Sqn, Bruggen
	ZD790	Panavia Tornado GR1 [DL]	RAF No 31 Sqn, Bruggen
	ZD792	Panavia Tornado GR1 [CF]	RAF No 17 Sqn, Bruggen
	ZD793	Panavia Tornado GR1 [CA]	RAF No 17 Sqn, Bruggen
	ZD809	Panavia Tornado GR1 [BA]	RAF No 14 Sqn, Bruggen
	ZD810	Panavia Tornado GR1 [AA]	RAF No 9 Sqn, Bruggen
	ZD811	Panavia Tornado GR1 [DF]	RAF No 31 Sqn, Bruggen
	ZD812	Panavia Tornado GR1T [DX]	RAF No 31 Sqn, Bruggen
	ZD842	Panavia Tornado GR1T [CY]	RAF No 17 Sqn, Bruggen
	ZD843	Panavia Tornado GR1 [DH]	RAF No 31 Sqn, Bruggen
	ZD844	Panavia Tornado GR1 [DE]	RAF No 31 Sqn, Bruggen
	ZD845	Panavia Tornado GR1 [AF]	RAF No 9 Sqn, Bruggen
	ZD846	Panavia Tornado GR1 [BL]	RAF No 14 Sqn, Bruggen
	ZD847	Panavia Tornado GR1 [CH]	RAF No 17 Sqn, Bruggen
	ZD848	Panavia Tornado GR1	RAF No 14 Sqn, Bruggen
	ZD849	Panavia Tornado GR1 [AJ-F]	RAF No 617 Sqn, Marham
	ZD850	Panavia Tornado GR1 [CL]	RAF No 17 Sqn, Bruggen
	ZD851	Panavia Tornado GR1 [AJ]	RAF No 9 Sqn, Bruggen
	ZD890	Panavia Tornado GR1 [AE]	RAF No 9 Sqn, Bruggen
	ZD892	Panavia Tornado GR1 [BJ]	RAF No 14 Sqn, Bruggen
	ZD895	Panavia Tornado GR1 [BF]	RAF No 14 Sqn, Bruggen
	ZD899	Panavia Tornado F2T	MoD(PE), BAe Warton
	ZD900	Panavia Tornado F2T	MoD(PE), A&AEE Boscombe Down
	ZD901	Panavia Tornado F2T [AA]	RAF, stored St Athan
	ZD902	Panavia Tornado F2T	MoD(PE), DRA Farnborough
	ZD903	Panavia Tornado F2T [AB]	RAF, stored St Athan
	ZD904	Panavia Tornado F2T [AE]	RAF, stored St Athan
	ZD905	Panavia Tornado F2 [AV]	RAF, stored St Athan
	ZD906	Panavia Tornado F2 [AN]	RAF, stored St Athan
	ZD932	Panavia Tornado F2 [AM]	RAF, stored St Athan
	ZD933	Panavia Tornado F2 [AO]	RAF, stored St Athan
	ZD934	Panavia Tornado F2T [AD]	RAF, stored St Athan
	ZD935	Panavia Tornado F2T	RAF, stored Coningsby
	ZD936	Panavia Tornado F2 [AP]	RAF, stored St Athan
	ZD937	Panavia Tornado F2 [AQ]	RAF, stored St Athan
	ZD938	Panavia Tornado F2 [AR]	RAF, stored St Athan
	ZD939	Panavia Tornado F2 [AS]	MoD(PE), BAe Warton
	ZD940	Panavia Tornado F2	RAF, stored St Athan
	ZD941	Panavia Tornado F2 [AU]	RAF, stored St Athan
	ZD948	Lockheed TriStar KC1 (G-BFCA)	RAF No 216 Sqn, Brize Norton
	ZD949	Lockheed TriStar K1 (G-BFCB)	RAF No 216 Sqn, Brize Norton
	ZD950	Lockheed TriStar KC1 (G-BFCC)	RAF No 216 Sqn, Brize Norton
	ZD951	Lockheed TriStar K1 (G-BFCD)	RAF No 216 Sqn, Brize Norton
	ZD952	Lockheed TriStar KC1 (G-BFCE)	RAF No 216 Sqn, Brize Norton
	ZD953	Lockheed TriStar KC1 (G-BFCF)	RAF No 216 Sqn, Brize Norton
	ZD974	Schempp-Hirth Kestrel TX1 (BGA2875)	RAF, stored Syerston
	ZD975	Schempp-Hirth Kestrel TX1 (BGA2876)	RAF, stored Syerston
	ZD980	B-V Chinook HC1	RAF No 240 OCU, Odiham
	ZD981	B-V Chinook HC1 [BD]	RAF No 18 Sqn, Laarbruch
	ZD982	B-V Chinook HC1 [BI]	RAF No 18 Sqn, Laarbruch
	ZD983	B-V Chinook HC1 [EF]	RAF No 7 Sqn, Odiham
	ZD984	B-V Chinook HC1 [EE]	RAF No 7 Sqn, Odiham
	ZD990	BAe Harrier T4A [Q]	RAF HOCU/No 20(R) Sqn, Wittering
	ZD991	BAe Harrier T4 [V]	RAF Wittering

Serial	Type	Owner or Operator	Notes
ZD992	BAe Harrier T4 [Y]	RAF Station Flight Laarbruch	
ZD993	BAe Harrier T4 [U]	RAF HOCU/No 20(R) Sqn, Wittering	
ZD996	Panavia Tornado GR1A [I]	RAF No 2 Sqn, Marham	
ZE116	Panavia Tornado GR1A [DG]	RAF No 31 Sqn, Bruggen	
ZE154	Panavia Tornado F3T [AD]	RAF F3 OCU/No 56(R) Sqn, Coningsby	
ZE155	Panavia Tornado F3	MoD(PE) A&AEE Boscombe Down	
ZE156	Panavia Tornado F3 [HE]	RAF No 111 Sqn, Leuchars	
ZE157	Panavia Tornado F3T [AB]	RAF F3 OCU/No 56(R) Sqn, Coningsby	
ZE158	Panavia Tornado F3 [DC]	RAF Leeming	
ZE159	Panavia Tornado F3 [EC]	RAF No 23 Sqn,Leeming	
ZE160	Panavia Tornado F3T [BY]	RAF No 29 Sqn, Coningsby	
ZE161	Panavia Tornado F3 [FG]	RAF No 25 Sqn, Leeming	
ZE162	Panavia Tornado F3 [FK]	RAF No 25 Sqn, Leeming	
ZE163	Panavia Tornado F3T [AA]	RAF F3 OCU/No 56(R) Sqn, Coningsby	
ZE164	Panavia Tornado F3 [DA]	RAF No 11 Sqn, Leeming	
ZE165	Panavia Tornado F3 [BJ]	RAF No 29 Sqn, Coningsby	
ZE166	Panavia Tornado F3T [AF]	RAF F3 OCU/No 56(R) Sqn, Coningsby	
ZE167	Panavia Tornado F3 [HM]	RAF No 111 Sqn, Leuchars	
ZE168	Panavia Tornado F3 [EB]	RAF No 23 Sqn, Leeming	
ZE199	Panavia Tornado F3T [FL]	RAF No 25 Sqn, Leeming	
ZE200	Panavia Tornado F3 [DB]	RAF No 11 Sqn, Leeming	
ZE201	Panavia Tornado F3 [ED]	RAF No 23 Sqn, Leeming	
ZE202	Panavia Tornado F3T [AH]	RAF F3 OCU/No 56(R) Sqn, Coningsby	
ZE203	Panavia Tornado F3 [FI]	RAF No 25 Sqn, Leeming	
ZE204	Panavia Tornado F3 [DD]	RAF No 11 Sqn, Leeming	
ZE205	Panavia Tornado F3T	RAF F3 OCU/No 56(R) Sqn, Coningsby	
ZE206	Panavia Tornado F3 [EW]	RAF No 23 Sqn, Leeming	
ZE207	Panavia Tornado F3 [GC]	RAF No 43 Sqn, Leuchars	
ZE208	Panavia Tornado F3T	RAF F3 OCU/No 56(R) Sqn, Coningsby	
ZE209	Panavia Tornado F3	RAF 1435 Flt, Mount Pleasant FI	
ZE210	Panavia Tornado F3 [FB]	RAF No 25 Sqn, Leeming	
ZE250	Panavia Tornado F3T [HZ]	RAF, stored Leuchars	
ZE251	Panavia Tornado F3 [DE]	RAF No 11 Sqn, Leeming	
ZE252	Panavia Tornado F3 [HH]	RAF No 111 Sqn, Leuchars	
ZE253	Panavia Tornado F3T	RAF F3 OCU/No 56(R) Sqn, Coningsby	
ZE254	Panavia Tornado F3 [CA]	RAF No 25 Sqn, Coningsby	
ZE255	Panavia Tornado F3 [HI]	RAF No 111 Sqn, Leuchars	
ZE256	Panavia Tornado F3T [AJ]	RAF F3 OCU/No 56(R) Sqn, Coningsby	
ZE257	Panavia Tornado F3 [BD]	RAF No 29 Sqn, Coningsby	
ZE258	Panavia Tornado F3 [GA]	RAF No 43 Sqn, Leuchars	
ZE287	Panavia Tornado F3T [AH]	RAF F3 OCU/No 56(R) Sqn, Coningsby	
ZE288	Panavia Tornado F3 [GG]	RAF No 43 Sqn, Leuchars	
ZE289	Panavia Tornado F3 [HF]	RAF No 111 Sqn, Leuchars	
ZE290	Panavia Tornado F3T [AD]	RAF F3 OCU/No 56(R) Sqn, Coningsby	
ZE291	Panavia Tornado F3 [AZ]	RAF F3 OCU/No 56(R) Sqn, Coningsby	
ZE292	Panavia Tornado F3 [AU]	RAF F3 OCU/No 56(R) Sqn, Coningsby	
ZE293	Panavia Tornado F3T [AC]	RAF No 111 Sqn, Leuchars	
ZE294	Panavia Tornado F3 [AQ]	RAF F3 OCU/No 56(R) Sqn, Coningsby	
ZE295	Panavia Tornado F3 [AR]	RAF F3 OCU/No 56(R) Sqn, Coningsby	
ZE296	Panavia Tornado F3T [AM]	RAF F3 OCU/No 56(R) Sqn, Coningsby	
ZE338	Panavia Tornado F3 [HG]	RAF No 111 Sqn, Leuchars	
ZE339	Panavia Tornado F3	RAF No 25 Sqn, Leeming	
ZE340	Panavia Tornado F3T [AE]	RAF F3 OCU/No 56(R) Sqn, Coningsby	
ZE341	Panavia Tornado F3 [HD]	RAF No 111 Sqn, Leuchars	
ZE342	Panavia Tornado F3	RAF No 29 Sqn, Coningsby	
ZE343	Panavia Tornado F3T [AI]	RAF F3 OCU/No 56(R) Sqn, Coningsby	
ZE351	McD Phantom F-4J(UK) [I] (9058M)	RAF Finningley Fire Section	
ZE353	McD Phantom F-4J(UK) [E] (9083M)	RAF Manston on display	
ZE354	McD Phantom F-4J(UK)	RAF Coningsby Fire Section	
ZE356	McD Phantom F-4J(UK) [Q] (9060M)	RAF Waddington Fire Section	
ZE359	McD Phantom F-4J(UK) [J]	Imperial War Museum, Duxford	
ZE360	McD Phantom F-4J(UK) [O] (9059M)	RAF FSCTE, Manston	
ZE361	McD Phantom F-4J(UK) [P] (9057M)	RAF Honington Fire Section	
ZE364	McD Phantom F-4J(UK) [Z] (9085M)	RAF Coltishall Fire Section	
ZE368	WS61 Sea King HAR3	RAF No 202 Sqn, D Flt, Lossiemouth	
ZE369	WS61 Sea King HAR3	RAF No 202 Sqn, St Mawgan	

Notes	Serial	Type	Owner or Operator
	ZE370	WS61 Sea King HAR3	RAF No 202 Sqn, E Flt, Leconfield
	ZE375	WS Lynx AH7 [L]	AAC No 665 Sqn, Aldergrove
	ZE376	WS Lynx AH7	MoD(PE) Westlands, Yeovil
	ZE378	WS Lynx AH7	MoD(PE) A&AEE Boscombe Down
	ZE379	WS Lynx AH7	AAC No 665 Sqn, Aldergrove
	ZE380	WS Lynx AH7	AAC No 665 Sqn, Aldergrove
	ZE381	WS Lynx AH7	AAC No 665 Sqn, Aldergrove
	ZE382	WS Lynx AH9	Westland, Yeovil (conversion)
	ZE395	BAe 125 CC3	RAF No 32 Sqn, Northolt
	ZE396	BAe 125 CC3	RAF No 32 Sqn, Northolt
	ZE410	Agusta A109A (AE-334)	AAC No 8 Flight, Netheravon
	ZE411	Agusta A109A (AE-331)	AAC No 8 Flight, Netheravon
	ZE412	Agusta A109A	AAC No 8 Flight, Netheravon
	ZE413	Agusta A109A	AAC No 8 Flight, Netheravon
	ZE418	WS61 Sea King HAS5 [701/PW]	RN No 819 Sqn, Prestwick
	ZE419	WS61 Sea King HAS6 [269]	RN No 814 Sqn, Culdrose
	ZE420	WS61 Sea King HAS6 [138]	RN No 826 Sqn, Culdrose
	ZE422	WS61 Sea King HAS6 [130]	RN No 826 Sqn, Culdrose
	ZE425	WS61 Sea King HC4 [VJ]	RN No 846 Sqn, Yeovilton
	ZE426	WS61 Sea King HC4	RN, AMG Yeovilton
	ZE427	WS61 Sea King HC4 [VH]	RN No 846 Sqn, Yeovilton
	ZE428	WS61 Sea King HC4 [G]	RN No 845 Sqn, Yeovilton
	ZE432	BAC 1-11/479 (DQ-FBV)	MoD(PE) ETPS, Boscombe Down
	ZE433	BAC 1-11/479 (DQ-FBQ)	MoD(PE) DRA Bedford
	ZE438	BAe Jetstream T3 [576]	RN FONA/Heron Flight, Yeovilton
	ZE439	BAe Jetstream T3 [577]	RN FONA/Heron Flight, Yeovilton
	ZE440	BAe Jetstream T3 [578]	RN, stored St Athan
	ZE441	BAe Jetstream T3 [579]	RN FONA/Heron Flight, Yeovilton
	ZE477	WS Lynx 3	IHM, Weston-super-Mare
	ZE495	Grob Viking T1 (BGA3000)	RAF No 622 VGS, Upavon
	ZE496	Grob Viking T1 (BGA3001)	RAF CGMF, Syerston
	ZE497	Grob Viking T1 (BGA3002)	RAF No 662 VGS, Arbroath
	ZE498	Grob Viking T1 (BGA3003)	RAF CGMF, Syerston
	ZE499	Grob Viking T1 (BGA3004)	RAF ACCGS, Syerston
	ZE501	Grob Viking T1 (BGA3006)	RAF ACCGS, Syerston
	ZE502	Grob Viking T1 (BGA3007)	RAF No 645 VGS, Catterick
	ZE503	Grob Viking T1 (BGA3008)	RAF No 625 VGS, South Cerney
	ZE504	Grob Viking T1 (BGA3009)	RAF No 645 VGS, Catterick
	ZE520	Grob Viking T1 (BGA3010)	RAF No 622 VGS, Upavon
	ZE521	Grob Viking T1 (BGA3011)	RAF CGMF, Syerston
	ZE522	Grob Viking T1 (BGA3012)	RAF No 618 VGS, West Malling
	ZE524	Grob Viking T1 (BGA3014)	RAF CGMF, Syerston
	ZE525	Grob Viking T1 (BGA3015)	ATC, Congleton, Cheshire
	ZE526	Grob Viking T1 (BGA3016)	RAF No 626 VGS, Predannack
	ZE527	Grob Viking T1 (BGA3017)	RAF CGMF, Syerston
	ZE528	Grob Viking T1 (BGA3018)	RAF No 618 VGS, West Malling
	ZE529	Grob Viking T1 (BGA3019)	RAF ACCGS, Syerston
	ZE530	Grob Viking T1 (BGA3020)	RAF No 645 VGS, Catterick
	ZE531	Grob Viking T1 (BGA3021)	RAF No 631 VGS, Samlesbury
	ZE532	Grob Viking T1 (BGA3022)	RAF No 618 VGS, West Malling
	ZE533	Grob Viking T1 (BGA3023)	RAF No 622 VGS, Upavon
	ZE534	Grob Viking T1 (BGA3024)	RAF No 662 VGS, Arbroath
	ZE550	Grob Viking T1 (BGA3025)	RAF No 622 VGS, Upavon
	ZE551	Grob Viking T1 (BGA3026)	RAF No 611 VGS, Swanton Morley
	ZE552	Grob Viking T1 (BGA3027)	RAF No 625 VGS, South Cerney
	ZE553	Grob Viking T1 (BGA3028)	RAF No 611 VGS, Swanton Morley
	ZE554	Grob Viking T1 (BGA3029)	RAF RGE, Chipping, Lancs
	ZE555	Grob Viking T1 (BGA3030)	RAF No 631 VGS, Sealand
	ZE556	Grob Viking T1 (BGA3031)	RAF RGE, Chipping, Lancs
	ZE557	Grob Viking T1 (BGA3032)	RAF No 661 VGS, Kirknewton
	ZE558	Grob Viking T1 (BGA3033)	RAF No 634 VGS, St Athan
	ZE559	Grob Viking T1 (BGA3034)	RAF No 645 VGS, Catterick
	ZE560	Grob Viking T1 (BGA3035)	RAF No 611 VGS, Swanton Morley
	ZE561	Grob Viking T1 (BGA3036)	RAF No 621 VGS, Weston-super-Mare
	ZE562	Grob Viking T1 (BGA3037)	RAF No 631 VGS, Sealand
	ZE563	Grob Viking T1 (BGA3038)	RAF No 634 Sqn, St Athan
	ZE564	Grob Viking T1 (BGA3039)	RAF No 661 VGS, Kirknewton
	ZE584	Grob Viking T1 (BGA3040)	RAF No 618 VGS, West Malling
	ZE585	Grob Viking T1 (BGA3041)	RAF No 614 VGS, Wethersfield
	ZE586	Grob Viking T1 (BGA3042)	RAF No 625 VGS, South Cerney
	ZE587	Grob Viking T1 (BGA3043)	RAF No 622 VGS, Upavon
	ZE589	Grob Viking T1 (BGA3045)	RAF No 634 VGS, St Athan
	ZE590	Grob Viking T1 (BGA3046)	RAF No 634 VGS, St Athan

Serial	Type	Owner or Operator	Notes
ZE591	Grob Viking T1 (BGA3047)	RAF No 662 VGS, Arbroath	
ZE592	Grob Viking T1 (BGA3048)	RAF No 625 VGS, South Cerney	
ZE593	Grob Viking T1 (BGA3049)	RAF No 631 VGS, Sealand	
ZE594	Grob Viking T1 (BGA3050)	RAF No 621 VGS, Weston-super-Mare	
ZE595	Grob Viking T1 (BGA3051)	RAF No 662 VGS, Arbroath	
ZE600	Grob Viking T1 (BGA3052)	RAF No 618 VGS, West Malling	
ZE601	Grob Viking T1 (BGA3053)	RAF No 625 VGS, South Cerney	
ZE602	Grob Viking T1 (BGA3054)	RAF No 617 VGS, Manston	
ZE603	Grob Viking T1 (BGA3055)	RAF No 617 VGS, Manston	
ZE604	Grob Viking T1 (BGA3056)	RAF ACCGS, Syerston	
ZE605	Grob Viking T1 (BGA3057)	RAF No 618 VGS, West Malling	
ZE606	Grob Viking T1 (BGA3058)	RAF No 614 VGS, Wethersfield	
ZE607	Grob Viking T1 (BGA3059)	RAF No 614 VGS, Wethersfield	
ZE608	Grob Viking T1 (BGA3060)	RAF No 625 VGS, South Cerney	
ZE609	Grob Viking T1 (BGA3061)	RAF No 622 VGS, Upavon	
ZE610	Grob Viking T1 (BGA3062)	RAF No 615 VGS, Kenley	
ZE611	Grob Viking T1 (BGA3063)	RAF No 621 VGS, Weston-super-Mare	
ZE612	Grob Viking T1 (BGA3064)	*Scrapped at Syerston*	
ZE613	Grob Viking T1 (BGA3065)	RAF No 621 VGS, Weston-super-Mare	
ZE614	Grob Viking T1 (BGA3066)	RAF No 661 VGS, Kirknewton	
ZE625	Grob Viking T1 (BGA3067)	RAF No 643 VGS, Binbrook	
ZE626	Grob Viking T1 (BGA3068)	RAF ACCGS, Syerston	
ZE627	Grob Viking T1 (BGA3069)	RAF ACCGS, Syerston	
ZE628	Grob Viking T1 (BGA3070)	RAF No 625 VGS, South Cerney	
ZE629	Grob Viking T1 (BGA3071)	RAF ACCGS, Syerston	
ZE630	Grob Viking T1 (BGA3072)	RAF No 636 VGS, Swansea	
ZE631	Grob Viking T1 (BGA3073)	RAF No 643 VGS, Binbrook	
ZE632	Grob Viking T1 (BGA3074)	RAF No 614 VGS, Wethersfield	
ZE633	Grob Viking T1 (BGA3075)	RAF No 615 VGS, Kenley	
ZE634	Grob Viking T1 (BGA3076)	*Scrapped at Syerston*	
ZE635	Grob Viking T1 (BGA3077)	RAF No 631 VGS, Sealand	
ZE636	Grob Viking T1 (BGA3078)	RAF No 622 VGS, Upavon	
ZE637	Grob Viking T1 (BGA3079)	RAF No 614 VGS, Wethersfield	
ZE650	Grob Viking T1 (BGA3080)	RAF No 626 VGS, Predannack	
ZE651	Grob Viking T1 (BGA3081)	RAF CGMF, Syerston	
ZE652	Grob Viking T1 (BGA3082)	RAF No 614 VGS, Wethersfield	
ZE653	Grob Viking T1 (BGA3083)	RAF No 614 VGS, Wethersfield	
ZE654	Grob Viking T1 (BGA3084)	RAF No 618 VGS, West Malling	
ZE655	Grob Viking T1 (BGA3085)	RAF No 618 VGS, West Malling	
ZE656	Grob Viking T1 (BGA3086)	RAF No 617 VGS, Manston	
ZE657	Grob Viking T1 (BGA3087)	RAF No 618 VGS, West Malling	
ZE658	Grob Viking T1 (BGA3088)	RAF No 621 VGS, Weston-super-Mare	
ZE659	Grob Viking T1 (BGA3089)	RAF No 611 VGS, Swanton Morley	
ZE677	Grob Viking T1 (BGA3090)	RAF No 615 VGS, Kenley	
ZE678	Grob Viking T1 (BGA3091)	RAF No 645 VGS, Catterick	
ZE679	Grob Viking T1 (BGA3092)	RAF No 631 VGS, Samlesbury	
ZE680	Grob Viking T1 (BGA3093)	RAF No 617 VGS, Manston	
ZE681	Grob Viking T1 (BGA3094)	RAF No 617 VGS, Manston	
ZE682	Grob Viking T1 (BGA3095)	RAF No 636 VGS, Swansea	
ZE683	Grob Viking T1 (BGA3096)	RAF No 662 VGS, Arbroath	
ZE684	Grob Viking T1 (BGA3097)	RAF No 661 VGS, Kirknewton	
ZE685	Grob Viking T1 (BGA3098)	RAF No 661 VGS, Kirknewton	
ZE686	Grob Viking T1 (BGA3099)	MoD(PE), Slingsby, Kirkbymoorside	
ZE690	BAe Sea Harrier FRS1 [711]	RN No 899 Sqn, Yeovilton	
ZE691	BAe Sea Harrier FRS1 [124/N]	RN No 800 Sqn, Yeovilton	
ZE692	BAe Sea Harrier FRS1 [713]	RN No 899 Sqn, Yeovilton	
ZE693	BAe Sea Harrier FRS1 [127]	RN No 800 Sqn, Yeovilton	
ZE694	BAe Sea Harrier FRS1	RN, AMG Yeovilton	
ZE695	BAe Sea Harrier FRS2	MoD(PE), BAe Dunsfold	
ZE696	BAe Sea Harrier FRS1 [001/R]	RN No 801 Sqn, Yeovilton	
ZE697	BAe Sea Harrier FRS1	RN, St Athan	
ZE698	BAe Sea Harrier FRS1 [716]	RN No 899 Sqn, Yeovilton	
ZE700	BAe 146 CC2	RAF Queen's Flight, Benson	
ZE701	BAe 146 CC2	RAF Queen's Flight, Benson	
ZE702	BAe 146 CC2	RAF Queen's Flight, Benson	
ZE704	Lockheed Tristar C2 (N508PA)	RAF No 216 Sqn, Brize Norton	
ZE705	Lockheed Tristar C2 (N509PA)	RAF No 216 Sqn, Brize Norton	
ZE706	Lockheed Tristar C2 (N503PA)	MoD(PE), Marshall, Cambridge	
ZE728	Panavia Tornado F3T [AN]	RAF F3 OCU/No 56(R) Sqn, Coningsby	
ZE729	Panavia Tornado F3 [BF]	RAF No 29 Sqn, Coningsby	
ZE730	Panavia Tornado F3	RAF No 43 Sqn, Leuchars	
ZE731	Panavia Tornado F3 [GK]	RAF No 43 Sqn, Leuchars	
ZE732	Panavia Tornado F3 [GI]	RAF No 43 Sqn, Leuchars	

Notes	Serial	Type	Owner or Operator
	ZE733	Panavia Tornado F3 [GE]	RAF No 43 Sqn, Leuchars
	ZE734	Panavia Tornado F3 [CX]	RAF No 5 Sqn, Coningsby
	ZE735	Panavia Tornado F3T [AL]	RAF F3 OCU/No 56(R) Sqn, Coningsby
	ZE736	Panavia Tornado F3 [HA]	RAF No 111 Sqn, Leuchars
	ZE737	Panavia Tornado F3 [FF]	RAF No 25 Sqn, Leeming
	ZE755	Panavia Tornado F3 [GB]	RAF No 43 Sqn, Leuchars
	ZE756	Panavia Tornado F3	RAF F3 OEU, Coningsby
	ZE757	Panavia Tornado F3 [GF]	RAF No 43 Sqn, Leuchars
	ZE758	Panavia Tornado F3 [C]	RAF No 1435 Flt, Mount Pleasant, FI
	ZE759	Panavia Tornado F3T [AG]	RAF F3 OCU/No 56(R) Sqn, Coningsby
	ZE760	Panavia Tornado F3 [P]	RAF No 43 Sqn, Leuchars
	ZE761	Panavia Tornado F3 [CB]	RAF No 5 Sqn, Coningsby
	ZE762	Panavia Tornado F3	RAF No 29 Sqn, Coningsby
	ZE763	Panavia Tornado F3 [BA]	RAF No 29 Sqn, Coningsby
	ZE764	Panavia Tornado F3 [DH]	RAF No 11 Sqn, Leeming
	ZE785	Panavia Tornado F3 [AO]	RAF F3 OCU/No 56(R) Sqn, Coningsby
	ZE786	Panavia Tornado F3T [CT]	RAF No 5 Sqn, Coningsby
	ZE787	Panavia Tornado F3 [EX]	RAF No 25 Sqn, Leeming
	ZE788	Panavia Tornado F3 [FH]	RAF No 25 Sqn, Leeming
	ZE789	Panavia Tornado F3 [AW]	RAF F3 OCU/No 56(R) Sqn, Coningsby
	ZE790	Panavia Tornado F3 [D]	RAF No 1435 Flt, Mount Pleasant, FI
	ZE791	Panavia Tornado F3 [A9]	RAF F3 OCU/No 56(R) Sqn, Coningsby
	ZE792	Panavia Tornado F3 [CU]	RAF No 5 Sqn, Coningsby
	ZE793	Panavia Tornado F3T [AK]	RAF F3 OCU/No 56(R) Sqn, Coningsby
	ZE794	Panavia Tornado F3 [HQ]	RAF No 111 Sqn, Leuchars
	ZE808	Panavia Tornado F3 [FA]	RAF No 25 Sqn, Leeming
	ZE809	Panavia Tornado F3 [HP]	RAF No 111 Sqn, Leuchars
	ZE810	Panavia Tornado F3	RAF F3 OCU/No 56(R) Sqn, Coningsby
	ZE811	Panavia Tornado F3 [HB]	RAF No 111 Sqn, Leuchars
	ZE812	Panavia Tornado F3 [F]	RAF No 1435 Flt, Mount Pleasant, FI
	ZE830	Panavia Tornado F3T [GD]	RAF No 43 Sqn, Leuchars
	ZE831	Panavia Tornado F3 [HJ]	RAF No 111 Sqn, Leuchars
	ZE832	Panavia Tornado F3 [A7]	RAF F3 OCU/No 56(R) Sqn, Coningsby
	ZE834	Panavia Tornado F3	RAF, on repair Leeming
	ZE835	Panavia Tornado F3 [HK]	RAF No 111 Sqn, Leuchars
	ZE836	Panavia Tornado F3 [AS]	RAF F3 OCU/No 56(R) Sqn, Coningsby
	ZE837	Panavia Tornado F3 [HY]	RAF No 111 Sqn, Leuchars
	ZE838	Panavia Tornado F3 [GH]	RAF No 43 Sqn, Leuchars
	ZE839	Panavia Tornado F3 [A6]	RAF F3 OCU/No 56(R) Sqn, Coningsby
	ZE858	Panavia Tornado F3 [A5]	RAF No 43 Sqn, Leuchars
	ZE862	Panavia Tornado F3T	RAF F3 OEU, Coningsby
	ZE887	Panavia Tornado F3 [DJ]	RAF No 11 Sqn, Leeming
	ZE888	Panavia Tornado F3T [EV]	RAF No 23 Sqn, Leeming
	ZE889	Panavia Tornado F3	RAF F3 OEU, Coningsby
	ZE907	Panavia Tornado F3 [EN]	RAF No 23 Sqn, Leeming
	ZE908	Panavia Tornado F3T [FC]	RAF No 25 Sqn, Leeming
	ZE911	Panavia Tornado F3[BE]	RAF No 29 Sqn, Coningsby
	ZE934	Panavia Tornado F3T [DX]	RAF No 11 Sqn, Leeming
	ZE936	Panavia Tornado F3 [EE]	RAF No 23 Sqn, Leeming
	ZE941	Panavia Tornado F3T [DY]	RAF No 11 Sqn, Leeming
	ZE942	Panavia Tornado F3 [DK]	RAF No 11 Sqn, Leeming
	ZE961	Panavia Tornado F3 [FD]	RAF No 25 Sqn, Leeming
	ZE962	Panavia Tornado F3 [FJ]	RAF No 25 Sqn, Leeming
	ZE963	Panavia Tornado F3 [ET]	RAF No 23 Sqn, Leeming
	ZE964	Panavia Tornado F3T [EU]	RAF No 23 Sqn, Leeming
	ZE965	Panavia Tornado F3 [BZ]	RAF No 29 Sqn, Coningsby
	ZE966	Panavia Tornado F3 [DZ]	RAF No 11 Sqn, Leeming
	ZE967	Panavia Tornado F3 [FE]	RAF No 25 Sqn, Leeming
	ZE968	Panavia Tornado F3	RAF F3 OEU, Coningsby
	ZE969	Panavia Tornado F3 [EA]	RAF No 23 Sqn, Leeming
	ZE982	Panavia Tornado F3 [DM]	RAF No 11 Sqn, Leeming
	ZE983	Panavia Tornado F3 [EZ]	RAF No 23 Sqn, Leeming
	ZF115	WS61 Sea King HC4	MoD(PE), ETPS, Boscombe Down
	ZF116	WS61 Sea King HC4	MoD(PE), A&AEE Boscombe Down
	ZF117	WS61 Sea King HC4 [VK]	RN No 846 Sqn, Yeovilton
	ZF118	WS61 Sea King HC4 [ZT]	RN No 707 Sqn, Yeovilton
	ZF119	WS61 Sea King HC4 [VO]	RNAY Fleetlands
	ZF120	WS61 Sea King HC4 [20/PO]	RN No 772 Sqn, Portland
	ZF121	WS61 Sea King HC4 [21]	RN No 772 Sqn, Portland
	ZF122	WS61 Sea King HC4 [22/PO]	RN No 772 Sqn, Portland
	ZF123	WS61 Sea King HC4 [23]	RN, AMG Yeovilton
	ZF124	WS61 Sea King HC4 [24/PO]	RN No 772 Sqn, Portland

Serial	Type	Owner or Operator	Notes
ZF130	BAe 125-600B (G-BLUW)	MoD(PE) BAe Dunsfold	
ZF135	Shorts Tucano T1	RAF CFS, Scampton	
ZF136	Shorts Tucano T1	MoD(PE) Shorts, Belfast	
ZF137	Shorts Tucano T1	MoD(PE) Shorts, Belfast	
ZF138	Shorts Tucano T1	RAF, stored Shawbury	
ZF139	Shorts Tucano T1	RAF No 3 FTS, Cranwell	
ZF140	Shorts Tucano T1	RAF No 3 FTS, Cranwell	
ZF141	Shorts Tucano T1	RAF No 3 FTS, Cranwell	
ZF142	Shorts Tucano T1	RAF No 3 FTS, Cranwell	
ZF143	Shorts Tucano T1	RAF No 3 FTS, Cranwell	
ZF144	Shorts Tucano T1	RAF No 3 FTS, Cranwell	
ZF145	Shorts Tucano T1	RAF No 1 FTS, Linton-on-Ouse	
ZF160	Shorts Tucano T1	RAF No 3 FTS, Cranwell	
ZF161	Shorts Tucano T1	RAF No 6 FTS, Finningley	
ZF162	Shorts Tucano T1	RAF No 6 FTS, Finningley	
ZF163	Shorts Tucano T1	RAF No 1 FTS, Linton-on-Ouse	
ZF164	Shorts Tucano T1	RAF No 3 FTS, Cranwell	
ZF165	Shorts Tucano T1	RAF No 3 FTS, Cranwell	
ZF166	Shorts Tucano T1	RAF No CFS, Scampton	
ZF167	Shorts Tucano T1	RAF No 3 FTS, Cranwell	
ZF168	Shorts Tucano T1	RAF CFS, Scampton	
ZF169	Shorts Tucano T1	RAF CFS, Scampton	
ZF170	Shorts Tucano T1	RAF No 1 FTS, Linton-on-Ouse	
ZF171	Shorts Tucano T1	RAF No 3 FTS, Cranwell	
ZF172	Shorts Tucano T1	RAF CFS, Scampton	
ZF200	Shorts Tucano T1	RAF No 3 FTS, Cranwell	
ZF201	Shorts Tucano T1	RAF No 3 FTS, Cranwell	
ZF202	Shorts Tucano T1	RAF No 3 FTS, Cranwell	
ZF203	Shorts Tucano T1	RAF CFS, Scampton	
ZF204	Shorts Tucano T1	RAF CFS, Scampton	
ZF205	Shorts Tucano T1	RAF No 3 FTS, Cranwell	
ZF206	Shorts Tucano T1	RAF CFS, Scampton	
ZF207	Shorts Tucano T1	RAF No 3 FTS, Cranwell	
ZF208	Shorts Tucano T1	RAF No 3 FTS, Cranwell	
ZF209	Shorts Tucano T1	RAF No 1 FTS, Linton-on-Ouse	
ZF210	Shorts Tucano T1	RAF No 3 FTS, Cranwell	
ZF211	Shorts Tucano T1	RAF No 1 FTS, Linton-on-Ouse	
ZF212	Shorts Tucano T1	RAF No 3 FTS, Cranwell	
ZF238	Shorts Tucano T1	RAF No 3 FTS, Cranwell	
ZF239	Shorts Tucano T1	RAF No 3 FTS, Cranwell	
ZF240	Shorts Tucano T1	RAF No 3 FTS, Cranwell	
ZF241	Shorts Tucano T1	RAF No 3 FTS, Cranwell	
ZF242	Shorts Tucano T1	RAF No 6 FTS, Finningley	
ZF243	Shorts Tucano T1	RAF No 1 FTS, Linton-on-Ouse	
ZF244	Shorts Tucano T1	RAF, stored Shawbury	
ZF245	Shorts Tucano T1	RAF No 3 FTS, Cranwell	
ZF263	Shorts Tucano T1	RAF No 3 FTS, Cranwell	
ZF264	Shorts Tucano T1	RAF No 3 FTS, Cranwell	
ZF265	Shorts Tucano T1	RAF CFS, Scampton	
ZF266	Shorts Tucano T1	RAF CFS, Scampton	
ZF267	Shorts Tucano T1	RAF No 3 FTS, Cranwell	
ZF268	Shorts Tucano T1	RAF No 3 FTS, Cranwell	
ZF269	Shorts Tucano T1	RAF CFS, Scampton	
ZF270	Shorts Tucano T1	RAF No 3 FTS, Cranwell	
ZF284	Shorts Tucano T1	RAF No 1 FTS, Linton-on-Ouse	
ZF285	Shorts Tucano T1	RAF No 3 FTS, Cranwell	
ZF286	Shorts Tucano T1	RAF No 3 FTS, Cranwell	
ZF287	Shorts Tucano T1	RAF No 3 FTS, Cranwell	
ZF288	Shorts Tucano T1	RAF CFS, Scampton	
ZF289	Shorts Tucano T1	RAF No 3 FTS, Cranwell	
ZF290	Shorts Tucano T1	RAF No 3 FTS, Cranwell	
ZF291	Shorts Tucano T1	RAF No 3 FTS, Cranwell	
ZF292	Shorts Tucano T1	RAF No 3 FTS, Cranwell	
ZF293	Shorts Tucano T1	RAF No 3 FTS, Cranwell	
ZF294	Shorts Tucano T1	RAF No 3 FTS, Cranwell	
ZF295	Shorts Tucano T1	RAF No 3 FTS, Cranwell	
ZF315	Shorts Tucano T1	RAF No 1 FTS, Linton-on-Ouse	
ZF316	Shorts Tucano T1	*Crashed nr Fochabers, Scotland, 12 May 1992*	
ZF317	Shorts Tucano T1	RAF No 3 FTS, Cranwell	
ZF318	Shorts Tucano T1	RAF CFS, Scampton	
ZF319	Shorts Tucano T1	RAF No 3 FTS, Cranwell	
ZF320	Shorts Tucano T1	RAF No 3 FTS, Cranwell	

Notes	Serial	Type	Owner or Operator
	ZF338	Shorts Tucano T1	RAF No 3 FTS, Cranwell
	ZF339	Shorts Tucano T1	RAF No 6 FTS, Finningley
	ZF340	Shorts Tucano T1	RAF No 3 FTS, Cranwell
	ZF341	Shorts Tucano T1	RAF No 3 FTS, Cranwell
	ZF342	Shorts Tucano T1	RAF No 3 FTS, Cranwell
	ZF343	Shorts Tucano T1	RAF CFS, Scampton
	ZF344	Shorts Tucano T1	RAF No 3 FTS, Cranwell
	ZF345	Shorts Tucano T1	RAF CFS, Scampton
	ZF346	Shorts Tucano T1	RAF No 3 FTS, Cranwell
	ZF347	Shorts Tucano T1	RAF No 3 FTS, Cranwell
	ZF348	Shorts Tucano T1	RAF No 3 FTS, Cranwell
	ZF349	Shorts Tucano T1	RAF No 3 FTS, Cranwell
	ZF350	Shorts Tucano T1	RAF No 3 FTS, Cranwell
	ZF372	Shorts Tucano T1	RAF CFS, Scampton
	ZF373	Shorts Tucano T1	RAF No 3 FTS, Cranwell
	ZF374	Shorts Tucano T1	RAF No 6 FTS, Finningley
	ZF375	Shorts Tucano T1	RAF No 3 FTS, Cranwell
	ZF376	Shorts Tucano T1	RAF No 1 FTS, Linton-on-Ouse
	ZF377	Shorts Tucano T1	RAF No 3 FTS, Cranwell
	ZF378	Shorts Tucano T1	RAF CFS, Scampton
	ZF379	Shorts Tucano T1	RAF No 6 FTS, Finningley
	ZF380	Shorts Tucano T1	RAF CFS, Scampton
	ZF405	Shorts Tucano T1	RAF No 6 FTS, Finningley
	ZF406	Shorts Tucano T1	RAF CFS, Scampton
	ZF407	Shorts Tucano T1	RAF No 1 FTS, Linton-on-Ouse
	ZF408	Shorts Tucano T1	RAF CFS, Scampton
	ZF409	Shorts Tucano T1	RAF No 3 FTS, Cranwell
	ZF410	Shorts Tucano T1	RAF No 1 FTS, Linton-on-Ouse
	ZF411	Shorts Tucano T1	RAF No 1 FTS, Linton-on-Ouse
	ZF412	Shorts Tucano T1	RAF No 1 FTS, Linton-on-Ouse
	ZF413	Shorts Tucano T1	RAF No 6 FTS, Finningley
	ZF414	Shorts Tucano T1	RAF No 3 FTS, Cranwell
	ZF415	Shorts Tucano T1	RAF No 3 FTS, Cranwell
	ZF416	Shorts Tucano T1	RAF No 1 FTS, Linton-on-Ouse
	ZF417	Shorts Tucano T1	RAF No 3 FTS, Cranwell
	ZF418	Shorts Tucano T1	RAF No 6 FTS, Finningley
	ZF444	BN2A Islander (G-WOTG)	RAF Parachute Association, Weston-on-the-Green
	ZF445	Shorts Tucano T1	RAF No 3 FTS, Cranwell
	ZF446	Shorts Tucano T1	RAF No 6 FTS, Finningley
	ZF447	Shorts Tucano T1	RAF No 3 FTS, Cranwell
	ZF448	Shorts Tucano T1	RAF CFS, Scampton
	ZF449	Shorts Tucano T1	RAF No 3 FTS, Cranwell
	ZF450	Shorts Tucano T1	RAF No 1 FTS, Linton-on-Ouse
	ZF483	Shorts Tucano T1	RAF No 3 FTS, Cranwell
	ZF484	Shorts Tucano T1	RAF No 1 FTS, Linton-on-Ouse
	ZF485	Short Tucano T1 (G-BULU)	RAF No 3 FTS, Cranwell
	ZF486	Shorts Tucano T1	RAF CFS, Scampton
	ZF487	Shorts Tucano T1	RAF No 3 FTS, Cranwell
	ZF488	Shorts Tucano T1	RAF No 3 FTS, Cranwell
	ZF489	Shorts Tucano T1	RAF
	ZF490	Shorts Tucano T1	RAF
	ZF491	Shorts Tucano T1	RAF No 3 FTS, Cranwell
	ZF492	Shorts Tucano T1	RAF
	ZF510	Shorts Tucano T1	RAF No 3 FTS, Cranwell
	ZF511	Shorts Tucano T1	RAF
	ZF512	Shorts Tucano T1	RAF
	ZF513	Shorts Tucano T1	RAF
	ZF514	Shorts Tucano T1	RAF
	ZF515	Shorts Tucano T1	RAF
	ZF516	Shorts Tucano T1	RAF
	ZF520	Piper PA-31 Navajo Chieftain 350 (N35823/G-BLZK)	MoD(PE) DRA Farnborough
	ZF521	Piper PA-31 Navajo Chieftain 350 (N27509)	MoD(PE) DRA Farnborough
	ZF522	Piper PA-31 Navajo Chieftain 350 (N4261A/G-RNAV/N27728)	MoD(PE) DRA Farnborough
	ZF534	BAe EAP	MoD(PE), stored BAe Warton
	ZF537	WS Lynx AH7 [F]	AAC No 671 Sqn, Middle Wallop
	ZF538	WS Lynx AH9	MoD(PE), Westland, Yeovil
	ZF539	WS Lynx AH9	MoD(PE), Westland, Yeovil
	ZF540	WS Lynx AH7	AAC No 667 Sqn, Middle Wallop
	ZF557	WS Lynx HAS3CTS [670]	RN No 815 Sqn OEU, Portland

Serial	Type	Owner or Operator	Notes
ZF558	WS Lynx HAS3CTS [672]	RN No 815 Sqn OEU, Portland	
ZF560	WS Lynx HAS3CTS [341/AG]	RN No 815 Sqn, Portland	
ZF562	WS Lynx HAS3CTS [320/AZ]	RN No 815 Sqn, Portland	
ZF563	WS Lynx HAS3CTS [671]	RN No 815 Sqn OEU, Portland	
ZF573	BN2T Islander (G-SRAY)	MoD(PE) PBN, Bembridge	
ZF577	BAC Lightning F53 (668)	*To USA, 1992*	
ZF578	BAC Lightning F53 (670)	Wales Aircraft Museum, Cardiff Airport	
ZF579	BAC Lightning F53 (671)	*To USA, 1992*	
ZF580	BAC Lightning F53 (672)	BAe Samlesbury gate	
ZF581	BAC Lightning F53 (675)	*To USA, 1992*	
ZF582	BAC Lightning F53 (676)	*To USA, 1992*	
ZF583	BAC Lightning F53 (681)	Solway Aviation Society, Carlisle	
ZF584	BAC Lightning F53 (682)	Ferranti Ltd, South Gyle, Edinburgh	
ZF585	BAC Lightning F53 (683)	*To USA, 1992*	
ZF586	BAC Lightning F53 (688)	*To USA, 1992*	
ZF587	BAC Lightning F53 (691)	*To USA, 1992*	
ZF588	BAC Lightning F53 (693)	East Midlands Airport Aero Park	
ZF589	BAC Lightning F53 (700)	*To USA, 1992*	
ZF590	BAC Lightning F53 (679)	*To USA, 1992*	
ZF591	BAC Lightning F53 (685)	*To USA, 1992*	
ZF592	BAC Lightning F53 (686)	*To USA, 1992*	
ZF594	BAC Lightning F53 (696)	North-East Aircraft Museum, Usworth	
ZF595	BAC Lightning T55 (714)	Stratford Aircraft Collection, Long Marston	
ZF596	BAC Lightning T55 (715)	*To USA, 1992*	
ZF597	BAC Lightning T55 (711)	*To USA, 1992*	
ZF598	BAC Lightning T55 (713)	Midland Air Museum, Coventry	
ZF622	Piper PA-31 Navajo Chieftain 350	MoD(PE) A&AEE Boscombe Down	
ZF641	WS/Agusta EH-101 [PP1]	MoD(PE) Westland, Yeovil	
ZF644	WS EH-101 Merlin [PP4]	MoD(PE) Westlands, Yeovil	
ZF649	WS EH-101 Merlin HAS1 [PP5]	MoD(PE) Westlands, Yeovil	
ZG101	WS EH-101 (mock-up) [GB]	Westland/Agusta, Yeovil	
ZG468	WS70 Blackhawk	Westland Helicopters, Yeovil	
ZG471	BAe Harrier GR7 [AB]	RAF No 3 Sqn, Laarbruch	
ZG472	BAe Harrier GR7 [O]	RAF SAOEU, Boscombe Down	
ZG474	BAe Harrier GR7 [AL]	RAF No 3 Sqn, Laarbruch	
ZG475	BAe Harrier GR7 [U]	RAF SAOEU, Boscombe Down	
ZG476	BAe Harrier GR7 [AP]	RAF No 3 Sqn, Laarbruch	
ZG477	BAe Harrier GR7 [AK]	RAF No 3 Sqn, Laarbruch	
ZG478	BAe Harrier GR7 [CD]	RAF No 3 Sqn, Laarbruch	
ZG479	BAe Harrier GR7 [AI]	RAF No 3 Sqn, Laarbruch	
ZG480	BAe Harrier GR7 [AY]	RAF No 3 Sqn, Laarbruch	
ZG500	BAe Harrier GR7 [AU]	RAF No 3 Sqn, Laarbruch	
ZG501	BAe Harrier GR7 [E]	RAF SAOEU, Boscombe Down	
ZG502	BAe Harrier GR7 [AM]	RAF No 3 Sqn, Laarbruch	
ZG503	BAe Harrier GR7 [AG]	RAF No 3 Sqn, Laarbruch	
ZG504	BAe Harrier GR7 [AR]	RAF No 3 Sqn, Laarbruch	
ZG505	BAe Harrier GR7 [AJ]	RAF No 3 Sqn, Laarbruch	
ZG506	BAe Harrier GR7 [CM]	RAF No 4 Sqn, Laarbruch	
ZG507	BAe Harrier GR7 [CA]	RAF No 4 Sqn, Laarbruch	
ZG508	Bae Harrier GR7 [CG]	RAF No 4 Sqn, Laarbruch	
ZG509	BAe Harrier GR7 [CH]	RAF No 4 Sqn, Laarbruch	
ZG510	BAe Harrier GR7 [CE]	RAF No 4 Sqn, Laarbruch	
ZG511	BAe Harrier GR7 [CD]	RAF No 4 Sqn, Laarbruch	
ZG512	BAe Harrier GR7 [CI]	RAF No 4 Sqn, Laarbruch	
ZG530	BAe Harrier GR7 [CL]	RAF No 4 Sqn, Laarbruch	
ZG531	BAe Harrier GR7	RAF No 4 Sqn, Laarbruch	
ZG532	BAe Harrier GR7 [CC]	RAF No 4 Sqn, Laarbruch	
ZG533	BAe Harrier GR7 [CF]	RAF No 4 Sqn, Laarbruch	
ZG705	Panavia Tornado GR1A [A]	RAF No 13 Sqn, Honington	
ZG706	Panavia Tornado GR1A [E]	RAF SAOEU, Boscombe Down	
ZG707	Panavia Tornado GR1A [B]	RAF No 13 Sqn, Honington	
ZG708	Panavia Tornado GR1A [C]	RAF No 13 Sqn, Honington	
ZG709	Panavia Tornado GR1A [I]	RAF No 13 Sqn, Honington	
ZG710	Panavia Tornado GR1A [D]	RAF No 13 Sqn, Honington	
ZG711	Panavia Tornado GR1A [E]	RAF No 13 Sqn, Honington	
ZG712	Panavia Tornado GR1A [F]	RAF No 13 Sqn, Honington	
ZG713	Panavia Tornado GR1A [G]	RAF No 13 Sqn, Honington	
ZG714	Panavia Tornado GR1A [H]	RAF No 13 Sqn, Honington	
ZG725	Panavia Tornado GR1A [J]	RAF No 13 Sqn, Honington	
ZG726	Panavia Tornado GR1A [K[	RAF No 13 Sqn, Honington	
ZG727	Panavia Tornado GR1A [L]	RAF No 13 Sqn, Honington	

Notes	Serial	Type	Owner or Operator
	ZG728	Panavia Tornado F3 [CI]	RAF No 5 Sqn, Coningsby
	ZG729	Panavia Tornado GR1A [M]	RAF No 13 Sqn, Honington
	ZG730	Panavia Tornado F3 [CC]	RAF No 5 Sqn, Coningsby
	ZG731	Panavia Tornado F3 [CG]	RAF No 5 Sqn, Coningsby
	ZG732	Panavia Tornado F3 [BC]	RAF No 29 Sqn, Coningsby
	ZG733	Panavia Tornado F3 [BK]	RAF No 29 Sqn, Coningsby
	ZG734	Panavia Tornado F3 [BG]	RAF No 29 Sqn, Coningsby
	ZG735	Panavia Tornado F3 [CO]	RAF No 5 Sqn, Coningsby
	ZG750	Panavia Tornado GR1T [Y]	RAF No 13 Sqn, Honington
	ZG751	Panavia Tornado F3 [CW]	RAF No 5 Sqn, Coningsby
	ZG752	Panavia Tornado GR1T [Z]	RAF No 13 Sqn, Honington
	ZG753	Panavia Tornado F3 [CH]	RAF No 5 Sqn, Coningsby
	ZG754	Panavia Tornado GR1T [AW]	RAF No 9 Sqn, Bruggen
	ZG755	Panavia Tornado F3 [GM]	RAF No 43 Sqn, Leuchars
	ZG756	Panavia Tornado GR1T [AX]	RAF No 9 Sqn, Bruggen
	ZG757	Panavia Tornado F3 [HL]	RAF No 111 Sqn, Leuchars
	ZG768	Panavia Tornado F3 [AX]	RAF F3 OCU/No 56(R) Sqn, Coningsby
	ZG769	Panavia Tornado GR1T [AY]	RAF No 9 Sqn, Bruggen
	ZG770	Panavia Tornado F3 [AP]	RAF F3 OCU/No 56(R) Sqn, Coningsby
	ZG771	Panavia Tornado GR1T [DW]	RAF No 31 Sqn, Bruggen
	ZG772	Panavia Tornado F3 [CN]	RAF No 5 Sqn, Coningsby
	ZG773	Panavia Tornado GR1	RAF Bruggen, ASF
	ZG774	Panavia Tornado F3 [GN]	RAF No 43 Sqn, Leuchars
	ZG775	Panavia Tornado GR1 [CC]	RAF No 17 Sqn, Bruggen
	ZG776	Panavia Tornado F3 [HN]	RAF No 111 Sqn, Leuchars
	ZG777	Panavia Tornado GR1 [CK]	RAF No 17 Sqn, Bruggen
	ZG778	Panavia Tornado F3 [HC]	RAF No 111 Sqn, Leuchars
	ZG779	Panavia Tornado GR1 [CM]	RAF No 17 Sqn, Bruggen
	ZG780	Panavia Tornado F3 [BH]	RAF No 29 Sqn, Coningsby
	ZG791	Panavia Tornado GR1	RAF No 31 Sqn, Bruggen
	ZG792	Panavia Tornado GR1 [BH]	RAF No 14 Sqn, Bruggen
	ZG793	Panavia Tornado F3	RAF No 5 Sqn, Coningsby
	ZG794	Panavia Tornado GR1	RAF
	ZG795	Panavia Tornado F3 [AY]	RAF F3 OCU/No 56(R) Sqn, Coningsby
	ZG796	Panavia Tornado F3 [AV]	RAF F3 OCU/No 56(R) Sqn, Coningsby
	ZG797	Panavia Tornado F3 [AU]	RAF F3 OCU/No 56(R) Sqn, Coningsby
	ZG798	Panavia Tornado F3	RAF
	ZG799	Panavia Tornado F3	RAF
	ZG816	WS61 Sea King HAS6 [709/AJ]	RN No 819 Sqn, Prestwick
	ZG817	WS61 Sea King HAS6 [504/CU]	RN No 810 Sqn, Culdrose
	ZG818	WS61 Sea King HAS6 [271]	RN No 814 Sqn, Culdrose
	ZG819	WS61 Sea King HAS6 [018/R]	RN No 820 Sqn, Culdrose
	ZG820	WS61 Sea King HC4 [F]	RN No 845 Sqn, Yeovilton
	ZG821	WS61 Sea King HC4 [VL]	RN No 846 Sqn, Yeovilton
	ZG822	WS61 Sea King HC4	RN, AMG Yeovilton
	ZG829	WS61 Sea King HC4	*Written off at Boscombe Down, October 1992*
	ZG844	PBN 2T Islander AL1 (G-BLNE)	AAC Islander Flight, Middle Wallop
	ZG845	PBN 2T Islander AL1 (G-BLNT)	AAC Islander Flight, Middle Wallop
	ZG846	PBN 2T Islander AL1 (G-BLNU)	AAC No 1 Flt, Aldergrove
	ZG847	PBN 2T Islander AL1 (G-BLNV)	AAC No 1 Flt, Aldergrove
	ZG848	PBN 2T Islander AL1 (G-BLNY)	AAC No 1 Flt, Aldergrove
	ZG856	BAe Harrier GR7 [CJ]	RAF No 4 Sqn, Laarbruch
	ZG857	Bae Harrier GR7 [CK]	RAF No 4 Sqn, Laarbruch
	ZG858	BAe Harrier GR7 [CB]	RAF No 4 Sqn, Laarbruch
	ZG859	BAe Harrier GR7 [P]	RAF HOCU/No 20(R) Sqn, Wittering
	ZG860	BAe Harrier GR7 [T]	RAF SAOEU, Boscombe Down
	ZG861	BAe Harrier GR7 [AA]	RAF No 3 Sqn, Laarbruch
	ZG862	BAe Harrier GR7 [CO]	RAF No 4 Sqn, Laarbruch
	ZG875	WS61 Sea King HAS6 [014]	RN No 820 Sqn, Culdrose
	ZG879	Powerchute Raider Mk 1	MoD(PE)/Powerchute, Hereford
	ZG884	WS Lynx AH9	MoD(PE), A&AEE Boscombe Down
	ZG885	WS Lynx AH9	AAC No 672 Sqn, Dishforth
	ZG886	WS Lynx AH9	AAC No 672 Sqn, Dishforth
	ZG887	WS Lynx AH9	AAC No 672 Sqn, Dishforth
	ZG888	WS Lynx AH9	AAC No 672 Sqn, Dishforth
	ZG889	WS Lynx AH9	AAC No 672 Sqn, Dishforth
	ZG914	WS Lynx AH9 [G]	AAC No 664 Sqn, Dishforth
	ZG915	WS Lynx AH9 [H]	AAC No 664 Sqn, Dishforth
	ZG916	WS Lynx AH9 [I]	AAC No 664 Sqn, Dishforth
	ZG917	WS Lynx AH9 [J]	AAC No 664 Sqn, Dishforth
	ZG918	WS Lynx AH9 [K]	AAC No 664 Sqn, Dishforth
	ZG919	WS Lynx AH9	AAC No 672 Sqn, Dishforth

Serial	Type	Owner or Operator	Notes
ZG920	WS Lynx AH9	AAC No 672 Sqn, Dishforth	
ZG921	WS Lynx AH9 [L]	AAC No 664 Sqn, Dishforth	
ZG922	WS Lynx AH9	AAC No 672 Sqn, Dishforth	
ZG923	WS Lynx AH9	AAC No 672 Sqn, Dishforth	
ZG969	Pilatus PC-9 (HB-HQE)	BAe Brough	
ZG989	PBN 2T Islander Astor	MoD(PE), A&AEE Boscombe Down	
ZG993	PBN 2T Islander AL1 (G-BOMD)	AAC No 1 Flight, Aldergrove	
ZG994	PBN 2T Islander AL1 (G-BPLN)	AAC No 1 Flight, Aldergrove	
ZH101	Boeing E-3D Sentry AEW1	RAF No 8 Sqn, Waddington	
ZH102	Boeing E-3D Sentry AEW1	RAF No 8 Sqn, Waddington	
ZH103	Boeing E-3D Sentry AEW1	RAF No 8 Sqn, Waddington	
ZH104	Boeing E-3D Sentry AEW1	RAF No 8 Sqn, Waddington	
ZH105	Boeing E-3D Sentry AEW1	RAF No 8 Sqn, Waddington	
ZH106	Boeing E-3D Sentry AEW1	RAF No 8 Sqn, Waddington	
ZH107	Boeing E-3D Sentry AEW1	RAF No 8 Sqn, Waddington	
ZH115	Grob Vigilant T1	RAF No 635 VGS, Samlesbury	
ZH116	Grob Vigilant T1	RAF No 632 VGS, Ternhill	
ZH117	Grob Vigilant T1	RAF No 632 VGS, Ternhill	
ZH118	Grob Vigilant T1	RAF No 632 VGS, Ternhill	
ZH119	Grob Vigilant T1	RAF No 616 VGS, Henlow	
ZH120	Grob Vigilant T1	RAF No 616 VGS, Henlow	
ZH121	Grob Vigilant T1	RAF No 635 VGS, Samlesbury	
ZH122	Grob Vigilant T1	RAF No 616 VGS, Henlow	
ZH123	Grob Vigilant T1	RAF No 642 VGS, Linton-on-Ouse	
ZH124	Grob Vigilant T1	RAF No 642 VGS, Linton-on-Ouse	
ZH125	Grob Vigilant T1	RAF No 633 VGS, Cosford	
ZH126	Grob Vigilant T1	RAF No 642 VGS, Linton-on-Ouse	
ZH127	Grob Vigilant T1	RAF CGMF, Syerston	
ZH128	Grob Vigilant T1	RAF No 613 VGS, Halton	
ZH129	Grob Vigilant T1	RAF No 612 VGS, Benson	
ZH144	Grob Vigilant T1	RAF No 612 VGS, Benson	
ZH145	Grob Vigilant T1	RAF No 632 VGS, Ternhill	
ZH146	Grob Vigilant T1	RAF No 637 VGS, Little Rissington	
ZH147	Grob Vigilant T1	RAF No 637 VGS, Little Rissington	
ZH148	Grob Vigilant T1	RAF No 616 VGS, Henlow	
ZH184	Grob Vigilant T1	RAF No 633 VGS, Cosford	
ZH185	Grob Vigilant T1	RAF No 633 VGS, Cosford	
ZH186	Grob Vigilant T1	RAF No 633 VGS, Cosford	
ZH187	Grob Vigilant T1	RAF No 663 VGS, Kinloss	
ZH188	Grob Vigilant T1	RAF No 663 VGS, Kinloss	
ZH189	Grob Vigilant T1	RAF ACCGS, Syerston	
ZH190	Grob Vigilant T1	RAF ACCGS, Syerston	
ZH191	Grob Vigilant T1	RAF No 663 VGS, Kinloss	
ZH192	Grob Vigilant T1	RAF No 663 VGS, Kinloss	
ZH193	Grob Vigilant T1	RAF No 663 VGS, Kinloss	
ZH194	Grob Vigilant T1	RAF ACCGS, Syerston	
ZH195	Grob Vigilant T1	RAF CGMF, Syerston	
ZH196	Grob Vigilant T1	RAF No 633 VGS, Cosford	
ZH197	Grob Vigilant T1	RAF No 624 VGS, Chivenor	
ZH200	BAe Hawk 200	MoD(PE), BAe Warton	
ZH205	Grob Vigilant T1	RAF No 624 VGS, Chivenor	
ZH206	Grob Vigilant T1	RAF No 624 VGS, Chivenor	
ZH207	Grob Vigilant T1	RAF No 613 VGS, Halton	
ZH208	Grob Vigilant T1	RAF No 613 VGS, Halton	
ZH209	Grob Vigilant T1	RAF No 613 VGS, Halton	
ZH210	Grob Vigilant T1 (wreck)	RAF CGMF Syerston	
ZH211	Grob Vigilant T1	RAF Syerston	
ZH247	Grob Vigilant T1	RAF CGMF, Syerston	
ZH248	Grob Vigilant T1	RAF No 616 VGS, Henlow	
ZH249	Grob Vigilant T1	RAF ACCGS, Syerston	
ZH257	B-V CH-47C Chinook (AE-520)	RAF, stored Fleetlands	
ZH263	Grob Vigilant T1	RAF No 637 VGS, Little Rissington	
ZH264	Grob Vigilant T1	RAF No 642 VGS, Linton-on-Ouse	
ZH265	Grob Vigilant T1	RAF ACCGS, Syerston	
ZH266	Grob Vigilant T1	RAF No 635 VGS, Samlesbury	
ZH267	Grob Vigilant T1	RAF No 635 VGS, Samlesbury	
ZH268	Grob Vigilant T1	RAF No 616 VGS, Henlow	
ZH269	Grob Vigilant T1	RAF No 633 VGS, Cosford	
ZH270	Grob Vigilant T1	RAF No 612 VGS, Benson	
ZH271	Grob Vigilant T1	RAF ACCGS, Syerston	
ZH272	Grob Vigilant T1		
ZH536	PBN2T Islander CC 2 (G-BSAH)	RAF Northolt, Station Flight	

Notes	Serial	Type	Owner or Operator
	ZH552	Panavia Tornado F3 [AZ]	RAF F3 OCU/No 56(R) Sqn, Coningsby
	ZH553	Panavia Tornado F3	RAF
	ZH554	Panavia Tornado F3	RAF
	ZH555	Panavia Tornado F3	RAF
	ZH556	Panavia Tornado F3	RAF
	ZH557	Panavia Tornado F3	RAF
	ZH558	Panavia Tornado F3	RAF
	ZH559	Panavia Tornado F3	RAF
	ZH563	DH115 Vampire T55 (U-1216)	RAF Benevolent Fund, Boscombe Down
	ZH566	WS61 Sea King 43B	*To Royal Norwegian AF*
	ZH570	BAe Hawk T60	*To Zimbabwe AF as 608, 16 June 1992*
	ZH571	BAe Hawk T60	*To Zimbabwe AF as 609, 16 June 1992*
	ZH572	BAe Hawk T60	*To Zimbabwe AF as 610, 1 Sept 1992*
	ZH580	WS Lynx 95 (ZF559)	*To Portuguese Navy*
	ZH586	Eurofighter 2000	MoD(PE)/BAe Warton
	ZH588	Eurofighter 2000	MoD(PE)/BAe Warton
	ZH590	Eurofighter 2000	MoD(PE)/BAe Warton
	ZH593	BAe Hawk T67	*To South Korean AF as 67-496*
	ZH594	BAe Hawk T67	*To South Korean AF as 67-497*
	ZH595	BAe Hawk T67	*To South Korean AF as 67-498*
	ZH596	BAe Hawk T67	*To South Korean AF as 67-499*
	ZH597	BAe Hawk T67	*To South Korean AF as 67-500*
	ZH598	BAe Hawk T67	*To South Korean AF as 67-501*
	ZH599	BAe Hawk T67	*To South Korean AF as 67-502*
	ZH603	BAe Hawk T67	*To South Korean AF as 67-503*
	ZH604	BAe Hawk T67	*To South Korean AF as 67-504*
	ZH605	BAe Hawk T67	*To South Korean AF as 67-505*
	ZH606	BAe Hawk T67	*To South Korean AF as 67-506*
	ZH607	BAe Hawk T67	*To South Korean AF as 67-507*
	ZH608	BAe Hawk T67	*To South Korean AF as 67-508*
	ZH609	BAe Hawk T67	*To South Korean AF as 67-509*
	ZH610	BAe Hawk T67	*To South Korean AF as 67-510*
	ZH611	BAe Hawk T67	*To South Korean AF as 67-511*
	ZH612	BAe Hawk T67	*To South Korean AF as 67-512*
	ZH613	BAe Hawk T67	*To South Korean AF as 67-513*
	ZH614	BAe Hawk T67	*To South Korean AF as 67-514*
	ZH615	BAe Hawk T67	*To South Korean AF as 67-515*
	ZH621	BAe Hawk T61	*To Abu Dhabi AF as 1051*
	ZH622	BAe Hawk T61	*To Abu Dhabi AF as 1052*
	ZH623	BAe Hawk T61	*To Abu Dhabi AF as 1053*
	ZH624	BAe Hawk T61	*To Abu Dhabi AF as 1054*
	ZH625	BAe Hawk T61	*To Abu Dhabi AF as 1055*
	ZH626	BAe Hawk T61	*To Abu Dhabi AF as 1056*
	ZH627	BAe Hawk T61	*To Abu Dhabi AF as 1057*
	ZH628	BAe Hawk T61	*To Abu Dhabi AF as 1058*
	ZH629	BAe Hawk T61	*To Abu Dhabi AF as 1059*
	ZH634	BAe Hawk T61	*To Abu Dhabi AF as 1060*
	ZH635	BAe Hawk T61	*To Abu Dhabi AF as 1061*
	ZH636	BAe Hawk T61	*To Abu Dhabi AF as 1062*
	ZH637	BAe Hawk T61	*To Abu Dhabi AF as 1063*
	ZH638	BAe Hawk T61	*To Abu Dhabi AF as 1064*
	ZH639	BAe Hawk T61	*To Abu Dhabi AF as 1065*
	ZH640	BAe Hawk T61	*To Abu Dhabi AF as 1066*
	ZH641	BAe Hawk T61	*To Abu Dhabi AF as 1067*
	ZH642	BAe Hawk T61	*To Abu Dhabi AF as 1068*
	ZJ100	BAe Hawk 100	BAe Warton
	ZJ101	BAe Hawk 100	BAe Warton
	ZJ201	BAe Hawk 200	BAe Warton

WD955 the oldest EE Canberra remaining in service is a T17A operated by No 360 Sqn. *PRM*

WL419 Gloster Meteor T7/8 is used for ejector seat trials by Martin Baker Aircraft. *PRM*

XR241 Auster AOP9 (G-AXRR) is operated by the Museum of Army Flying at Middle Wallop. *PRM*

XR991/G-MOUR HS Gnat T1 carries the RAF *Yellowjacks* team colours. *PRM*

XV292 Lockheed Hercules C1P remains in the 25th anniversary colours. *PRM*

ZA683 B-V Chinook HC1 demonstrated here by No 7 Sqn. *PRM*

ZE339 Panavia Tornado F3 from No 25 Sqn in special display colours. *PRM*

ZH107 is the seventh Boeing E-3D Sentry AEW1 with No 8 Sqn at Waddington. *PRM*

RAF Maintenance Command/ Support Command 'M' number cross-reference

1764M/K4972
2365M/K6038
3858M/X7688
4354M/BL614
5377M/EP120
5405M/LF738
5466M/*BN230*/(LF751)
5690M/MK356
5718M/BM597
5758M/DG202
5854M/WH903
6457M/ML427
6490M/LA255
6850M/TE184
6944M/RW386
6946M/RW388
6948M/DE673
6960M/MT847
7000M/TE392
7008M/EE549
7014M/N6720
7015M/NL985
7035M/*K2567*/(DE306)
7060M/VF301
7090M/EE531
7118M/LA198
7119M/LA226
7150M/PK683
7154M/WB188
7174M/VX272
7175M/VV106
7200M/VT812
7241M/*X4474*
7243M/TE462
7244M/X4277
7245M/RW382
7246M/TD248
7256M/TB752
7257M/TB252
7279M/TB752
7281M/TB252
7285M/VV119
7288M/PK724
7293M/RW393
7323M/VV217
7325M/R5868
7326M/VN485
7362M/475081/(VP546)
7416M/WN907
7421M/WT660
7422M/WT684
7428M/WK198
7432M/WZ724
7443M/WX853
7458M/WX905
7464M/XA564
7467M/WP978
7470M/XA553
7473M/XE946
7491M/WT569
7496M/WT612
7499M/WT555
7510M/WT694
7525M/WT619
7530M/WT648
7532M/WT651
7533M/WT680
7544M/WN904
7548M/PS915
7554M/FS890
7564M/XE982
7570M/XD674
7582M/*WP180*/(WP190)
7583M/WP185
7602M/WE600
7605M/WS692
7606M/WV562
7607M/TJ138
7615M/WV679
7616M/WW388
7618M/WW442
7621M/WV686
7622M/WV606
7625M/WD356
7630M/VZ304
7631M/VX185
7641M/XA634
7645M/WD293
7646M/VX461
7648M/XF785
7663M/XA571
7673M/WV332
7688M/WW421
7693M/WV483
7696M/WV493
7697M/WV495
7698M/WV499
7703M/WG725
7704M/TW536
7705M/WL505
7706M/WB584
7709M/WT933
7711M/PS915
7712M/WK281
7715M/XK724
7716M/WS776
7718M/WA577
7719M/WK277
7722M/XA571
7728M/WZ458
7729M/WB758
7734M/XD536
7736M/WZ559
7737M/XD602
7739M/XA801
7741M/VZ477
7750M/*WK864*/(WL168)
7751M/WL131
7755M/WG760
7758M/PM651
7759M/PK664
7761M/XH318
7762M/XE670
7770M/WT746
7796M/WJ676
7805M/TW117
7806M/TA639
7809M/XA699
7816M/WG763
7817M/TX214
7822M/XP248
7825M/WK991
7827M/XA917
7829M/XH992
7839M/WV781
7840M/XK482
7841M/WV783
7847M/WV276
7849M/XF319
7851M/WZ706
7852M/XG506
7854M/XM191
7855M/XK416
7859M/XP283
7860M/XL738
7862M/XR246
7863M/WZ679
7864M/XP244
7865M/TX226
7866M/XH278
7867M/XH980
7868M/WZ736
7869M/WK935
7870M/XM556
7872M/*WZ826*/(XD826)
7881M/WD413
7882M/XD525
7883M/XT150
7886M/XR985
7887M/XD375
7890M/XD453
7891M/XM693
7894M/XD818
7895M/WF784
7898M/XP854
7899M/XG540
7900M/WA576
7902M/WZ550
7906M/WH132
7917M/WA591
7920M/WL360
7923M/XT133
7928M/XE849
7930M/WH301
7931M/RD253
7932M/WZ744
7933M/XR220
7937M/WS843
7938M/XH903
7939M/XD596
7940M/XL764
7949M/XF974
7955M/XH767
7957M/XF545
7959M/WS774
7960M/WS726
7961M/WS739
7964M/WS760
7965M/WS792
7967M/WS788
7969M/WS840
7970M/WP907
7971M/XK699
7973M/WS807
7976M/XK418
7979M/XM529
7980M/XM561
7982M/XH892
7983M/XD506
7986M/WG777
7988M/XL149
7990M/XD452
7997M/XG452
7998M/*XM515*/(XD515)
8005M/WG768
8009M/XG518
8010M/XG547
8012M/VS562
8016M/XT677
8017M/XL762
8018M/XN344
8019M/WZ869
8021M/XL824
8022M/XN341
8023M/XD463
8027M/XM555
8032M/XH837
8033M/XD382
8034M/XL703
8041M/XF690
8043M/XF836
8046M/XL770
8049M/WE168
8050M/XG329
8051M/XN929
8052M/WH166
8054AM/XM410
8054BM/XM417
8055AM/XM402
8055BM/XM404
8056M/XG337
8057M/XR243
8060M/WW397
8062M/XR669
8063M/WT536
8070M/EP120
8072M/PK624
8073M/TB252
8074M/TE392
8075M/RW382
8076M/XM386
8077M/XN594
8078M/XM351
8079M/XN492
8080M/XM480
8081M/XM468
8083M/XM367
8085M/XM467
8086M/TB752
8088M/XN602
8092M/WK654
8094M/WT520
8102M/WT486
8103M/WR985
8106M/WR982
8108M/WV703
8113M/WV753
8114M/WL798
8117M/WR974
8118M/WZ549

RAF Maintenance cross-reference

8119M/WR971
8121M/XM474
8128M/WH775
8131M/WT507
8133M/WT518
8139M/XJ582
8140M/XJ571
8141M/XN688
8142M/XJ560
8143M/XN691
8147M/XR526
8151M/WV795
8153M/WV903
8154M/WV908
8155M/WV797
8156M/XE339
8158M/XE369
8159M/XD528
8160M/XD622
8161M/XE993
8162M/WM913
8163M/XP919
8164M/*WN105*/(WF299)
8165M/WH791
8169M/WH364
8171M/XJ607
8173M/XN685
8176M/WH791
8177M/*WM311*
8179M/XN928
8180M/XN930
8182M/XN953
8183M/*XN972*/(XN962)
8184M/WT520
8186M/WR977
8187M/WH791
8189M/WD646
8190M/XJ918
8192M/XR658
8196M/XE920
8197M/WT346
8198M/WT339
8203M/XD377
8205M/XN819
8206M/WG419
8207M/WD318
8208M/WG303
8209M/WG418
8210M/WG471
8211M/WK570
8212M/WK587
8213M/WK626
8214M/WP864
8215M/WP869
8217M/WZ866
8224M/XN699
8226M/XP921
8229M/XM355
8230M/XM362
8231M/XM375
8232M/*ZD713*
8233M/XM408
8234M/XN458
8235M/XN549
8236M/XP573
8237M/XS179
8238M/XS180
8239M/XS210
8344M/WH960
8345M/XG540
8346M/XN734
8350M/WH840
8352M/XN632

8355M/*KG374*/(KN645)
8357M/WK576
8359M/WF825
8360M/WP863
8361M/WB670
8362M/WG477
8363M/WG463
8364M/WG464
8365M/XK421
8366M/XG454
8367M/XG474
8368M/XF926
8369M/WE139
8370M/N1671
8371M/XA847
8372M/K8042
8373M/P2617
8375M/NX611
8376M/RF398
8377M/R9125
8378M/*T9707*
8379M/DG590
8380M/Z7197
8382M/VR930
8383M/K9942
8384M/X4590
8385M/N5912
8386M/NV778
8387M/T6296
8388M/XL993
8389M/VX573
8390M/SL542
8392M/SL674
8393M/XK987
8394M/WG422
8395M/WF408
8396M/XK740
8399M/WR539
8401M/XP686
8402M/XN769
8403M/XK531
8406M/XP831
8407M/XP585
8408M/XS186
8409M/XS209
8410M/XR662
8413M/XM192
8414M/XM173
8417M/XM144
8422M/XM169
8427M/XM172
8428M/XH593
8429M/XH592
8431M/XR651
8434M/XM411
8435M/XN512
8436M/XN554
8437M/WG362
8439M/WZ846
8440M/WD935
8442M/XP411
8445M/XK968
8447M/XP359
8453M/XP745
8457M/XS871
8458M/XP672
8459M/XR650
8460M/XP680
8462M/XX477
8463M/XP355
8464M/XJ758
8465M/W1048
8466M/L-866

8467M/WP912
8468M/MM5701/(BT474)
8470M/584219
8471M/701152
8472M/120227/(VN679)
8473M/WP180/(WP190)
8478M/494083
8475M/360043/(PJ876)
8476M/24
8477M/4101/(DG200)
8478M/10639/(RN228)
8479M/730301
8482M/112372/(VK893)
8483M/420430
8484M/5439
8485M/997
8487M/J-1172
8488M/WL627
8491M/WJ880
8492M/WJ872
8493M/XR571
8494M/XP557
8495M/XR672
8498M/XR670
8501M/XP640
8502M/XP686
8503M/XS451
8505M/XL384
8506M/XR704
8507M/XS215
8508M/XS218
8509M/XT141
8510M/XP567
8513M/XN724
8514M/XS176
8515M/WH869
8516M/XR643
8535M/XN776
8538M/XN781
8546M/XN728
8548M/WT507
8549M/WT534
8550M/XT595
8551M/XN774
8554M/TG511
8556M/XN855
8559M/XN467
8560M/XR569
8561M/XS100
8565M/*WT720*/*(E-408)*
8566M/XV279
8568M/XP503
8569M/XR535
8570M/XR954
8572M/XM706
8573M/XM708
8575M/XP542
8576M/XP502
8578M/XR534
8579M/XR140
8580M/XP516
8581M/WJ775
8582M/XE874
8585M/XE670
8586M/XE643
8587M/XP677
8588M/XR681
8589M/XR700
8590M/XM191
8591M/XA813
8595M/XH278
8596M/LH208
8598M/WP270

8602M/*PF179*/(XR541
8606M/XP530
8608M/XP540
8609M/XR953
8610M/XL502
8611M/WF128
8615M/XP532
8617M/XM709
8619M/XP511
8620M/XP534
8621M/XR538
8622M/XR980
8623M/XR998
8624M/*XR991*/(XS102
8626M/XS109
8627M/XP558
8628M/XJ380
8630M/WG362
8631M/XR574
8634M/WP314
8638M/XS101
8640M/XR977
8642M/XR537
8645M/XD163
8647M/XP338
8648M/XK526
8653M/XS120
8654M/XL898
8655M/XN126
8656M/XP405
8657M/VZ634
8661M/XJ727
8662M/XR458
8666M/XE793
8667M/WP972
8668M/WJ821
8670M/XL384
8671M/XJ435
8672M/XP351
8673M/XD165
8674M/XP395
8676M/XL577
8678M/XE656
8679M/XF526
8680M/XF527
8681M/XG164
8682M/XP404
8684M/XJ634
8685M/XF516
8687M/XJ639
8691M/WT518
8693M/WH863
8695M/WJ817
8696M/WH773
8699M/ZD232
8702M/XG196
8703M/VW453
8704M/XN643
8705M/XT281
8706M/XF383
8708M/XF509
8709M/XG209
8710M/XG274
8711M/XG290
8713M/XG225
8714M/XK149
8715M/*XF445*/(XG264
8716M/XV155
8718M/XX396
8719M/XT257
8720M/XP353
8721M/XP354
8723M/XL567

8724M/XW923
8726M/XP299
8727M/XR486
8728M/WT532
8729M/WJ815
8730M/XD186
8731M/XP361
8732M/XJ729
8733M/XL318
8736M/XF375
8739M/XH170
8741M/XW329
8743M/WD790
8745M/XL392
8746M/XH171
8747M/WJ629
8749M/XH537
8751M/XT255
8752M/XR509
8753M/WL795
8755M/*WH699*/(WJ637)
8756M/XL427
8757M/XM656
8760M/XL386
8762M/WH740
8763M/WH665
8764M/XP344
8767M/XX635
8768M/A-522
8769M/A-528
8770M/XL623
8771M/XM602
8772M/WR960
8777M/XX914
8778M/XM598
8779M/XM607
8780M/WK102
8781M/WE982
8782M/XH136
8783M/XW272
8784M/VP976
8785M/XS642
8786M/XN495
8789M/XK970
8790M/XK986
8791M/XP329
8792M/XP345
8793M/XP346
8794M/XP398
8796M/XK943
8799M/WV787
8800M/XG226
8801M/XS650
8802M/XJ608
8805M/XT772
8806M/XP140
8807M/XL587
8810M/XJ825
8813M/VT260
8814M/XM927
8816M/XX734
8818M/XK527
8819M/XS479
8820M/VP952
8821M/XX115
8822M/VP957
8824M/VP971
8826M/XV638
8828M/XS587
8829M/XE653
8830M/XF515
8831M/XG160
8832M/XG172
8833M/XL569
8836M/XL592
8838M/*34037*/(429356)
8839M/XG194
8840M/XG252
8845M/XS572
8847M/XX344
8848M/XZ135
8851M/XT595
8853M/XT277
8854M/XV154
8855M/XT284
8857M/XW544
8860M/XW549
8861M/XW528
8862M/XN473
8863M/XG154
8864M/WJ678
8865M/XN641
8866M/XL609
8867M/XK532
8868M/WH775
8869M/WH957
8870M/WH964
8871M/WJ565
8873M/XR453
8874M/XE597
8875M/XE624
8876M/*VM791*/(XA312)
8877M/XP159
8879M/XX948
8880M/XF435
8881M/XG254
8882M/XR396
8883M/XX946
8884M/VX275
8886M/XA243
8887M/WK162
8888M/XA231
8889M/XN239
8890M/WT532
8892M/XL618
8894M/XT669
8895M/XX746
8896M/XX821
8897M/XX969
8898M/XX119
8899M/XX756
8900M/XZ368
8901M/XZ383
8902M/XX739
8903M/XX747
8904M/XX966
8905M/XX975
8906M/XX976
8907M/XZ371
8908M/XZ382
8909M/XV784
8910M/XL160
8911M/XH673
8913M/XT857
8915M/XH132
8917M/XM372
8918M/XX109
8919M/XT486
8920M/XT469
8921M/XT466
8923M/XX819
8924M/XP701
8925M/XP706
8931M/XV779
8932M/XR718
8933M/XX297
8934M/XR749
8935M/XR713
8937M/XX751
8938M/WV746
8939M/XP741
8940M/XR716
8941M/XT456
8942M/XN185
8943M/XE799
8944M/WZ791
8945M/XX818
8946M/XZ389
8947M/XX726
8948M/XX757
8949M/XX743
8950M/XX956
8951M/XX727
8952M/XX730
8953M/XX959
8954M/XZ384
8955M/XX110
8956M/XN577
8957M/XN582
8958M/XN501
8959M/XN472
8960M/XM455
8961M/XS925
8962M/XM371
8967M/XV263
8968M/XM471
8969M/XR753
8971M/XN548
8972M/XR754
8973M/XS922
8974M/XM473
8978M/XX837
8979M/XV747
8981M/XW764
8983M/XM478
8984M/XN551
8985M/WK127
8986M/XV261
8987M/XM358
8988M/XN593
8989M/XV740
8990M/XM419
8991M/XR679
8992M/XP547
8993M/XS219
8994M/XS178
8995M/XM425
8996M/XM414
8997M/XX669
8998M/XT864
9000M/XZ282
9001M/XV778
9003M/XZ390
9004M/XZ370
9005M/XZ374
9006M/XX967
9007M/XX968
9008M/XX140
9009M/XX763
9010M/XX764
9011M/XM412
9012M/XN494
9014M/XN584
9015M/XW320
9016M/XN640
9017M/ZE449
9018M/XW365
9019M/XX824
9020M/XX825
9021M/XX826
9022M/XX958
9023M/XX844
9024M/XL192
9025M/XR701
9026M/XP629
9027M/XP556
9028M/XP563
9029M/XS217
9030M/XR674
9031M/XP688
9023M/XR673
9033M/XS181
9034M/XP638
9035M/XR653
9036M/XM350
9037M/XN302
9038M/XV810
9039M/XN586
9040M/XZ138
9041M/
9042M/
9043M/
9044M/XS177
9045M/XN636
9046M/XM349
9047M/XW409
9048M/XM403
9049M/XW404
9050M/XG577
9051M/XM472
9052M/WJ717
9053M/XT755
9054M/XT766
9055M/XT770
9056M/XS488
9057M/ZE361
9058M/ZE351
9059M/ZE360
9060M/ZE356
9061M/XW335
9062M/XW351
9064M/XT867
9065M/XV577
9066M/XV582
9067M/XV586
9068M/XT874
9069M/XV570
9070M/XV581
9071M/XT853
9072M/XW768
9073M/XW924
9074M/XV738
9075M/XV753
9076M/XV808
9077M/XZ967
9078M/XV752
9079M/XZ130
9083M/ZE353
9084M/ZE354
9085M/ZE364
9087M/XX753
9088M/XV587
9090M/XW353
9091M/XW434
9092M/XH669
9093M/WK124
9095M/
9096M/WV322
9097M/XW366
9098M/XV406
9099M/XT900
9100M/XL188

9101M/WL756
9102M
9103M/XV411
9104M
9105M
9106M
9107M/XV482
9108M/XT475
9109M/XW312
9110M/XX736
9111M/XW421
9112M/XM475
9113M
9114M/XL162
9115M/XV863
9116M
9117M/XV161
9119M/XW303
9120M/XW419
9121M/XN606
9122M
9123M/XT773
9124M/XW427
9125M/XW410
9126M/XW413
9127M/XW432
9128M/XW292
9129M/XW294
9130M/XW327
9131M
9133M
9134M/XT288

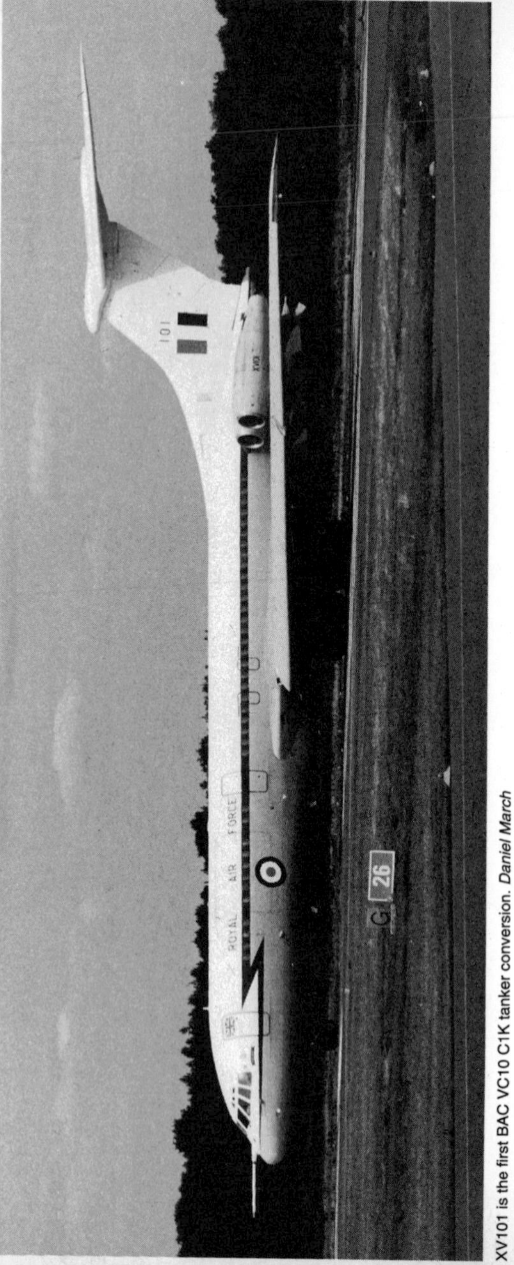

XV101 is the first BAC VC10 C1K tanker conversion. *Daniel March*

ZG885 Westland Lynx AH9 is based at Dishforth with No 672 Sqn. *PRM*

RN Landing Platform and Shore Station Code-letters

Notes	Serial	Type	Owner or Operator	
	323	AB	HMS *Ambuscade* (F172)	Type 21 (815 Sqn)
	341	AG	HMS *Avenger* (F185)	Type 21 (815 Sqn)
	327	AL	HMS *Alacrity* (F174)	Type 21 (815 Sqn)
	472	AM	HMS *Andromeda* (F57)	Leander (829 Sqn)
		AS	RFA *Argus* (A135)	Aviation Training ship
	466	AT	HMS *Argonaut* (F56)	Leander (829 Sqn)
	322	AV	HMS *Active* (F171)	Type 21 (815 Sqn)
	326	AW	HMS *Arrow* (F173)	Type 21 (815 Sqn)
	365/6	AY	HMS *Argyll* (F232)	Type 23
	320	AZ	HMS *Amazon* (F169)	Type 21 (815 Sqn)
	328/9	BA	HMS *Brave* (F94)	Type 22 (829 Sqn)
	—	BD	RFA *Sir Bedivere* (L3004)	Landing ship
	—	BE	RFA *Blue Rover* (A270)	Fleet tanker
	333	BM	HMS *Birmingham* (D86)	Type 42 (815 Sqn)
	342/3	BT	HMS *Brilliant* (F90)	Type 22 (829 Sqn)
	—	BV	HMS *Black Rover* (A273)	Fleet tanker
	346/7	BW	HMS *Broadsword* (F88)	Type 22 (829 Sqn)
	402/3	BX	HMS *Battleaxe* (F89)	Type 22 (829 Sqn)
	330/1	BZ	HMS *Brazen* (F91)	Type 22 (829 Sqn)
	335	CF	HMS *Cardiff* (D108)	Type 42 (815 Sqn)
	350/1	CL	HMS *Cumberland* (F85)	Type 22
	348/9	CM	HMS *Chatham* (F87)	Type 22 (829 Sqn)
	338/9	CT	HMS *Campbeltown* (F86)	Type 22 (829 Sqn)
	—	CU	RNAS Culdrose (HMS *Seahawk*)	
	336/7	CV	HMS *Coventry* (F98)	Type 22 (829 Sqn)
	412/3	CW	HMS *Cornwall* (F99)	Type 22 (829 Sqn)
	—	DC	HMS *Dumbarton Castle* (P268)	Fishery protection
	—	DG	RFA *Diligence* (A132)	Maintenance
	411	EB	HMS *Edinburgh* (D97)	Type 42 (815 Sqn)
	434/5	ED	HMS *Endurance* (A176)	Ice Patrol
	420	EX	HMS *Exeter* (D89)	Type 42 (815 Sqn)
	—	FA	RFA *Fort Austin* (A386)	Support ship
	—	FG	RFA *Fort Grange* (A385)	Support ship
	—	FL	RNAY Fleetlands	
	—	FS	HMS *Fearless* (L10)	Assault
	410	GC	HMS *Gloucester* (D96)	Type 42 (815 Sqn)
	321	GIB	Gibraltar Airport	(815 Sqn)
	—	GD	RFA *Sir Galahad* (L3005)	Landing ship
	—	GN	RFA *Green Rover* (A268)	Fleet tanker
	—	GR	RFA *Sir Geraint* (L3027)	Landing ship
	—	GV	RFA *Gold Rover* (A271)	Fleet tanker
	344	GW	HMS *Glasgow* (D88)	Type 42 (815 Sqn)
	—	GY	RFA *Grey Rover* (A269)	Fleet tanker
	—	HC	HMS *Hecla* (A133)	Survey ship
	—	HR	HMS *Herald*	Survey ship
	—	ID	HMS *Intrepid* (L11)	Assault
	—	L	HMS *Illustrious* (R06)	Carrier
	—	LA	HMS *Lancaster* (F229)	Type 23
	—	LC	HMS *Leeds Castle* (P258)	Fishery protection
	406	LN	HMS *London* (F95)	Type 22 (829 Sqn)
	332	LP	HMS *Liverpool* (D92)	Type 42 (815 Sqn)
	—	LS	RNAS Lee-on-Solent (HMS *Daedalus*)	
	363/4	MA	HMS *Marlborough* (F233)	Type 23 (829 Sqn)
	360	MC	HMS *Manchester* (D95)	Type 42 (815 Sqn)
	—	N	HMS *Invincible* (R05)	Carrier
	345	NC	HMS *Newcastle* (D87)	Type 42 (815 Sqn)
	361/2	NF	HMS *Norfolk* (F230)	Type 23 (829 Sqn)
	417	NM	HMS *Nottingham* (D91)	Type 42 (815 Sqn)
	—	OD	RFA *Olmeda* (A124)	Fleet tanker
	—	ON	RFA *Olna* (A123)	Fleet tanker
	—	OW	RFA *Olwen* (A122)	Fleet tanker
	—	PO	RNAS Portland (HMS *Osprey*)	
	—	PV	RFA *Sir Percival* (L3036)	Landing ship
	—	PW	Prestwick Airport (HMS *Gannet*)	
	—	R	HMS *Ark Royal* (R09)	Carrier
	—	RG	RFA *Regent* (A486)	Support ship

Serial	Type	Owner or Operator	Notes
—	RS	RFA *Resource* (A480)	Support ship
432	SC	HMS *Scylla* (F71)	Leander (829 Sqn)
352/3	SD	HMS *Sheffield* (F96)	Type 23 (829 Sqn)
334	SN	HMS *Southampton* (D90)	Type 42 (815 Sqn)
450	SS	HMS *Sirius* (F40)	Leander (815 Sqn)
—	TM	RFA *Sir Tristram* (L3505)	Landing ship
374/5	VB	HMS *Beaver* (F93)	Type 22 (829 Sqn)
—	VL	RNAS Yeovilton (HMS *Heron*)	
—	WU	RNAY Wroughton	
376	XB	HMS *Boxer* (F92)	Type 22 (829 Sqn)
407	YK	HMS *York* (D98)	Type 42 (815 Sqn)
—	—	HMS *Iron Duke* (F234)	Type 23
—	—	HMS *Monmouth* (F235)	Type 23
—	—	HMS *Montrose* (F236)	Type 23
—	—	HMS *Westminster* (F237)	Type 23
—	—	HMS *Northumberland* (F238)	Type 23
—	—	HMS *Richmond* (F239)	Type 23
—	—	HMS *Somerset* (F240)	Type 23
—	—	HMS *Grafton* (F241)	Type 23
—	—	HMS *Sutherland* (F242)	Type 23
—	—	RFA *Fort Victoria* (A387)	Auxiliary Oiler
—	—	RFA *Fort George* (A388)	Auxiliary Oiler

RN Code – Squadron – Base – Aircraft Cross-Check

Code Numbers	Deck/Base Letters	Unit	Location	Aircraft Type(s)
000 — 005	R	801 Sqn	Yeovilton	Sea Harrier FRS1
010 — 020	R	820 Sqn	Culdrose	Sea King HAS5
122 — 129	N	800 Sqn	Yeovilton	Sea Harrier FRS1
130 — 139	—	826 Sqn	Culdrose	Sea King HAS5
180 — 187	—	849 Sqn	Culdrose	Sea King AEW2A
264 — 274	N	814 Sqn	Culdrose	Sea King HAS5
300 — 306	PO	815 Sqn	Portland	Lynx HAS2/HAS3
320 — 479	*	815/829 Sqns	Portland	Lynx HAS2/HAS3
500 — 510	—	810 Sqn	Culdrose	Sea King HAS5
538 — 559	CU	705 Sqn	Culdrose	Gazelle HT2
560 — 575	CU	750 Sqn	Culdrose	Jetstream T2
576 — 579	CU	750 Sqn	Culdrose	Jetstream T3
580 — 599	—	706 Sqn	Culdrose	Sea King HAS5
600 — 605	PO	829 Sqn	Portland	Lynx HAS2/HAS3
606 — 612	PO	829 Sqn	Portland	Lynx HAS2/HAS3
620 — 628	PO	772 Sqn	Portland	Sea King HC4
630 — 638	PO	702 Sqn	Portland	Lynx HAS2/HAS3
640 — 648	PO	702 Sqn	Portland	Lynx HAS2/HAS3
670 — 672	—	700L Sqn	Portland	Lynx HAS3 CTS
701 — 708	PW	819 Sqn	Prestwick	Sea King HAS6
710 — 716	VL	899 Sqn	Yeovilton	Sea Harrier FRS1
717-718, 723	VL	899 Sqn	Yeovilton	Harrier T4N
719 — 720	VL	899 Sqn	Yeovilton	Hunter T8M
721 — 722	VL	899 Sqn	Yeovilton	Sea Harrier FRS1
738 — 739	VL	Station Flt	Yeovilton	Chipmunk T10
816 — 817	—	771 Sqn	Culdrose	Chipmunk T10
820 — 826	CU	771 Sqn	Culdrose	Sea King HAR5

Code Numbers	Deck/Base Letters	Unit	Location	Aircraft Type(s)
830 — 838	VL	FRADU	Yeovilton	Hunter GA11
860 — 868	VL	FRADU	Yeovilton	Hunter GA11
869 — 880	VL	FRADU	Yeovilton	Hunter T7/T8
901 — 912	—	FGF	Plymouth	Chipmunk T10

*See foregoing separate ships' Deck Letter Analysis

Note that only the 'last two' digits of the code are worn by some aircraft types, especially helicopters.

Ships' Numeric Code — Deck Letters Analysis

	0	1	2	3	4	5	6	7	8	9
32	AZ	GIB	AV	AB			AW	AL	BA	BA
33	BZ	BZ	LP	BM	SN	CF	CV	CV	CT	CT
34		AG	BT	BT	GW	NC	BW	BW	CM	CM
35	CL	CL	SD							
36	MC	NF	NF	MA	MA	AY	AY			
37					VB	VB	XB			
40			BX	BX		LN	LN	YK		
41	GC	EB	CW	CW				NM		
42	EX									
43			SC		ED	ED				
45	SS									
46							AT			
47						AM				

British-based Historic Aircraft in Overseas Markings

Some 'Historic' aircraft carry the markings of overseas air arms and can be seen in the UK, mainly preserved in museums and collections or taking part in air shows.

Serial	*Type*	*Owner or Operator*	*Notes*
Argentina			
A-515	FMA IA58 Pucara (ZD485)	RAF Cosford Aerospace Museum	
A-517	FMA IA58 Pucara (G-BLRP)	Privately owned, Channel Islands	
A-522	FMA IA58 Pucara (8768M)	FAA Museum, RNAS Yeovilton	
A-528	FMA IA58 Pucara (8769M)	Museum of Army Flying, Middle Wallop	
A-533	FMA IA58 Pucara	Museum of Army Flying, Middle Wallop	
A-549	FMA IA58 Pucara	Imperial War Museum, Duxford	
AE-406	Bell UH-1H Iroquois	Museum of Army Flying, Middle Wallop	
AE-409	Bell UH-1H [656] Iroquois	Museum of Army Flying, Middle Wallop	
AE-422	Bell UH-1H	FAA Museum, RNAS Yeovilton	
PA-12	SA330L Puma HC1 (ZE449/9017M)	RAF Odiham, BDRT	
0729	Beech T-34C Turbo Mentor	FAA Museum, RNAS Yeovilton	
0767	Macchi MB339AA [4-A-116]	Rolls-Royce, Filton	
Australia			
A2-4	Supermarine Seagull V (VH-ALB)	RAF Museum, Hendon	
A17-199	Lockheed Hudson IIIA (G-BEOX) (FH174) [SF-R]	RAF Museum, Hendon	
A17-48	DH Tiger Moth (N48DH) (G-BPHR)	Privately owned, Norfolk	
Belgium			
FT-36	Lockheed T-33A	Dumfries & Galloway Aviation Museum, Tinwald Downs	
HD-75	Hanriot HD1 (OO-APJ/ G-AFDX/N75)	RAF Museum, Hendon	
MT-11	Fouga CM-170R Magister (G-BRFU)	Vintage Aircraft Team, Cranfield	
Canada			
232	Hawker Sea Fury FB11 (N232J)	Privately owned, North Weald	
671	DHC Chipmunk 22 (G-BNZC)	British Aerial Museum, Duxford	
920	VS Stranraer (CF-BXO) [Q-N]	RAF Museum, Hendon	
9059	Bristol Bolingbroke IVT	Privately owned, Portsmouth	
9893	Bristol Bolingbroke IVT	Imperial War Museum store	
9940	Bristol Bolingbroke IVT	Royal Scottish Museum of Flight, East Fortune	
10201	Bristol Bolingbroke IVT (G-BPIV)	British Aerial Museum, Duxford	
16693	Auster J/1N Alpha (G-BLPG) [693]	Privately owned, Headcorn	
18013	DHC Chipmunk 22 [013]	Privately owned, Little Gransden	
18393	Avro Canada CF-100 (G-BCYK)	Imperial War Museum, Duxford	
18671	DHC Chipmunk 22 (G-BNZC)	Privately owned, Duxford	
20310	CCF Harvard IV (G-BSBG)	Privately owned, Shoreham	
20385	CCF T-6J Harvard IV (G-BGPB)	Harvard Formation Team, North Weald	
23140	NA Sabre [AX] (fuselage)	Midland Air Museum, Coventry	
51-16622	Piasecki HUP-3 Retriever (N6699D)	IHM, Weston-super-Mare	
Denmark			
E-402	Hawker Hunter F51	Privately owned, Bournemouth	
E-419	Hawker Hunter F51	North-East Aircraft Museum, Usworth	
E-420	Hawker Hunter F51 (G-9-442)	Privately owned, Marlow	
E-421	Hawker Hunter F51	Brooklands Museum, Weybridge	
E-423	Hawker Hunter F51 (G-9-444)	Second World War Aircraft Preservation Society, Lasham	
E-424	Hawker Hunter F51 (G-9-445)	S Yorks Air Museum, Firbeck	
E-425	Hawker Hunter F51	Midland Air Museum, Coventry	
E-427	Hawker Hunter F51 (G-9-447)	Privately owned, Bruntingthorpe	
E-430	Hawker Hunter F51	Privately owned, Charlwood, Surrey	
ET-272	Hawker Hunter T7 (cockpit only)	Jet Heritage Ltd, Bournemouth	
ET-273	Hawker Hunter T7	Historical Aviation Society, Chelford	
L-866	Consolidated Catalina (8466M)	RAF Cosford Aerospace Museum	
R-756	Lockheed F-104G Starfighter	Midland Air Museum, Coventry	

Notes	Serial	Type	Owner or Operator
	Egypt		
	705	Yak 18 (G-OYAK)	Privately owned, Earls Colne
	France		
	9	Dassault Mystère IVA [004]	RAF Bentwaters
	16	Dassault Mystère IVA	RAF Lakenheath
	25	Dassault Mystère IVA	RAF Woodbridge
	36	Dassault Mystère IVA [EABDR 8]	RAF Upper Heyford BDRT
	37	Nord 3400 [MAB] (G-ZARA)	Privately owned, Boston
	45	SNCAN Stampe SV4C (G-BHFG)	Privately owned, Enstone
	45	CM-170 Fouga Magister (G-FUGA)	Privately owned, East Midlands
	46	Dassault Mystère IVA [EABDR 9]	RAF Upper Heyford
	50	Dassault Mystère IVA	RAF Woodbridge
	57	Dassault Mystère IVA [8-MT]	Imperial War Museum, Duxford
	59	Dassault Mystère IVA [314-TH]	Wales Aircraft Museum, Cardif
	65	Nord 3202 (G-BMBF)	Privately owned, Stamford
	68	Nord 3400 [MHA]	Privately owned, Coventry
	70	Dassault Mystère IVA [8-NV]	Midland Air Museum, Coventry
	75	Dassault Mystère IVA [11]	RAF Lakenheath
	79	Dassault Mystère IVA [8-NB]	Norfolk & Suffolk Aviation Museum, Flixton
	83	Dassault Mystère IVA [8-MS]	Newark Air Museum, Winthorpe
	84	Dassault Mystère IVA [8-NF]	Robertsbridge Aviation Society, Headcorn
	85	Nord 3202B [AJG] (G-BRVA)	Privately owned, Breighton
	85	Dassault Mystère IVA	Privately owned, Bruntingthorpe
	99	Dassault Mystère IVA	RAF Lakenheath
	101	Dassault Mystère IVA [8-MN]	Bomber County Aviation Museum, Hemswell
	104	Dassault Mystère IVA	RAF Bentwaters, BDRT
	FR108	SO1221 Djinn	International Helicopter Museum, Weston-super-Mare
	113	Dassault Mystère IVA	RAF Lakenheath
	120	SNCAN Stampe SV4C (G-AZGC)	Privately owned, Booker
	121	Dassault Mystère IVA [8-MY]	City of Norwich Aviation Museum
	126	Dassault Mystère IVA	RAF Lakenheath
	127	Dassault Mystère IVA [EABDR 7]	RAF Upper Heyford
	129	Dassault Mystère IVA [EABDR 6]	RAF Upper Heyford
	133	Dassault Mystère IVA	RAF Woodbridge
	145	Dassault Mystère IVA	RAF Lakenheath
	FR145	SO1221 Djinn [CDL]	Privately owned, Luton Airport
	146	Dassualt Mystère IVA [8-MC]	North East Aircraft Museum, Usworth
	157	Morane MS230 (G-AVEB)	Privately owned, Booker
	192	MH1521M Broussard (G-BKPT)	Privately owned, Chiseldon
	241	Dassault Mystère IVA [2]	RAF Lakenheath
	276	Dassault Mystère IVA	RAF Woodbridge
	285	Dassault Mystère IVA	RAF Lakenheath
	300	Dassault Mystère IVA [5]	RAF Lakenheath
	309	Dassault Mystère IVA [8]	RAF Lakenheath
	318	Dassault Mystère IVA [8-NY]	Dumfries & Galloway Aviation Museum, Tinwald Downs
	319	Dassault Mystère IVA [8-ND]	Rebel Air Museum, Andrewsfield
	851	Nord 3202 (G-BRVA)	Privately owned, Liverpool
	133722	CV F4U-7 Corsair (NX1337A) [15F-22]	Privately owned, Duxford
	14286	Lockheed T-33A [WK]	Imperial War Museum, Duxford
	16718	Lockheed T-33A [314-UJ]	City of Norwich Aviation Museum
	42163	NA F-100D Super Sabre [11-YG]	Dumfries & Galloway Aviation Museum, Tinwald Downs
	42165	NA F-100D Super Sabre [11-ML]	Imperial War Museum, Duxford
	54433	Lockheed T-33A [WD]	Norfolk & Suffolk Aviation Museum, Flixton
	63938	NA F-100F Super Sabre [11-MU]	Lashenden Air Warfare Museum, Headcorn
	S3398	Spad XIII Replica (G-BFYO) [2]	FAA Museum, RNAS Yeovilton
	S4523	Spad S-XIII (N4727V) [1]	Imperial War Museum, Duxford
	Germany		
	C19/18	Albatros Replica (BAPC 118)	South Yorks Air Museum, Firbeck
	D5397/17	Albatros D.VA Replica (G-BFXL)	FAA Museum, RNAS Yeovilton
	1Z+NK	Amiot AAC1 (Part.AF6316)	IWM, Duxford
	LG+01	Bucker Bu.133C Jungmeister	Privately owned, Duxford

Serial	Type	Owner or Operator	Notes
	(G-AYSJ)		
BU+CK	CASA 1-131E Jungmann (G-BUCK)	Privately owned, White Waltham	
BU+EM	CASA 1-131E Jungmann (G-BUEM)	Privately owned, Shoreham	
6J+PR	CASA 2.111D (G-AWHB)	Aces High Ltd, North Weald	
	Fieseler Fi103 (V-1) (8583M)	RAF Cosford Aerospace Museum	
442795	Fieseler Fi103 (V-1) (BAPC 199)	Science Museum, South Kensington	
475081	Fieseler Fi156C-7 Storch [GM+AK] (VP546/7362M)	RAF Cosford Aerospace Museum	
28368	Flettner Fl282/B-V20 Kolibri (frame only)	Midland Air Museum, Coventry	
100143	Focke-Achgelis Fa330A	Imperial War Museum, Duxford	
100502	Focke-Achgelis Fa330A	Lincolnshire Aviation Museum, East Kirkby	
100509	Focke-Achgelis Fa330A	Science Museum, stored South Kensington	
100545	Focke-Achgelis Fa330A	Fleet Air Arm Museum, Yeovilton	
100549	Focke-Achgelis Fa330A	Greater Manchester Museum of Science and Industry	
2100	Focke Wulf FW189A-1	Privately owned, Lancing	
2+1	Focke Wulf FW190 Replica (G-SYFW) [7334]	Privately owned, Guernsey, CI	
4	Focke Wulf FW190 Replica (G-BSLX)	Privately owned, Shoreham	
8	Focke Wulf FW190 Replica (G-WULF)	Privately owned, Elstree	
1227	Focke Wulf FW190A-5 [OG+HO]	Warbirds of GB, Bournemouth	
584219/39	Focke Wulf FW190F-8/U1 (8470M)	RAF Museum, Hendon	
733682	Focke Wulf FW190A-8/R7	Imperial War Museum, Lambeth	
4253/18	Fokker D.VII Replica (G-BFPL)	Privately owned, Duxford	
626/18	Fokker D.VII Replica (N6268)	Blue Max Movie Aircraft Museum, Booker	
8417/18	Fokker D.VII	RAF Museum store Cardington	
102/18	Fokker Dr.1 Dreidekker Replica (BAPC 88)	FAA Museum, RNAS Yeovilton	
152/17	Fokker Dr.1 Dreidekker Replica (G-ATJM)	Privately owned, Rendcomb, Glos	
152/17	Fokker Dr.1 Dreidekker Replica (G-BTYV/N152JS)	Privately owned, North Weald	
425/17	Fokker Dr.1 Dreidekker Replica (BAPC 133)	Newark Air Museum	
425/17	Fokker Dr.1 Dreidekker Replica (G-BEFR)	Privately owned, Dunkeswell	
210/16	Fokker EIII (BAPC56)	Science Museum, Kensington	
422/15	Fokker EIII replica (G-AVJO)	Privately owned, Booker	
701152	Heinkel He111H-23 (8471M) [NT+SL]	RAF Museum, Hendon	
120227	Heinkel He162A-2 Salamander (VN679/8472M) [2]	RAF Museum, Hendon	
120235	Heinkel He162A Salamander	Imperial War Museum, Lambeth	
—	Hispano HA 1112 (G-BOML)	Privately owned, Duxford	
494083	Junkers Ju87D-5 (8474M) [RI+JK]	RAF Museum, Hendon	
360043	Junkers Ju88R-1 (PJ876/8475M) [D5+EV]	RAF Museum, Hendon	
6234	Junkers Ju87R-4 [L1+FW]	Privately owned, Shoreham	
22+35	Lockheed F104G Starfighter	Second World War Aircraft Preservation Society, Lasham	
22+57	Lockheed F104G Starfighter	Starfighter Preservation Group, New Waltham	
7198/18	LVG C.VI (G-AANJ)	Shuttleworth Collection, Old Warden	
6357	Messerschmitt Bf109 Replica (BAPC 74)	Kent Battle of Britain Museum, Hawkinge	
14	Messerschmitt Bf109 Replica (BAPC 67)	Kent Battle of Britain Museum, Hawkinge	
1190	Messerschmitt Bf109E-3	Privately owned, Bournemouth	
1480	Messerschmitt Bf109 [6] (BAPC 66)	Kent Battle of Britain Museum, Hawkinge	
4101	Messerschmitt Bf109E-3 (DG200/8477M) [12]	RAF Museum, Hendon	
10639	Messerschmitt Bf109G-2 (Trop) (RN228/8478M) (G-USTV)	Imperial War Museum, Duxford	
4502	Messerschmitt Bf110E-2 [M8+ZE]	Privately owned, Lancing	

Notes	Serial	Type	Owner or Operator
	730301	Messerschmitt Bf110G-4 (8479M) [D5+RL]	RAF Museum, Hendon
	191316	Messerschmitt Me163B Komet	Science Museum, South Kensington
	191614	Messerschmitt Me163B Komet (8481M)	RAF Cosford Aerospace Museum
	191659	Messerschmitt Me163B Komet [15] (8481M)	Royal Scottish Museum of Flight, East Fortune
	191660	Messerschmitt Me163B Komet [3]	Imperial War Museum, Duxford
	112372	Messerschmitt Me262A-2a (VK893/8482M) [9K-XK]	RAF Cosford Aerospace Museum
	420430	Messerschmitt Me410A-1/U2 (8483M) [3U+CC]	RAF Cosford Aerospace Museum
	7A+WN	Morane-Saulnier MS500 (G-AZMH)	Privately owned, Chalmington, Dorset
	FI+S	Morane-Saulnier MS505 (G-BIRW) Criquet	Royal Scottish Museum of Flight, East Fortune
	TA+RC	Morane-Saulnier MS 505 Criquet (G-BPHZ)	Privately owned, Duxford
	NJ+C11	Nord 1002 (G-ATBG)	Privately owned, Sutton Bridge
	14	Pilatus P-2 (J-108/G-BJAX)	Privately owned, Andrewsfield
	CC+43	Pilatus P-2 (U-143/G-CJCI)	Privately owned, Micheldever
	97+04	Putzer Elster B (G-APVF)	Privately owned, Duxford
	Greece		
	52-6541	Republic F-84F Thunderflash	North-East Aircraft Museum, Usworth
	51-6171	Canadair F-86D Sabre (really 51-6171)	North-East Aircraft Museum, Usworth
	Hungary		
	501	Mikoyan MiG-21PF	Privately owned, St Athan
	503	Mikoyan MiG-21SMT (G-BRAM)	Aces High, North Weald
	India		
	Q497	EE Canberra T4 (WH847)	BAe Warton Fire Service
	Iraq		
	333	DH Vampire T55 (pod only)	Military Aircraft Preservation Group, Hadfield, Derbys
	Israel		
	28	NA P-51D Mustang	Privately owned, Fowlmere, Cambs
	41	NA P-51D Mustang (472028)	Privately owned, Teeside Airport
	Italy		
	MM5701	Fiat CR42 (BT474/8468M) [13-95]	RAF Museum, Hendon
	MM53211	Fiat G.46-4 (BAPC 79)	Privately owned, Duxford
	MM53432	NA T-6D Texan [RM-11]	Privately owned, South Wales
	MM53692	CCF T-6G Texan	RAeS Medway Branch, Rochester
	MM53795	CCF Harvard IV (G-BJST) [SC-66]	Privately owned, South Wales
	MM53797	CCF Harvard IV [SC-52]	Privately owned, Coventry
	MM54099	CCF T-6G Harvard [RR-56] (G-BRBC)	Privately owned, Chigwell
	Japan		
	24	Kawasaki Ki100-1B (8476M) (BAPC 83)	RAF Cosford Aerospace Museum
	5439	Mitsubishi Ki46-III (8484M) (BAPC 84)	RAF St Athan
	997	Yokosuka MXY 7 Ohka 11 (8485M) (BAPC 98)	Greater Manchester Museum of Science and Industry
	15-1585	Yokosuka MXY 7 Ohka 11 (BAPC 58)	FAA Museum, RNAS Yeovilton
	—	Yokosuka MXY 7 Ohka 11(8486M) (BAPC 99)	RAF Cosford Aerospace Museum
	—	Yokosuka MXY 7 Ohka 11 (BAPC 159)	Defence School, Chattenden
	Netherlands		
	115	Hawker Sea Fury FB.10(G-BTTA)	Privately owned, Duxford
	361	Hawker Sea Fury FB.10 (N36SF)	Privately owned, Benson
	B-163	NA Harvard 11B	Privately owned, Windsor
	B-168	NA Harvard 11B	Privately owned, Duxford (spares use)
	204/V	Lockheed SP-2H Neptune	RAF Cosford Aerospace Museum
	E-15	Fokker S-11 Instructor (G-BIYU)	Privately owned, Chessington

Serial	Type	Owner or Operator	Notes
E-31	Fokker S-11 Instructor (G-BEPV)	Strathallan Aircraft Collection	
N-202	Hawker Hunter F6 (nose only) [10]	Pinewood Studios, Elstree	
N-250	Hawker Hunter F6 (nose only) [G-9-185]	Science Museum, South Kensington	
N-315	Hawker Hunter T7	Privately owned, Batley	
R-163	Piper L-21B Super Cub (G-BIRH)	Privately owned, Lee-on-Solent	
R-167	Piper L-21B Super Cub (G-LION)	Privately owned, Nayland	
NZ5648	Goodyear FG-1D Corsair (NX55JP)	Privately owned, Duxford	
Norway			
56321	Saab S91B Safir (G-BKPY) [U-AB]	Newark Air Museum, Winthorpe	
Poland			
05	WSK SM-2 (Mi-Z)	IHM, Weston-super-Mare	
1120	MiG-15bis	RAF Museum, Hendon	
1420	MiG-15 (G-BMZF)	FAA Museum, RNAS Yeovilton, stored	
6247	WSK SBLim-2A (MiG-I5UTI)	Privately owned, Shoreham	
Portugal			
1513	NA Harvard II	Privately owned, Cranfield	
1662	NA Harvard III(G-ELMH)	Privately owned, Sudbury	
1681	NA T-6G Texan (G-BSBD)	Privately owned, Shoreham	
1730	CCF Harvard IV (G-BSBE)	Privately owned, Thruxton	
1736	CCF Harvard IV (G-BSBF)	Privately owned, Shoreham	
1766	CCF Harvard IV (G-ELMH)	Privately owned, Sudbury	
1780	CCF Harvard IV (G-BSBC)	Privately owned, Thruxton	
1788	CCF Harvard IV (G-BSBB)	Privately owned, Thruxton	
3460	Dornier 27-1 (G-BMFG)	Privately owned, Old Buckenham	
Qatar			
QA10	Hawker Hunter FGA 78	Yorkshire Air Museum, Elvington	
QA12	Hawker Hunter FGA 78	Lovaux Ltd, Bournemouth (dismantled)	
Russia			
07	Yak-18M (G-BMJY)	Privately owned, North Weald	
165221	WSK-Mielec An-2T (G-BTCU)[77]	Privately owned, Henstridge	
South Africa			
6130	Lockheed Ventura II (AJ469)	RAF Cosford Aerospace Museum	
Spain			
E3B-143	CASA 1.131E Jungmann (G-JUNG)	Privately owned, White Waltham	
E3B-153	CASA 1.131E Jungmann (G-BPTS)	Privately owned, Duxford	
E3B-369	CASA 1.131E (G-BPDM) [781-32]	Privately owned, Shoreham	
EM-01	DH60G Moth (G-AAOR) [30-76]	Privately owned, Shoreham	
T2-124	Messerschmitt Bf-109K [FE-124]	Warbirds of GB, Bournemouth	
C4E-88	Messerschmitt Bf-109E	Tangmere Military Aviation Museum	
Sweden			
16105	NA Harvard IIB(G-BTXI)	Privately owned, Duxford	
29640	Saab J-29F [20-08]	Midland Air Museum, Coventry	
32028	SAAB 32A Lansen (G-BMSG)	Privately owned, Cranfield	
35075	Saab J-35J Draken [40]	Imperial War Museum, Duxford	
Switzerland			
A-806	Pilatus P3-03 (G-BTLL)	Privately owned, Biggin Hill	
A-867	Pilatus P3-05 (G-BUKM)	Privately owned, Shipdham	
J-1008	DH Vampire FB6	Mosquito Aircraft Museum, London Colney	
J-1149	DH Vampire FB6 (G-SWIS)	Jet Heritage, Bournemouth	
J-1167	DH Vampire FB6 (G-MKVI)	Vintage Aircraft Team, Cranfield	
J-1172	DH Vampire FB6 (8487M)	Greater Manchester Museum of Science and Industry	
J-1173	DH Vampire FB6 (G-DHXX)	Privately owned, Southampton	
J-1523	DH Venom FB50 (G-VENI)	Privately owned, Cranfield	
J-1605	DH Venom FB50 (G-BLID)	Privately owned, Charlwood, Surrey	
J-1614	DH Venom FB50 (G-BLIE)	Privately owned, Glasgow	
J-1632	DH Venom FB50 (G-VNOM)	Privately owned, Cranfield	
J-1704	DH Venom FB54	RAF Cosford Aerospace Museum	
J-1758	DH Venom FB54 (G-BLSD/ N203DM)	Aces High Ltd, North Weald	
U-80	Bucker Bu.133D Jungmeister [G-BUKK]	Privately owned, White Waltham	

Notes	Serial	Type	Owner or Operator
	U-110	Pilatus P-2 (G-PTWO)	Privately owned, North Weald
	U-142	Pilatus P-2 (G-BONE)	Privately owned, Southend
	U-1214	DH Vampire T55 (G-DHVV)	Privately owned, Southampton
	U-1215	DH Vampire T55 (G-HELV)	Jet Heritage, Bournemouth
	U-1219	DH Vampire T55 (G-DHWW)	Privately owned, Southampton
	U-1230	DH Vampire T55 (G-DHZZ)	Privately owned, Southampton
	USA		
	I-492	Ryan PT-22 (G-BPUD)	Privately owned, Swanton Morley
	01532	Northrop F-5E Tiger II (Replica)	RAF Alconbury on display
	O-17899	Convair VT-29B	Imperial War Museum, Duxford
	112	Boeing-Stearman E75 (PT-130) Kaydet (G-BSWC)	Privately owned, Redcamb
	11042	Wolf W-II Boredom Fighter (Replica) (G-BMZX) [7]	Privately owned, Haverfordwest
	11083	Wolf W-II Boredom Fighter (Replica) (G-BNAI) [5]	Privately owned, Haverfordwest
	11989	Cessna L-19A Bird Dog (N33600)	Museum of Army Flying, Middle Wallop
	114526	NA T-6G Texan (G-BRWB)	Privately owned, Duxford
	114700	NA T-6G Texan	Aces High, North Weald
	115042	NA T-6G Texan (G-BGHU) [TA-042]	Privately owned, Headcorn
	115302	Piper L-18C Super Cub (G-BJTP) [TP]	Privately owned, Winterbourne, Bristol
	1164	Beech C-45 (G-BKGL) (really RCAF 5193)	British Aerial Museum, Duxford
	118	Boeing Stearman (G-BSDS)	Privately owned, Norwich
	1180	Boeing Stearman (G-BRSK)	Privately owned, Old Buckenham
	121821	Vultee BT-15 Valiant (N513L)	Aces High, North Weald
	122095	Grumman F8F-1B Bearcat (G-BUCF)	Privately owned, Duxford
	122179	CV F4U-5NL Corsair [NP-9] (N179NP)	Warbirds of GB, Bournemouth
	122351	Beech C-45G (G-BKRG)	Aces High, North Weald
	124485	Boeing B-17G (G-BEDF/485784) [DF-A]	B-17 Preservation, Duxford
	126922	Douglas AD-4NA Skyraider (F-AZED) [937-JS]	The Fighter Collection, Duxford
	1411	Grumman Widgeon (N444M)	Privately owned, Biggin Hill
	14419	Lockheed T-33A	Midland Air Museum, Coventry
	146289	NA T-28C Trojan (N99153)	Norfolk & Suffolk Aviation Museum, Flixton
	14700	NA T-6G Texan	Privately owned, Coventry
	150225	Westland Wessex (G-AWOX) [123]	International Helicopter Museum, Weston-super-Mare
	151632	NA TB-25N Mitchell (NL9494Z) (really 430925)	Privately owned, Coventry
	15195	Fairchild PT-19A Cornell	RAF Museum, stored Cardington
	153008	McD F-4N Phantom	RAF Alconbury, BDRT
	155848	McD F-4S Phantom (WT-11/ VMFA-232]	FAA Museum, RNAS Yeovilton
	159233	AV-8A Harrier [CG-33] (VMA-231)	FAA Museum, RNAS Yeovilton
	160810	Bell AH-1T Sea Cobra (fuselage)	GEC, Rochester
	16579	Bell UH-1H Iroquois (FY66)	IHM, Weston-super-Mare
	17473	Lockheed T-33A	RAF Cosford Aerospace Museum
	17657	Douglas A-26K Invader (FY64) (nose only)	Booker Aircraft Museum
	18-2001	Piper L-18C Super Cub (G-BIZV) (really 52-2401)	Privately owned, White Waltham
	19252	Lockheed T-33A	Tangmere Military Aviation Museum
	208	Boeing Stearman N2S (N75664)	Privately owned, Spanhoe
	217448	Boeing Stearman (N3922B)	Privately owned, Old Buckenham
	217786	Boeing Stearman (G-BRTK)[177]	Privately owned, Old Buckenham
	224211	Douglas C-47A Dakota (G-BPMP)	Privately owned, Coventry
	226671	Republic P-47M Thunderbolt [MX-X] (NX47DD)	The Fighter Collection, Duxford
	23	Fairchild PT-23 (N49272)(43-437)	Privately owned, Cosford
	231983	Boeing B-17G (F-BDRS) [IY-G]	Imperial War Museum, Duxford
	232	Hawker Sea Fury FB11 (N232J) (really 44-83735)	Privately owned, North Weald
	236800	Piper L-4A Cub (G-BHPK) [44-A] (really 42-38410)	Privately owned, Tibenham

Serial	Type	Owner or Operator	Notes
243809	Waco CG-4A Hadrian (BAPC 185)	Museum of Army Flying, Middle Wallop	
24518	Kaman HH-43F Huskie (24535)	Midland Air Museum, Coventry	
26	Boeing-Stearman N2S (G-BAVO)	Privately owned, Tatenhill	
27	NA SNJ-7 Texan (G-BRVG)	Intrepid Aviation, North Weald	
2807	NA T-6G Texan (G-BHTH) [V-103]	Privately owned, Kidlington	
295	Ryan PT-22 (N56028)	Privately owned, Rendcomb	
29963	Lockheed T-33A	Wales Aircraft Museum, Cardiff	
30861	NA TB-25J Mitchell (N9089Z)	Privately owned, North Weald	
315509	Douglas C-47A (G-BHUB)	Imperial War Museum, Duxford	
314887	Fairchild Argus III (G-AJPI)	Privately owned, Liverpool	
31923	Aeronca O-58B (G-BRHP)	Privately owned, White Waltham	
320	Boeing Stearman (G-BPMD)	Privately owned, White Waltham	
329417	Piper L-4A Cub (G-BDHK) (really 42-38400)	Privately owned, Coleford	
32947	Piper L-4H Cub (G-BGXA) [44-F]	Privately owned, Martley, Worcs	
329601	Piper L-4H Cub (G-AXHR) [D-44]	Privately owned, Nayland	
329854	Piper L-4H Cub (G-BMKC) [R-44]	Privately owned, Booker	
329934	Piper L-4H Cub (G-BCPH) [72-B]	Privately owned, Booker	
330485	Piper L-4H Cub (G-AJES) [44-C]	Privately owned, Dunkeswell	
34037	NA TB-25N Mitchell (N9115Z/ 8838M) (really 429366)	RAF Museum, Hendon	
343251	Boeing Stearman (G-NZSS)	Privately owned, Cumbernauld	
361	Boeing Stearman (G-ELAN)	Privately owned, White Waltham	
37414	McD F-4C Phantom	Midland Air Museum, Coventry	
37699	McD F-4C Phantom (FY63)	RAF Upper Heyford, BDRT	
379	Boeing Stearman (G-ILLE)	Privately owned, Ipswich	
390	Boeing PT-17 Stearman (G-BTRJ)	Privately owned, Coventry	
40707	McD F-4C Phantom (FY64)	RAF Lakenheath, BDRT	
41	NA T-6G Texan (G-DDMV)	Privately owned, Sywell	
413048	Piper L-4J Cub (G-BCXJ) [39-E] (really 44-80752)	Privately owned, Compton Abbas	
41-33275	NA AT6D Harvard (G-BICE) [CE]	Privately owned, Ipswich	
41386	Thomas-Morse S4 Scout Replica (G-MJTD)	Privately owned, Hitchin	
42-1241	NA AT-16 Harvard	Thameside Aviation Museum	
42157	NA F-100D Super Sabre	North-East Aviation Museum, Usworth	
42174	NA F-100D Super Sabre [UH]	Midland Air Museum, Coventry	
42-6183	Taylorcraft DF-65 [IY] (G-BRIY)	Privately owned, Elstree	
42-17786	Boeing PT-13D Stearman [177] (G-BTZM)	Privately owned, Little Gransden	
42196	NA F-100D Super Sabre [LT]	Norfolk & Suffolk Aviation Museum, Flixton	
42223	NA F-100D Super Sabre	Newark Air Museum, Winthorpe	
42265	NA F-100D Super Sabre	RAF Lakenheath	
42-67543	Lockheed P-38J Lightning (N3145X)	The Fighter Collection, Duxford	
42-69047	Bell P-63A Kingcobra (G-BTWR)	The Fighter Collection, Duxford	
42-69097	Bell P-63A Kingcobra (G-BTWR)	The Fighter Collection, Duxford	
42-78044	Aeronca IIAC Chief (G-BRXL)	Privately owned, Denham	
43-437	Ryan PT-23 (N49272)	Privately owned, Cosford	
430823	TB-25J Mitchell (N1042B) [69]	Aces High Ltd, North Weald	
431171	NA B-25J Mitchell (N7614C)	Imperial War Museum, Duxford	
43-1952	Aeronca O-58B (G-BRPR)	Privately owned, Empingham	
44	Piper PA-18-95 Super Cub (G-BJLH) [K-33]	Privately owned, Felthorpe	
441	Boeing Stearman (G-BTFG)	Privately owned, Rhyl	
44-14574	NA P-51D Mustang (fuselage)	East Essex Aviation Museum, Clacton	
442	Boeing A75 Stearman (G-BPTB)	Privately owned, Paddock Wood	
44-30861	NA TB-25J Mitchell (N9089Z)	Aces High North Weald	
445562	Douglas A-26 Invader (N7079G)	Privately owned, North Weald	
44-79609	Piper L-4H Cub (G-BHXY) [PR]	Privately owned, Bryngwyn Bach, Clwyd	
44-80594	Piper L-4 Cub (G-BEDJ)	Privately owned, Overton	
44-83184	Fairchild 24R Argus III [7] (G-RGUS)	Privately owned, Tongham	
454467	Piper J-3C-65 Cub (G-BILI) [44-J]	Privately owned, Bristol	
454537	Piper L-4J Cub (G-BFDL) [44-J]	Privately owned, Meppershall	
45-49192	Republic P-47D Thunderbolt (47DD)	Imperial War Museum, Duxford	
461748	Boeing B-29A Superfortress (G-BHDK) [Y]	Imperial War Museum, Duxford	
463221	NA P-51D Mustang (G-BTCD) (really 473149) [G4-S]	The Fighter Collection, Duxford	
472216	NA P-51D Mustang (G-BIXL) [AJ-L]	Privately owned, North Weald	

Notes	Serial	Type	Owner or Operator
	472258	NA P-51D Mustang (really 473979) [WZ-I]	Imperial War Museum, Lambeth
	472773	NA P-51D Mustang [AJ-C] (G-SUSY)	Privately owned, North Weald
	472917	CAC-18 Mustang 23 [AJ-A] (G-HAEC)	Privately owned, Duxford
	473415	NA P-51D Mustang (N6526D) [B6-V]	RAF Museum, Hendon
	473877	NA P-51D Mustang (N167F) [B6-S]	Privately owned, Duxford
	474008	NA P-51D Mustang (N51RR)	Privately owned, North Weald
	479766	Piper L-4H Cub (G-BKHG) [63-D]	Privately owned Goldcliff
	480015	Piper L-4H Cub (G-AKIB)	Privately owned, Bodmin
	480133	Piper L-4J Cub (G-BDCD) [44-B]	Privately owned, Slinfold
	480321	Piper L-4J Cub (G-FRAN) [44-H]	Privately owned, Stapleford
	480480	Piper L-4J Cub (G-BECN) [44-E]	Privately owned, Milden
	480594	Piper L-4J Cub (G-BEDJ)	Privately owned, Ashford Hill
	480752	Piper L-4 Cub (G-BCXJ) [E-39]	Privately owned, Compton Abbas
	483009	NA AT-6D Texan (really 244450) (G-BPSE)	Aces High, North Weald
	483868	Boeing B-17G Fortress (N5237V) [A-N]	RAF Museum, Hendon
	511371	NA P-51D Mustang (N1051S) [VF-S]	Privately owned, Southend
	51-15227	NA T-6G Harvard (G-BKRA) [10]	Privately owned, Shoreham
	51-15673	Piper L-18C Super Cub (G-CUBI)	Privately owned, Felixkirk
	53319	Grumman TBM-3R Avenger (G-BTDP) [319-RB]	Privately owned, North Weald
	540	Piper L-4H Cub (G-BCNX) (really 43-29877)	Privately owned, Monewden
	54137	CCF Harvard IV (G-CTKL) [69]	Privately owned, Shoreham
	54-21261	Lockheed T-33A (N33VC)	Privately owned, Duxford
	542447	Piper L-21B Super Cub (G-SCUB)	Privately owned, Anwick
	542457	Piper L-21B Super Cub (G-LION/R-167)	Privately owned, Nayland
	54-2474	Piper L-21B Super Cub (G-PCUB)	Privately owned, Overton
	54439	Lockheed T-33A	North East Aviation Museum, Usworth
	5547	Lockheed T-33A (really 19036)	Newark Air Museum, Winthorpe
	60312	McDonnell TF-101B Voodoo [AR]	Midland Air Museum, Coventry
	60689	Boeing B-52D Stratofortress	Imperial War Museum, Duxford
	607327	Piper L-21B Super Cub [09-L] (G-ARAO)	Privately owned, Lambley
	612414	Boeing CH-47A Chinook	RAF Odiham, instructional use
	63000	NA F-100D Super Sabre (really 42160)	Wales Aircraft Museum, Cardiff
	63000	NA F-100D Super Sabre [FW-000] (really 42212)	RAF Upper Heyford, at gate
	63319	NA F-100D Super Sabre (really 42269) [319-FW]	RAF Lakenheath, at gate
	63-419	McD F-4C Phantom (37419) [SA]	RAF Alconbury, BDRT
	63-428	Republic F-105G Thunderchief (really 24428)	RAF Upper Heyford
	63449	McD F-4C Phantom (37449)	RAF Upper Heyford, BDRT
	63-471	McD F-4C Phantom (37471)	RAF Lakenheath, BDRT
	63-610	McD F-4C Phantom (37610)	RAF Lakenheath, BDRT
	80141	Grumman F6F-5K Hellcat (G-BTCC)	The Fighter Collection, Duxford
	6692	Lockheed U-2C	IWM, Duxford
	67543	Lockheed P-38J Lightning (NX3145X) [MC-O]	The Fighter Collection, Duxford
	6771	Republic F-84F Thunderstreak (really 52-7133) (ex-FU-6)	Cosford Aerospace Museum, store
	68-011	GD F-111E [UH]	RAF Lakenheath, for display
	68-060	GD F-111E (pod)	Dumfries & Galloway Aviation Museum, Tinwald Downs
	70270	McDonnell F-101B Voodoo (fuselage)	Midland Air Museum, Coventry
	70-494	Republic F-105G Thunderchief [LN] (really 62-4434)	RAF Lakenheath, BDRT
	70524	Lockheed C-130A Hercules (FY57)	RAF Mildenhall, BDRT
	74-131	McD F-15A Eagle (40131) [LN]	RAF Lakenheath, BDRT
	77-259	Fairchild A-10A (70259) [AR]	IWM, Duxford

Serial	Type	Owner or Operator	Notes
7797	Aeronca L-16A (G-BFAF)	Privately owned, Finmere	
80-219	Fairchild GA-10A (00219) [AR]	RAF Alconbury, on display	
80260	McDonnell F-101B Voodoo	RAF Molesworth	
80483	Grumman F7F-3 Tigercat [483-JW] (N6178C)	Privately owned, Duxford	
8178	NA F-86A Sabre (48-178/G-SABR)	Privately owned, Bournemouth	
82062	DHC U-6A Beaver	Midland Air Museum, Coventry	
85	WAR P-47 Thunderbolt Replica (N47DL/G-BTBI)	Privately owned, Manchester	
854	Ryan PT-22 (G-BTBH)	Privately owned, Woodbridge	
855	Ryan PT-22 (N56421)	Privately owned, Cosford	
88297	Goodyear FG-1D Corsair (N8297/G-FGID) [29]	Privately owned, Duxford	
88-9696	NA AT-6C-1NT Harvard IIA (G-TEAC)	Harvard Formation Team, North Weald	
897	Aeronca 11AC Chief (G-BJEV)	Privately owned, Little Gransden	
91007	Lockheed T-33A (G-TJET/ G-NASA) (really 51-8566) [TR-007]	Privately owned, Cranfield	
93542	CCF Harvard IV (G-BRLV) [LTA-542]	Privately owned, North Weald	
985	Boeing A75N-1 Stearman (G-ERIX) (really 42-16930)	Privately owned, Sutton Bridge	
Yugoslavia			
5	Nord 1203 (G-BEDB)	Privately owned, Chirk	
13064	Republic P-47D Thunderbolt	RAF Museum Restoration Centre, Cardington	
30146	Soko P-2 Kraguj [146] (G-BSXD)	Privately owned, Shoreham	
30149	Soko P-2 Kraguj [149] (G-BRXK)	Privately owned, Liverpoool	

8178/G-SABR is the world's only airworthy F-86A Sabre. *PRM*

Irish Army Air Corps Military Aircraft Markings

Notes	Serial	Type	Owner or Operator
	34	Miles Magister	Irish Aviation Museum Store, Castlemoate House, Dublin
	141	Avro Anson XIX	Irish Aviation Museum Store, Castlemoate House, Dublin
	157	VS Seafire III (RX158)	Privately owned, Battle
	164	DH Chipmunk T20	Engineering Wing, Baldonnel (stored)
	168	DH Chipmunk T20	Training Wing, Gormanston (stored)
	172	DH Chipmunk T20	Training Wing, Gormanston (stored)
	173	DH Chipmunk T20	South East Aviation Enthusiasts, Waterford
	176	DH Dove 4 (VP-YKF)	South East Aviation Enthusiasts, Waterford
	177	Percival Provost T51 (G-BLIW)	Privately owned, Shoreham
	181	Percival Provost T51	Privately owned, Thatcham
	183	Percival Provost T51	Irish Aviation Museum Store, Castlemoate House, Dublin
	184	Percival Provost T51	South East Aviation Enthusiasts, Waterford
	187	DH Vampire T55	Aviation Society of Ireland, stored, Waterford
	189	Percival Provost T51	Baldonnel, Fire Section
	191	DH Vampire T55	Irish Aviation Museum Store, Castlemoate House, Dublin
	192	DH Vampire T55	South East Aviation Enthusiasts, Waterford
	193	DH Vampire T55 (pod)	Baldonnel Fire Section
	195	Sud Alouette III	No 1 Support Wing, Baldonnel
	196	Sud Alouette III	No 1 Support Wing, Baldonnel
	197	Sud Alouette III	No 1 Support Wing, Baldonnel
	198	DH Vampire T11 (XE977)	On display, Baldonnel
	199	DH Chipmunk T22	Training Wing store, Gormanston (spares)
	202	Sud Alouette III	No 1 Support Wing, Baldonnel
	203	Cessna FR172H	No 2 Support Wing, Gormanston
	205	Cessna FR172H	No 2 Support Wing, Gormanston
	206	Cessna FR172H	No 2 Support Wing, Gormanston
	207	Cessna FR172H	No 2 Support Wing, Gormanston
	208	Cessna FR172H	No 2 Support Wing, Gormanston
	209	Cessna FR172H	No 2 Support Wing, Gormanston
	210	Cessna FR172H	No 2 Support Wing, Gormanston
	211	Sud Alouette III	No 1 Support Wing, Baldonnel
	212	Sud Alouette III	No 1 Support Wing, Baldonnel
	213	Sud Alouette III	No 1 Support Wing, Baldonnel
	214	Sud Alouette III	No 1 Support Wing, Baldonnel
	215	Fouga Super Magister	No 1 Support Wing, Baldonnel
	216	Fouga Super Magister	No 1 Support Wing, Baldonnel
	217	Fouga Super Magister	No 1 Support Wing, Baldonnel
	218	Fouga Super Magister	No 1 Support Wing, Baldonnel
	219	Fouga Super Magister	No 1 Support Wing, Baldonnel
	220	Fouga Super Magister	No 1 Support Wing, Baldonnel
	221	Fouga Super Magister [3-KE]	Engineering Wing, Baldonnel
	222	SIAI SF-260WE Warrior	Training Wing, Baldonnel
	225	SIAI SF-260WE Warrior	Training Wing, Baldonnel
	226	SIAI SF-260WE Warrior	Training Wing, Baldonnel
	227	SIAI SF-260WE Warrior	Training Wing, Baldonnel
	228	SIAI SF-260WE Warrior	Training Wing, Baldonnel
	229	SIAI SF-260WE Warrior	Training Wing, Baldonnel
	230	SIAI SF-260WE Warrior	Training Wing, Baldonnel
	231	SIAI SF-260WE Warrior	Training Wing, Baldonnel
	233	SIAI SF-260MC	Engineering Wing, Baldonnel (stored)
	235	SIAI SF-260WE Warrior	Training Wing, Baldonnel
	237	Aerospatiale SA342L Gazelle	Advanced Flying Training School, Baldonnel
	238	BAe125/700B	Transport & Training Squadron, Baldonnel
	240	Beech King Air 200	Transport & Training Squadron, Baldonnel
	241	Aerospatiale Gazelle	Advanced Flying Training School, Baldonnel
	243	Cessna FR172K	No 2 Support Wing, Gormanston
	244	SA365F Dauphin II	No 3 Support Wing, Baldonnel
	245	SA365F Dauphin II	No 3 Support Wing, Baldonnel
	246	SA365F Dauphin II	No 3 Support Wing, Baldonnel
	247	SA365F Dauphin II	No 3 Support Wing, Baldonnel

Serial	Type	Owner or Operator	Notes
248	SA365F Dauphin II	No 3 Support Wing, Baldonnel	
250	Airtech CN.235	Transport & Training Squadron, Baldonnel	
251	Grumman Gulfstream IV	Transport & Training Squadron, Baldonnel	
252	Airtech CN.235 MP		
253	Airtech CN.235 MP		

XV239 HS Nimrod MR2P from the Kinloss MR Wing. *PRM*

XX185 one of No 7 FTS/19(R) Sqn's Hawk T1s based at Chivenor. *PRM*

XX497 SA Jetstream T1 wears the markings of No 45 (R) Sqn. *PRM*

Overseas Military Aircraft Markings

Aircraft included in this section are a selection of those likely to be seen visiting UK civil and military airfields on transport flights, exchange visits, exercises and for air shows. It is not a comprehensive list of *all* aircraft operated by the air arms concerned.

Serial Unit

ALGERIA
Force Aerienne Algerienne
Lockheed C-130H Hercules
4911 (7T-WHT)
4912 (7T-WHS)
4913 (7T-WHY)
4914 (7T-WHZ)
4924 (7T-WHR)
4926 (7T-WHQ)
4928 (7T-WHJ)
4930 (7T-WHI)
4934 (7T-WHF)
4935 (7T-WHE)

Lockheed C-130H-30 Hercules
4987 (7T-WHD)
4989 (7T-WHL)
4997 (7T-WHA)
5224 (7T-WHB)

AUSTRALIA
Royal Australian Air Force
Boeing 707-338C
33 Sqn, Richmond, NSW
A20-623
A20-624
A20-627
A20-629

Boeing 707-368C
33 Sqn, Richmond, NSW
A20-261

Lockheed C-130H Hercules
36 Sqn, Richmond, NSW
A97-001
A97-002
A97-003
A97-004
A97-005
A97-006
A97-007
A97-008
A97-009
A97-010
A97-011
A97-012

Lockheed C-130E Hercules
37 Sqn, Richmond, NSW
A97-159
A97-160
A97-167
A97-168
A97-171
A97-172

Serial Unit

A97-177
A97-178
A97-180
A97-181
A97-189
A97-190

Lockheed P-3C Orion
Edinburgh, NSW
10/11 Sqns
A9-656 11 Sqn
A9-657 11 Sqn
A9-658 11 Sqn
A9-659 11 Sqn
A9-660 11 Sqn
A9-661 11 Sqn
A9-662 11 Sqn
A9-663 11 Sqn
A9-664 11 Sqn
A9-665 11 Sqn
A9-751 10 Sqn
A9-752 10 Sqn
A9-753 10 Sqn
A9-755 10 Sqn
A9-756 10 Sqn
A9-757 10 Sqn
A9-758 10 Sqn
A9-759 10 Sqn
A9-760 10 Sqn

AUSTRIA
Oesterreichische Luftstreitkrafte
Saab 105ÖE
JbG, Linz
1/Uberwg, Zeltweg
2/Uberwg, Graz
1101/A JbG
1102/B JbG
1104/D JbG
1105/E JbG
1106/F
1107/G JbG
1108/H
1109/I
1110/J JbG
1111/A
1112/B
1114/D
1116/F JbG
1117/G JbG
1119/I
1120/J JbG
1122/B
1123/C JbG
1124/D
1125/E JbG
1126/F
1127/G

Serial Unit

1128/H
1129/I 1/Uberwg
1130/J
1131/A
1132/B JbG
1133/C JbG
1134/D 2/Uberwg
1135/E 1/Uberwg
1136/F
1137/G
1139/I JbG
1140/J 2/Uberwg

Short SC7 Skyvan 3M
Flachenstaffel, Tulln
5S-TA
5S-TB

BELGIUM
Force Aérienne Belge/ Belgische Luchtmacht
D-BD Alpha Jet
7/11 Smaldeel
Brustem (9Wg)
AT01
AT02
AT03
AT05
AT06
AT08
AT09
AT10
AT11
AT12
AT13
AT14
AT15
AT16
AT17
AT18
AT19
AT20
AT21
AT22
AT23
AT24
AT25
AT26
AT27
AT28
AT29
AT30
AT31
AT32
AT33

Boeing 727-29C
21 Smaldeel, Melsbroek
CB01

Overseas Serials

Serial	Unit
CB02	

Swearingen Merlin IIIA
21 Smaldeel, Melsbroek

CF01
CF02
CF04
CF05
CF06

Lockheed C-130H Hercules
20 Smaldeel, Melsbroek

CH01
CH02
CH03
CH04
CH05
CH06
CH07
CH08
CH09
CH10
CH11
CH12

Dassault Falcon 20E
21 Smaldeel, Melsbroek

CM01
CM02

Hawker-Siddeley HS748 Srs 2A
21 Smaldeel, Melsbroek

CS01
CS02
CS03

General Dynamics F-16A
349/350 Smaldeel, Bevekom (1 Wg);
1/2 Smaldeel, Florennes (2Wg);
23/31 Smaldeel, Kleine Brogel (10 Wg)

Serial	Unit
FA01	349 Sm
FA02	350 Sm
FA03	349 Sm
FA04	350 Sm
FA05	349 Sm
FA09	349 Sm
FA10	349 Sm
FA16	349 Sm
FA17	349 Sm
FA18	350 Sm
FA19	350 Sm
FA20	349 Sm
FA21	349 Sm
FA22	350 Sm
FA23	349 Sm
FA25	349 Sm
FA26	349 Sm
FA27	349 Sm
FA28	350 Sm
FA30	350 Sm
FA31	349 Sm
FA32	350 Sm
FA34	349 Sm
FA36	350 Sm
FA37	349 Sm
FA38	350 Sm
FA39	350 Sm
FA40	349 Sm
FA43	349 Sm
FA44	350 Sm
FA45	349 Sm
FA46	349 Sm
FA47	349 Sm
FA48	349 Sm
FA49	350 Sm
FA50	350 Sm
FA51	350 Sm
FA53	350 Sm
FA55	349 Sm
FA56	31 Sm
FA57	23 Sm
FA58	31 Sm
FA60	31 Sm
FA61	23 Sm
FA64	31 Sm
FA65	23 Sm
FA66	31 Sm
FA67	23 Sm
FA68	31 Sm
FA69	23 Sm
FA70	31 Sm
FA71	23 Sm
FA72	31 Sm
FA73	23 Sm
FA74	31 Sm
FA75	23 Sm
FA76	31 Sm
FA77	23 Sm
FA78	31 Sm
FA80	31 Sm
FA81	23 Sm
FA82	31 Sm
FA83	23 Sm
FA84	31 Sm
FA86	31 Sm
FA87	31 Sm
FA88	31 Sm
FA89	23 Sm
FA90	31 Sm
FA91	350 Sm
FA92	31 Sm
FA93	23 Sm
FA94	31 Sm
FA95	23 Sm
FA96	31 Sm
FA97	1 Sm
FA98	2 Sm
FA99	1 Sm
FA100	2 Sm
FA101	1 Sm
FA102	2 Sm
FA103	1 Sm
FA104	2 Sm
FA106	2 Sm
FA107	1 Sm
FA108	2 Sm
FA109	1 Sm
FA110	2 Sm
FA111	1 Sm
FA112	2 Sm
FA113	23 Sm
FA114	2 Sm
FA115	1 Sm
FA116	2 Sm
FA117	1 Sm
FA118	2 Sm
FA119	1 Sm
FA120	2 Sm
FA121	1 Sm
FA122	2 Sm
FA123	1 Sm
FA124	23 Sm
FA125	1 Sm
FA126	23 Sm
FA127	1 Sm
FA128	2 Sm
FA129	1 Sm
FA130	2 Sm
FA131	1 Sm
FA132	2 Sm
FA133	1 Sm
FA134	2 Sm
FA135	1 Sm
FA136	2 Sm

General Dynamics F-16B
349/350 Sm, Bevekom (1Wg);
1/2 Sm, Florennes (2Wg);
OCS, Bevekom;
23/31 Sm, Kleine Brogel (10 Wg)

Serial	Unit
FB01	OCS
FB02	349 Sm
FB03	OCS
FB04	OCS
FB05	OCS
FB07	OCS
FB08	OCS
FB09	OCS
FB10	OCS
FB11	OCS
FB12	OCS
FB14	10 Wg
FB15	10 Wg
FB17	10 Wg
FB18	10 Wg
FB19	10 Wg
FB20	10 Wg
FB21	1 Sm
FB22	2 Sm
FB23	1 Sm
FB24	2 Sm

Fouga CM170 Magister
33 Sm (9Wg), Brustem

MT3
MT04
MT14
MT29
MT30
MT31
MT33
MT34
MT36
MT37
MT40
MT44
MT46
MT48
MT49

Siai Marchetti SF.260MB/SF.260MD*
Ecole de Pilotage Elementaire,
(5 Sm) Gossoncourt

ST01
ST03

Serial	Unit
ST04	
ST05	
ST06	
ST08	
ST09	
ST10	
ST11	
ST12	
ST13	
ST14	
ST15	
ST16	
ST17	
ST18	
ST19	
ST20	
ST21	
ST22	
ST23	
ST24	
ST25	
ST26	
ST27	
ST28	
ST29	
ST30	
ST31	
ST32	
ST33	
ST34	
ST35	
ST36	
ST37*	
ST38*	
ST39*	
ST40*	
ST41*	
ST42*	
ST43*	
ST44*	
ST45*	
ST46*	
ST47*	
ST48*	
ST49*	
ST50*	
ST51*	

Belgische Landmacht

Britten-Norman BN-2A Islander

15/16 Smaldeel, Brasschaat;
SvHLV, Brasschaat

Serial	Unit
B01/LA	SvHLV
B02/LB	15/16 Sm
B03/LC	15/16 Sm
B04/LD	SvHLV
B07/LG	SvHLV
B08/LH	SvHLV
B09/LI	15/16 Sm
B10/LJ	SvHLV
B11/LK	SvHLV
B12/LL	SvHLV

Agusta A109HO

Serial	Unit
H01	
H02	
H03	
H04	
H05	
H06	
H07	
H08	
H09	
H10	
H11	
H12	

Sud Alouette II

16 Sm, Butzweilerhof;
17 Sm, Werl;
18 Sm, Merzbrück;
SvHLV, Brasschaat

Serial	Unit
A04	16 Sm
A05	SvHLV
A09	17 Sm
A12	17 Sm
A14	16 Sm
A15	17 Sm
A16	SvHLV
A18	16 Sm
A22	16 Sm
A23	18 Sm
A24	SvHLV
A26	SvHLV
A31	SvHLV
A32	18 Sm
A34	17 Sm
A35	SvHLV
A37	18 Sm
A38	17 Sm
A40	17Sm
A41	SvHLV
A42	18 Sm
A43	SvHLV
A44	18 Sm
A45	18 Sm
A46	18 Sm
A47	18 Sm
A48	17 Sm
A49	16 Sm
A50	16 Sm
A53	18 Sm
A54	SvHLV
A55	SvHLV
A57	17 Sm
A59	16 Sm
A61	16 Sm
A62	17 Sm
A64	16 Sm
A65	18 Sm
A66	SvHLV
A67	17 Sm
A68	17 Sm
A69	SvHLV
A70	17 Sm
A72	SvHLV
A73	SvHLV
A74	16 Sm
A75	18 Sm
A76	16 Sm
A77	SvHLV
A78	16 Sm
A79	16 Sm
A80	18 Sm
A81	16 Sm
A90	SvHLV
A92	16 Sm
A93	SvHLV
A94	SvHLV
A95	SvHLV

Force Navale Belge/Belgische Zeemacht

SA316B Alouette III

Koksijde Heli Flight

Serial	Unit
M1	(OT-ZPA)
M2	(OT-ZPB)
M3	(OT-ZPC)

Westland Sea King Mk48

40 Smaldeel, Koksijde

Serial	Unit
RS01	
RS02	
RS03	
RS04	
RS05	

BRAZIL

Forca Aerea Brazileira

Boeing KC-137

2 GT 2 Esq Afonsos

Serial	Unit
2401	
2402	
2403	
2404	

Lockheed C-130E Hercules

1 GT Afonsos

Serial	Unit
C-130 2451	
C-130 2453	
C-130 2454	
C-130 2455	
C-130 2456	
C-130 2458	
C-130 2460	

Lockheed KC-130H Hercules

1 GT 1 Esq Afonsos

Serial	Unit
C-130 2461	
C-130 2462	

Lockheed C-130H Hercules

1 GT 1 Esq Afonsos

Serial	Unit
C-130 2463	
C-130 2464	
C-130 2465	
C-130 2466	
C-130 2467	

Lockheed RC-130E Hercules

6 GAV 1 Esq Recife

Serial	Unit
C-130 2459	

CANADA

Canadian Forces

McDonnell Douglas CF-18A/CF-18B* Hornet

439 Squadron,
Sollingen, Germany

Serial	Unit
188709	
188742	
188743	
188745	
188750	
188760	
188763	
188764	
188766	

Serial	Unit
188769	
188770	
188782	
188784	
188795	
188923*	
188926*	
188927*	
188928*	

Canadair
CC-109 Cosmopolitan
412 Sqn, Lahr, Germany

Serial	Unit
109151	
109152	
109160	

Lockheed
CC-130E Hercules
413 Sqn, Greenwood
429 Sqn, Trenton
435 Sqn, Edmonton
436 Sqn, Trenton

Serial	Unit
130305	429 Sqn
130306	435 Sqn
130307	429 Sqn
130308	436 Sqn
130310	436 Sqn
130311	435 Sqn
130313	435 Sqn
130314	436 Sqn
130315	436 Sqn
130316	429 Sqn
130317	436 Sqn
130319	436 Sqn
130320	436 Sqn
130321	435 Sqn
130323	436 Sqn
130324	436 Sqn
130325	436 Sqn
130326	435 Sqn
130327	436 Sqn
130328	436 Sqn

Lockheed
CC-130H/CC-130T*
Hercules

Serial	Unit
130332	436 Sqn
130333	436 Sqn
130334	436 Sqn
130335	435 Sqn
130336	435 Sqn
130337	435 Sqn
130338	435 Sqn
130339*	435 Sqn
130340	435 Sqn
130341	435 Sqn
130342	435 Sqn

Boeing CC-137
(B.707-374C)
437 Sqn, Trenton

Serial	Unit
13701	
13702	
13703	
13704	
13705	

Lockheed
CP-140 Aurora
404/405/415 Sqns,
Greenwood;
407 Sqn, Comox

Serial	Unit
140101	
140102	407 Sqn
140103	
140104	
140105	407 Sqn
140106	
140107	
140108	407 Sqn
140109	
140110	407 Sqn
140111	407 Sqn
140112	
140113	
140114	
140115	
140116	
140117	407 Sqn
140118	407 Sqn

Lockheed
CP-140A Arctura

Serial	Unit
140119	
140120	
140121	

Canadair CC-144/CE-144A*
Challenger
412 Sqn, Ottawa-Uplands
414 Sqn, North Bay

Serial	Unit
144601	412 Sqn
144602	412 Sqn
144603*	414 Sqn
144604	412 Sqn
144605	412 Sqn
144606	412 Sqn
144607	414 Sqn
144608	412 Sqn
144609	412 Sqn
144610	412 Sqn
144611	414 Sqn
144613	412 Sqn
144614	412 Sqn
144615	412 Sqn
144616	412 Sqn

CHILE
Fuerza A_rea de Chile

Boeing 707-321B/330/351C

Serial	Unit
902	351C
903	330B
904	321B

Lockheed
C-130B/H Hercules
Grupo 10, Santiago

Serial	Unit
995	C-130H
996	C-130H
997	C-130B

DENMARK
Kongelige Danske
Flyvevaabnet

Saab A-35XD
Draken
Eskadrille 729, Karup

Serial	Unit
A001	
A002	
A004	
A005	
A006	
A007	
A008	
A009	
A010	
A011	
A012	
A014	
A017	
A018	
A019	
A020	

Saab S-35XD
Draken
Eskadrille 729, Karup

Serial	Unit
AR102	
AR104	
AR105	
AR106	
AR107	
AR108	
AR109	
AR110	
AR111	
AR112	
AR113	
AR114	
AR115	
AR116	
AR117	
AR118	
AR119	
AR120	

Saab Sk-35XD
Draken
Eskadrille 729, Karup

Serial	Unit
AT151	
AT152	
AT153	
AT154	
AT155	
AT156	
AT157	
AT158	
AT160	

Lockheed
C-130H Hercules
Eskadrille 721, Vaerl_se

Serial	Unit
B678	
B679	
B680	

General Dynamics F-16
Eskadrille 723, Aalborg;
Eskadrille 726, Aalborg;
Eskadrille 727, Skrydstrup;
Eskadrille 730, Skrydstrup

F-16A

Serial	Unit
E004	Esk 726
E005	Esk 726
E006	Esk 726
E007	Esk 726
E008	Esk 726
E016	Esk 726
E017	Esk 726
E018	Esk 726
E174	Esk 727
E176	Esk 723
E177	Esk 723
E178	Esk 730

Serial	Unit
E180	Esk 723
E181	Esk 727
E182	Esk 730
E183	Esk 723
E184	Esk 723
E187	Esk 727
E188	Esk 723
E189	Esk 723
E190	Esk 723
E191	Esk 730
E192	Esk 730
E193	Esk 727
E194	Esk 723
E195	Esk 723
E196	Esk 723
E197	Esk 730
E198	Esk 730
E199	Esk 727
E200	Esk 723
E202	Esk 730
E203	Esk 723
E596	Esk 726
E597	Esk 730
E598	Esk 730
E599	Esk 730
E600	Esk 727
E601	Esk 727
E602	Esk 727
E603	Esk 727
E604	Esk 726
E605	Esk 730
E606	Esk 730
E607	Esk 726
E608	Esk 723
E609	Esk 727
E610	Esk 727
E611	Esk 727

F-16B

Serial	Unit
ET022	Esk 726
ET197	Esk 726
ET198	Esk 726
ET199	Esk 726
ET204	Esk 727
ET205	Esk 730
ET206	Esk 730
ET207	Esk 727
ET208	Esk 730
ET210	Esk 730
ET612	Esk 730
ET613	Esk 727
ET614	Esk 723
ET615	Esk 727

Grumman
Gulfstream III
Eskadrille 721, Vaerløse
F249
F313
F330

Saab T-17
Supporter
Flyveskolen, Avnø (FLSK);
Haerens Flyvetjaeneste
(Danish Army), Vandel;
Eskadrille 721, Vaerløse

Serial	Unit
T401	Aalborg Stn Flt
T402	Skrydstrup Stn Flt
T403	Karup Stn Flt
T404	Karup Stn Flt
T405	Karup Stn Flt
T407	Karup Stn Flt
T408	Esk 721
T409	Esk 721
T410	Karup Stn Flt
T411	Esk 721
T413	Army
T414	Army
T415	Army
T417	Army
T418	FLSK
T419	Esk 721
T420	Esk 721
T421	FLSK
T422	FLSK
T423	FLSK
T425	Aalborg Stn Flt
T426	FLSK
T427	FLSK
T428	FLSK
T429	Aalborg Stn Flt
T430	FLSK
T431	FLSK
T432	FLSK

Sikorsky S-61A Sea King
Eskadrille 722, Vaerløse
Detachments at:
Aalborg, Ronne, Skrydstrup
U240
U275
U276
U277
U278
U279
U280
U481

Sovaernets
Flyvetjaeneste
(Navy)
Westland Lynx Mk 80/90*
Eskadrille 722, Vaerløse
S035
S134
S142
S170
S175
S181
S191
S249*
S256*

Haerens
Flyvetjaeneste (Army)
Hughes 500M
Vandel
H201
H202
H203
H205
H206
H207
H209
H210
H211
H212
H213
H244
H245
H246

Aerospatiale AS550C-2
Fennec
P-090
P-234
P-254
P-275
P-276
P-287
P-288
P-319
P-320
P-339
P-352
P-369

ECUADOR
Lockheed
C-130H Hercules
FAE-812
FAE-893

EGYPT
Al Quwwat al-Jawwiya
Ilmisriya
Lockheed
C-130H Hercules
***C-130H-30 Hercules**
16 Sqn, Cairo West
1271/SU-BAB
1272/SU-BAC
1273/SU-BAD
1274/SU-BAE
1275/SU-BAF
1277/SU-BAI
1278/SU-BAJ
1279/SU-BAK
1280/SU-BAL
1281/SU-BAM
1282/SU-BAN
1283/SU-BAP
1284/SU-BAQ
1285/SU-BAR
1286/SU-BAS
1287/SU-BAT
1288/SU-BAU
1289/SU-BAV
1290/SU-BEW
1291/SU-BEX
1292/SU-BEY
1293/SU-BKS*
1294/SU-BKT*
1295/SU-BKU*

FRANCE
Armée de l'Air
Aerospatiale SN601
Corvette
CEV, Bretigny
1 MV
2 MW
Aérospatiale TB-30
Epsilon
GE315, Cognac

1	315-UA
2	315-UB
3	315-FZ
4	315-UC
5	315-UD
6	315-UE
7	315-UF
8	315-UG
9	315-UH

Serial	Unit
10	315-UI
11	315-UJ
12	315-UK
13	315-UL
14	315-UM
15	315-UN
16	315-UO
17	315-UP
18	315-UQ
19	315-UR
20	315-US
21	315-UT
23	315-UV
24	315-UW
25	315-UX
26	315-UY
27	315-UZ
28	315-VA
29	315-VB
30	315-VC
31	315-VD
32	315-VE
33	315-VF
34	315-VG
35	315-VH
36	315-VI
37	315-VJ
38	315-VK
39	315-VL
40	315-VM
41	315-VN
42	315-VO
43	315-VP
44	315-VQ
45	315-VR
46	315-VS
47	315-VT
48	315-VU
49	315-VV
50	315-VW
51	2-BD
52	315-VX
53	315-VY
54	315-VZ
56	315-WA
57	F-ZVLB
58	315-WB
60	315-WC
61	315-WD
62	315-WE
63	315-WF
64	315-WG
65	315-WH
66	315-WI
67	315-WJ
68	315-WK
69	315-WL
70	315-WM
71	315-WN
72	315-WO
73	315-WP
74	315-WQ
75	315-WR
76	315-WS
77	315-WT
78	315-WU
79	315-WV
80	315-WW
81	315-WX
82	315-WY
83	315-WZ
84	315-XA
85	315-XB
86	315-XC
87	315-XD
88	315-XE
89	315-XF
90	315-XG
91	315-XH
92	315-XI
93	315-XJ
94	315-XK
95	315-XL
96	315-XM
97	315-XN
98	315-XO
99	315-XP
100	315-XQ
101	315-XR
102	315-XS
103	315-XT
104	315-XU
105	315-XV
106	315-XW
107	315-XX
108	315-XY
109	315-XZ
110	315-YA
111	315-YB
112	315-YC
113	315-YD
114	315-YE
115	315-YF
116	315-YG
117	315-YH
118	315-YI
119	315-YJ
120	315-YK
121	315-YL
122	315-YM
123	315-YN
124	315-YO
125	315-YP
126	315-YQ
127	315-YR
128	315-YS
129	315-YT
130	315-YU
131	315-YV
132	315-YW
133	315-YX
134	315-YY
135	315-YZ
136	315-ZA
137	315-ZB
138	315-ZC
139	315-ZD
140	315-ZE
141	315-ZF
142	315-ZG
143	315-ZH
144	315-ZI
145	315-ZJ
146	315-ZK
148	315-ZL
149	315-ZM
150	315-ZN
152	315-ZO
153	315-ZP
154	315-ZQ
155	315-ZR
158	315-ZS
159	315-ZT

Airtech CN-235
CEAM, Mont de Marsan

Serial	Unit
043	330-ID
045	330-IE

Boeing C-135FR
ERV 93, Avord, Istres and Mont de Marsan

Serial	Unit
38470	93-CA
38471	93-CB
38472	93-CC
38474	93-CE
38475	93-CF
12735	93-CG
12736	93-CH
12737	93-CI
12738	93-CJ
12739	93-CK
12740	93-CL

Boeing KC-135R
ERV93
23516
38033
71439

Boeing E-3F Sentry
EDA-36, Avord

Serial	Unit
201	36-CA
202	36-CB
203	36-CC
204	36-CD

CASA 212
CEV

Serial	Unit
377	MO
378	MP
386	MQ
387	MR
388	MS

Cessna 310
CEV

Serial	Unit
045	AU
46	AV
185	AU
186	BI
187	BJ
188	BK
190	BL
192	BM
193	BG
0194	BH
242	AW
244	AX
513	BE
820	CL
981	BF

Cessna 411
CEV

Serial	Unit
6	AD
8	AE
185	AC
248	AB

D-BD Alpha Jet
Patrouille de France (PDF);
EC 1/8, EC 2/8 Cazaux;
GE 314, Tours;
CEAM (330), Mont de Marsan

Serial	*Unit*	
01	F-ZJTS	CEV
02	F-ZWRU	
E1		CEV
E3	8-NC	EC2/8
E4	CEV	
E5		
E7		
E8		
E9	314-TN	
E10	8-NM	EC2/8
E11	314-LO	
E12	CEV	
E13	314-LH	
E14		
E15	314-TT	
E17	8-NK	EC2/8
E18	8-MD	EC1/8
E19		
E20		
E21	314-UD	
E22	314-LD	
E23	314-TK	
E24	314-TW	
E25		
E26	314-LE	
E27	8-MU	EC1/8
E28		
E29	8-MF	EC1/8
E30	8-NR	EC2/8
E31	314-UA	
E32	8-MM	EC1/8
E33	8-NN	EC2/8
E34	314-TC	
E35	314-LJ	
E36	F-TERC (2)	
PDF		
E37	314-TE	
E38	314-TP	
E40	8-MD	EC1/8
E41	314-LC	
E42	314-LQ	
E43	314-TZ	
E44		CEV
E45	314-TF	
E46	F-ZJTJ	AMD-BA
E47		
E48		
E49	F-TERI (5)	
PDF		
E50	314-UH	
E51	314-UB	
E52	314-TX	
E53		
E55	314-UN	
E58	314-TD	
E59	314-LN	
E60		CEV
E61		
E63	314-LJ	
E64	8-NH	EC2/8
E65		
E66	8-M	EC1/8
E67	8-MR	EC1/8
E68	8-NA	EC2/8
E69	8-NX	EC2/8
E70		
E72	314-LR	
E73		
E74	314-LB	
E75	314-TU	
E76	314-TH	
E79		

Serial	*Unit*	
E80		CEV
E81	8-MP	EC1/8
E82		
E83	8-NSL	EC2/8
E84	8-MH	EC1/8
E85	330-BR	
E86	314-TJ	
E87		
E88	314-TF	
E89		
E90	8-NE	EC2/8
E91		
E92	F-TERE (6)	
PDF		
E93	8-NV	EC2/8
E94	8-NI	EC2/8
E95	314-LX	
E96		
E97	314-LV	
E98		
E99	314-LW	
E100		CEV
E101	8-ND	EC2/8
E102	8-ME	EC1/8
E103	F-TERO (8)	
PDF		
E104	314-TL	
E105	314-UI	
E106		
E107		
E108	8-NQ	EC2/8
E109		
E110		
E112	8-NO	EC2/8
E113		
E114	314-UC	
E115		
E116	8-NP	EC2/8
E117	314-UL	
E118	8-NT	EC2/8
E119	314-TR	
E120	314-LM	
E121		
E122	314-UG	
E123		
E124	8-MC	EC1/8
E125	F-TERH (4)	
PDF		
E126	F-TERA (7)	
PDF		
E127		
E128	314-LS	
E129	8-NB	EC2/8
E130	314-TI	
E131	314-TY	
E132	F-TERN (9)	
PDF		
E133		
E134		
E135		
E136	314-LY	
E137	8-MI	EC1/8
E138	314-UK	
E139	314-UF	
E140	314-LL	
E141	8-MQ	EC1/8
E142	8-MB	EC1/8
E143	314-UJ	
E144	8-NU	EC2/8
E145	314-TO	
E146	8-MG	EC1/8
E147		

Serial	*Unit*	
E148	8-NJ	EC2/8
E149	314-LI	
E150	8-NF	EC2/8
E151	314-LA	
E152	314-TA	
E153	314-TM	
E154	8-MA	EC1/8
E155	F-TERF (0)	
PDF		
E156	F-TERG (1)	
PDF		
E157	314-LG	
E158		
E159	314-LT	
E160	314-LZ	
E161	8-NG	EC2/8
E162	314-LF	
E163	314-TQ	
E164	8-MJ	EC1/8
E165	314-TB	
E166	8-MN	EC1/8
E167	8-MV	EC1/8
E168	8-MK	EC1/8
E169	314-LK	
E170	F-TERM (3)	
PDF		
E171	314-LP	
E173		
E174		
E175	314-LU	
E176	330-BT	

Dassault Mystère 20
CEV
SIET 98/120, Cazaux
ET 1/65, Villacoublay
CITAC 339, Luxeuil

22	CS (CEV)	
49	(120-FA 98/120)	
79	CT	(CEV)
86	CG	(CEV)
93		(ET1/65)
96	CB	(CEV)
104	CW	(CEV)
115	339-WL	
124	CC	(CEV)
131	CD	(CEV)
138	CR	(CEV)
145	CU	(CEV)
167	B	(ET1/65)
182		(CITAC)
186	339-JE	
188	CX	(CEV)
238		(ET1/65)
252	CA	(CEV)
260		(ET1/65)
263	CY	(CEV)
268		(ET1/65)
288	CV	(CEV)
291	P	(ET3/65)
342		(ET1/65)
375	CZ	(CEV)
422		(ET3/65)
451	339-JC	
483	339-WO	

Dassault Falcon 50
ET.60 Villacoublay

5	(F-RAFI)
27	(F-RAFK)
034	(F-RAFL)

Serial	Unit	
78	(F-RAFJ)	

Dassault Falcon 900
ET.60, Villacoublay

Serial	Unit	
1		
2	(F-RAFP)	
4	(F-RAFQ)	

Dassault Mirage IVA/IVP
EB 1/91, Mont-de-Marsan;
EB 2/91, Cazaux;
EB 1/94 Istres;

Serial	Unit	
1	AP	
2	AA	
4	AC	
5	AD	
6	AE	
7	AF	
8/01	AG	
11	AJ	1/91
12	AK	
13	AL	
14	AM	
15	AN	
19	AR	
20	AS	
21	AT	
23	AV	1/91
24	AW	
25	AX	1/91
26	AY	1/91
27	AZ	
31	BD	1/91
32	BE	
34	BG	
36	BI	1/91
37	BJ	
39	BL	
42	BO	
44	BQ	
45	BR	
46	BS	
47	BT	
48	BU	1/91
49	BV	
51	BX	
52	BY	1/91
53	BZ	1/91
54	CA	1/91
55	CB	
56	CC	
57	CD	1/91
59	CF	1/91
61	CH	2/91
62	CI	1/91

Dassault Mirage F.1C/F.1CT*
EC 12 Cambrai,
EC 13 Colmar,
EC4/30 Djibouti
EC 30, Rheims;
CEAM (330) Mont de Marsan

Serial	Unit	
2		
3		
4	330-AM	
5		
6	30-MA	2/30
9		
10	12-ZM	2/12
14	30-SJ	1/30
15		
16	12-ZC	2/12
17		
18	12-KF	3/12
19	12-ZQ	1/12
20		
21		
22		
23	12-ZE	2/12
24		
25	12-KC	3/12
26		
27		
29	12-ZM	2/12
30	12-ZJ	2/12
31		
32		
33		
35		
36	30-MP	2/30
37	12-KJ	3/12
38		
39	12-KI	3/12
40	30-FM	3/30
41		
42	30-MC	2/30
43	30-SA	1/30
44		
47		
49		
50		
52	30-MH	2/30
54	30-MJ	2/30
55	30-MN	2/30
60	30-MM	2/30
62	30-SC	1/30
63		
64		
67	30-SE	1/30
68	30-FN	3/30
69	30-SF	1/30
70	30-FG	3/30
71		
72		
73	30-SM	1/30
74		
75		
76		
77	30-SD	1/30
78		
79		
80		
81		
82		
83		
84	30-LP	4/30
85		
87	330-AP	
90		
100	30-LC	4/30
101	30-LI	4/30
102		
103	30-SO	1/30
201	30-SA	1/30
202	30-ML	2/30
203	330-AC	
205	12-KH	3/12
206	30-MA	2/30
207		
210	30-SN	1/30
211		
213	30-MO	2/30
214	12-KM	3/12
216	30-SG	1/30
217	330-AO	
218	30-SK 1/30	
219		
220		
221	30-SI	1/30
223		
224	30-MD	2/30
225	30-ME	2/30
226		
227*		CEV
228		
229*		
230	12-KR	3/12
231		
232		
233*		
234	30-LH	4/30
235		
236	12-KB	3/12
237*	30-LE	4/30
238	12-ZH	2/12
239*		
240		
241	12-2E	2/12
242*		
243	12-KL	3/12
244		
245	12-KO	3/12
246	30-SL	1/30
247	12-ZF	2/12
248	12-KQ	3/12
249	30-SQ	1/30
251	30-SR	1/30
252	30-MB	2/30
253	12-ZG	2/12
254*		
255	12-KD	3/12
256	12-KA	3/12
257	12-ZB	2/12
258	30-LA	4/30
259	30-SH	1/30
260	30-MI	2/30
261	12-ZI	2/12
262*		
264	12-KN	3/12
265		
267*	13-QB	1/13
268		
271	30-SF	1/30
272		
273*	330-AI	
274*	330-AJ	
275	12-ZA	2/12
277		
278*	13-QA 1/13	
279		
280*		
281	30-MR	2/30
282	12-KG	3/12
283	30-IMF	2/30

Dassault Mirage F.1CR
ER33 Strasbourg;
CEAM (330) Mont de Marsan;
CEV, Istres

Serial	Unit	
601		
CEV		
602		

Serial	Unit	
CEV		
603	33-CB	1/33
604	33-CE	1/33
605	33-NF	2/33
606	33-TS	3/33
607	33-NM	2/33
608	33-TN	3/33
609		
610	33-CH	1/33
611	330-AA	
612	33-NJ	2/33
613	33-NK	2/33
614		
615	33-CU	1/33
616		
617	33-CI	1/33
619	33-CC	1/33
620		
622	33-CL	1/33
623	33-CM	1/33
624	33-NY	2/33
625		
627	33-NI	2/33
628	33-TC	3/33
629	33-CG	1/33
630		
631	33-CD	1/33
632	33-TM	3/33
634	33-CK	1/33
635	33-TH	3/33
636	33-CS	1/33
637	33-CP	1/33
638	33-TP	3/33
640	33-TE	3/33
641	33-NT	2/33
642	33-NC	2/33
643	33-CO	1/33
645	33-NO	2/33
646	33-TU	2/33
647	33-TR	3/33
648	33-CF	1/33
649		
650	33-CJ	1/33
651	33-NB	2/33
653	33-TB	3/33
654	33-TF	3/33
655		
656	33-CN	1/33
657		
658	33-TI	3/33
659	33-CA	1/33
660	33-ND	2/33
661	33-TA	3/33
662	33-NA	2/33

Dassault Mirage 2000B

CEAM (330), Mont de Marsan (330);
ECT 2/2, Dijon;
EC 3/2, Dijon

Serial	Unit	
01		
CEV		
501	2-EQ	CEV
502	2-FL	2/2
504		
505	2-FB	2/2
506	2-FC	2/2
507	2-FD	2/2
508	2-FE	2/2
509	2-FF	2/2
510	2-FG	2/2
511	2-FH	2/2
512	2-FI	2/2
513	2-FJ	2/2
514	2-FK	2/2
515	330-AN	
516	2-FM	2/2
517	2-FO	2/2
518	5-OJ	2/5
519	5-AM	3/5
520	5-AN	3/5
521	5-OL	2/5
522	5-NA	1/5
523	12-YA	1/12

Dassault Mirage 2000C

EC2, Dijon;
EC 5, Orange; EC12 Cambrai;
CEAM (330), Mont de Marsan

Serial	Unit	
01		
03		CEV
04		
1	2-EP	CEV
2		CEV
3	2-EG	1/2
4	2-LB	3/2
5	2-LH	3/2
8	2-EC	1/2
9	2-LO	3/2
11	2-EF	1/2
12	2-EH	1/2
13	2-EI	1/2
14	2-EO	1/2
15	2-EK	1/2
16	2-EL	1/2
17	2-EM	1/2
18		
19	2-LA	3/2
20	2-LE	3/2
21	2-LF	3/2
22	2-LG	3/2
25	2-LK	3/2
27	2-LM	3/2
28	2-LN	3/2
29	2-ED	1/2
30		
32	2-EP	1/2
33	2-LQ	3/2
34	2-LI	3/2
35	2-EE	1/2
36	2-EN	1/2
37	2-LS	3/2
38	5-AR	3/5
39	5-OF	2/5
40	5-OI	2/5
41	5-AA	3/5
42	5-NC	1/5
43		
44		
45	5-OM	2/5
46	5-AD	3/5
47	5-AC	3/5
48	5-AF	3/5
49	5-AE	3/5
51	5-OG	2/5
52	5-OC	2/5
53	5-AH	3/5
54	5-AI	3/5
55	5-OH	2/5
56	5-OA	2/5
57	5-AJ	3/5
58	5-AK	3/5
59	5-OB	2/5
61	5-OD	2/5
62	5-AL	3/5
63	5-OK	2/5
64	330-AQ	
65		
66	5-ON	2/5
67		
68	5-AP	3/5
69		
70	5-AO	3/5
71		
72	5-OE	2/5
73		
74	5-OP	2/5
75		
76	5-NP	1/5
77	5-ND	1/5
78	5-NE	1/5
79	5-NF	1/5
80	330-AS	
81	330-AY	
82	330-AX	
83	5-NG	1/5
84	5-NH	1/5
85	5-NI	1/5
86	5-NJ	1/5
87	5-NK	1/5
88	5-NL	1/5
89	5-NM	1/5
90	5-NN	1/5
91	5-NO	1/5
92		
93	5-NR	1/5
94	12-YK	1/12
95		
96	12-YC	1/12
97	12-YF	1/12
98	12-YJ	1/12
99	12-YP	1/12
100	12-YH	1/12
101	12-YQ	1/12
102	12-YE	1/12
103	12-YN	1/12
104	12-YD	1/12
105	12-YL	1/12
106	12-YM	1/12
107	12-YR	1/12
108	330-AT	
109		
110		
111		
112	5-NB	1/5
113		
114		
115		
116		
117		
118		
119		
120		
121		
122		
123		
124		
125		
126		
127		
128		
129		
130		

Serial	Unit	
131		
132		
133		
134		
135		
136		
137		
138		
139		
140		
141		
142		
143		
144		
145		
146		

Dassault Mirage 2000N
CEAM (330), Mont de Marsan;
EC.1/4, EC.2/4 Luxeuil;
EC.2/3 Nancy
EC.3/4 Istres

Serial	Unit	
301		
302	4-CA	3/4
303	4-CP	3/4
304	3-CB	3/4
305	4-BF	2/4
306	4-CC	3/4
307	4-BC	2/4
308	4-CD	3/4
309	4-BA	2/4
310	4-CE	3/4
311	4-BD	2/4
312	4-CF	3/4
313	4-BE	2/4
314	4-CG	3/4
315	4-BG	2/4
316	4-CH	3/4
317	4-BH	2/4
318	4-CI	3/4
319	4-BI	2/4
320	4-CJ	3/4
322	4-CK	3/4
323	4-BK	2/4
324	4-CL	3/4
325	4-BL	2/4
326	4-CM	3/4
327	4-BM	2/4
329	4-BN	2/4
330	4-CO	3/4
331	4-BO	2/4
332	330-AR	
333	330-AG	
334	330-AV	
335	4-BJ	2/4
336	4-BP	2/4
337	4-AB	2/4
338	4-AC	1/4
339	4-AD	1/4
340	4-AA	1/4
341	4-AF	1/4
342	4-AG	1/4
343	4-AH	1/4
344	4-AJ	1/4
345	4-BE	2/4
346	4-AL	1/4
347	4-AM	1/4
348	4-AN	1/4
349	4-AO	1/4
350	4-AP	1/4
351	4-AQ	1/4
353	3-JB	2/3
354	3-JC	2/3
355	3-JD	2/3
356	3-JE	2/3
357	3-JF	2/3
358	3-JG	2/3
359	3-JH	2/3
361	3-JJ	2/3
362	3-JK	2/3
363	3-JL	2/3
364	3-JM	2/3
365	3-JN	2/3
366	3-JO	2/3
367	3-JP	2/3
360	3-JI	2/3
368	4-AR	1/4
369	4-AS	1/4
370	4-AT	1/4
371		
372		
373		
374		
375		

Dassault Rafale-A

Serial	Unit
01	CEV

Dassault Rafale-C

Serial	Unit
01	CEV
02	
03	

Dassault Rafale-M

Serial	Unit
01	AMD-BA

DHC6 Twin Otter
ET 65 Villacoublay;
GAM 56 Evreux

Serial	Unit	
292	OW	GAM 56
298	OY	GAM 56
300	OZ	GAM 56
603	CY	ET 65
730	CA	ET 65
742	CB	ET 65
743	CZ	ET 65
745	CV	ET 65
786		
790		

Douglas DC-8F
EE 51 Evreux
ET3/60 Charles de Gaulle

Serial	Unit	
45570	F-RAFE	EE 51
45819	F-RAFC	ET3/60
46013	F-RAFG	ET3/60
46043	F-RAFD	ET3/60
46130	F-RAFF	ET3/60

Embraer Xingu
GE 319 Avord;
ETE 43 Bordeaux;
ETE 44 Aix-en-Provence
CITAC339, Luxeuil,
CEAM, Mont de Marsan

Serial	Unit
054	YX
064	YY
072	YA
073	YB
075	YC
076	YD
078	YE
080	YF
082	YG
084	JA (CITAC)
086	YI
089	YJ
091	YK
092	YL
095	YM
096	YN
098	YO
099	YP
101	IB (CEAM)
102	YS
103	YT (ETE.44)
105	YU
107	YV
108	YW
111	YQ

Lockheed C-130H Hercules
***C-130H-30 Hercules**
ET-2/61, Orleans

Serial	Unit
5114	61-PA
5116	61-PB
5119	61-PC
5140	61-PD
5142*	61-PE
5144*	61-PF
5150*	61-PG
5151*	61-PH
5152*	61-PI
5153*	61-PJ
5226*	61-PK
5227*	61-PL

Morane Saulnier 760 Paris
ETE 41 Metz;
ETE 43 Bordeaux;
ETE 44 Aix-en-Provence;
ET 65 Villacoublay; CEAM
GI-312 Salon de Provence
(330) Mont de Marsan

Serial	Unit	
1	330-DA	
14		
19	330-DC	
23		
24	65-LA	
25	41-AP	
26	65-LN	
27		
29	65-LD	
30	65-LW	
34	43-BB	
35		
36	43-BD	
38	41-AS	
44	65-LM	
45	43-BC	
51		
53	DD	
54	330-DQ	
56	41-AD	
57	65-LK	
58	312-DG	
59		
60	41-AT	
61	312-DF	
62		
68	NB	(CEV)
70	65-LF	

Serial	Unit	
71	41-AC	
73		
75	330-DB	
77	330-DD	
78	65-LI	
80	DE	
81	65-LL	
83	NC	(CEV)
91	65-LU	
92	13-TB	
93		
94		
97	65-LH	
100	NG	(CEV)
113	NI	(CEV)
114	NJ	(CEV)
115	OV	(CEV)
116	ON	(CEV)
117	AZ	(CEV)
118	NQ	(CEV)
119	NL	(CEV)

Nord 262 Fregate
EdC 70 Chateaudun;
ET 65 Villacoublay;
ETE 41 Metz;
ETE 44, Aix-en-Provence;
GE 316 Toulouse;
CEAM (330) Mont de Marsan;
CEV, Istres

Serial	Unit	
01		CEV
3	OH	CEV
55	MH	CEV
58	MJ	CEV
64	AA	ET65
66	AB	ET65
67	MI	CEV
68	AC	ET65
76	316-DA	
77	AK	ET65
78	AF	ET65
80	AW	ET65
81	AH	ET65
83	316-DB	
86	316-DD	
87	316-DC	
88	AL	ET65
89	AZ	ET65
91	AT	ETE44
92	316-DE	
93	MB	CEV
94	AU	ETE 44
95	AR	ET65
105	AE	ET65
106	MA	EdC70
107	AX	ET65
108	AG	ET65
109	AM	ET65
110	AS	ET65

SEPECAT Jaguar A
EC 1/7, 2/7, 3/7 St Dizier;
EC 1/11, 2/11, 3/11 Toul;
CEAM (330) Mont de Marsan
CITAC, Luxeuil

Serial	Unit	
A1		
A2	11-EA	1/11
A3		CEV
A5		
A7	7-PG 2/7	
A8	7-IE 3/7	
A9	7-PB	2/7
A11	7-PA	
A12	11-MW	2/11
A13	11-EM	1/11
A14	11-MA	2/11
A15	7-IJ	3/7
A16		
A17	7-HF	1/7
A19	7-IB	3/7
A21	7-IA	3/7
A22	7-IN	3/7
A23	7-HK	1/7
A24		
A25		
A26	7-HM	1/7
A27	11-MQ	2/11
A28		
A29	7-HR	1/7
A31		
A32	7-IC	3/7
A33	7-HL	1/7
A34	7-ID	3/7
A35	7-IF	3/7
A36	7-IP	3/7
A37	7-HA	1/7
A38	7-HG	1/7
A39	7-IT	3/7
A40		
A41	7-HM	1/7
A43	7-HJ	1/7
A44		
A46	7-II	3/7
A47		
A48		
A49	7-IM	3/7
A50	11-MC	2/11
A53	11-RB	3/11
A54	7-HB	1/7
A55	7-IL	3/7
A58	11-RH	3/11
A59	7-IO	3/7
A60		
A61		
A64		
A65	7-IS	3/7
A66		
A67	7-HQ	1/7
A70	7-IH	3/7
A72	7-PD	2/7
A73	7-HH	1/7
A74		
A75	7-HE	1/7
A76	7-HI	1/7
A79		
A80	7-HN	1/7
A82		
A83		
A84	11-RQ	3/11
A86		
A87		
A88	11-MG	2/11
A89		
A90	11-MB	2/11
A92		
A93	7-HP	1/7
A94	11-EC	1/11
A96	11-MD	2/11
A97	11-RG	3/11
A98		
A99	11-EI	1/11
A100	11-EE	1/11
A101	11-EF	1/11
A103		
A104		
A107	11-MV	2/11
A108	11-MP	2/11
A112	11-EK	1/11
A113		
A115	11-RL	3/11
A117	11-MH	2/11
A118	11-MD	2/11
A119	11-RE	3/11
A120	11-MR	2/11
A121		
A122	11-EQ	1/11
A123		
A124	11-ME	2/11
A126		
A127	11-MM	2/11
A128	11-MF	2/11
A129		
A130		
A131	11-RO	3/11
A133	11-EL	1/11
A135	11-RJ	3/11
A137	11-EM	1/11
A138	11-RX	3/11
A139	11-RC	3/11
A140	11-EN	1/11
A141		
A142		
A144	11-RW	3/11
A145	11-ET	1/11
A148	11-EQ	1/11
A149	11-RK	3/11
A150		
A151	11-ER	1/11
A152		
A153	11-RA	3/11
A154	11-RP	3/11
A157	11-ES	1/11
A158	11-RM	3/11
A159	11-RV	3/11
A160	11-RT	3/11

SEPECAT Jaguar E

Serial	Unit	
E1		CEV
E2	7-HD	1/7
E3	339-WI	
E4	11-MO	2/11
E5		
E6	11-EZ	1/11
E7	7-PN	2/7
E8	339-WG	
E9	7-PQ	2/7
E10	339-WF	
E11	7-PE	2/7
E12		
E13	11-RD	3/11
E15	7-PF	2/7
E16	7-HO	1/7
E18	7-PI	2/7
E19	11-EG	1/11
E20	11-RI	3/11
E21		
E22	7-PC	2/7
E23		
E24		
E25	7-PK	2/7
E27	7-IG	3/7
E28	7-PR	2/7

Overseas Serials

Serial	Unit	
E29	7-PL	2/7
E30	7-PO	2/7
E32		
E33		
E35	339-WH	
E36	7-PP	2/7
E37		
E38	7-PM	2/7
E39	7-PH	2/7
E40		

SOCATA TBM 700
ET 2/65, Villacoublay

Serial	Unit
33	65-XA
35	65-XB

Transall C-160
Transall C-160H[1]
Transall C-160NG[2]
ETOM 55, Dakar
EE 59, Evreux (C160H);
ET 61 Orleans (C160A/F);
ET 64 Evreux (C160NG)

Serial	Unit	
A02	61-MI	
A04	61-BI	CEV
A06	61-ZB	
F1	61-MA	
F2	61-MB	
F3	61-MC	
F4	61-MD	
F5		
F11	61-MF	
F12	61-MG	
F13		
F15	61-MJ	
F16	61-MK	
F17		
F18	61-MM	
F42	61-MN	
F43	61-MO	
F44	61-MP	
F45	61-MQ	
F46	61-MR	
F48	61-MT	
F49	61-MU	
F51	61-MW	
F52	61-MX	
F53	61-MY	
F54	61-MZ	
F55	61-ZC	
F86	61-ZD	
F87	61-ZE	
F88	61-ZF	
F89	61-ZG	
F90	61-ZH	
F91	61-ZI	
F92	61-ZJ	
F93	61-ZK	
F94	61-ZL	
F95	61-ZM	
F96	61-ZN	
F97	61-ZO	
F98	61-ZP	
F99	61-ZQ	
F100	61-ZR	
F153	61-ZS	
F154	61-ZT	
F155	61-ZU	
F157	61-ZW	
F158	61-ZX	
F159	61-ZY	
F160	61-ZZ	
F201[2]	64-GA	
F202[2]	64-GB	
F203[2]	64-GC	
F204[2]	64-GD	
F205[2]	64-GE	
F206[2]	64-GF	
F207[2]	64-GG	
F208[2]	64-GH	
F210[2]	64-GJ	
F211[2]	64-GK	
F212[2]	64-GL	
F213[2]	64-GM	
F214[2]	64-GN	
F215[2]	64-GO	
F216[2]	64-GT	
F217[2]	64-GQ	
F218[2]	64-GR	
F221[2]	F-ZJUU	
F222[2]	64-GV	
F223[2]	64-GW	
F224[2]	64-GX	
F225[2]	64-GY	
F226[2]	64-GZ	
F227[2]	64-GP	
F230[1]	F-ZJUA	
F231[1]	F-ZJUB	
F232[1]		
H01[1]	59-BA	
H02[1]	59-BB	
H03[1]	59-BC	
H04[1]	59-BD	

Aéronavale/Marine
Aérospatiale SA.321G
Super Frelon
32 F, Lanveoc;
33 F, San Mandrier

Serial	Unit
101	(32F)
2	(32F)
105	(33F)
106	(32F)
118	(32F)
120	(32F)
22	(32F)
134	(32F)
137	(32F)
141	(32F)
144	(33F)
148	(33F)
149	(32F)
60	(32F)
162	(32F)
163	(33F)
164	(32F)
165	(32F)

Breguet 1050
Alizé
4F, Lann Bihoué;
6F, Nimes-Garons;
ES 59, Hyeres

Serial	Unit
11	(4F)
12	(4F)
17	(4F)
22	(4F)
24	(6F)
25	(6F)
26	(6F)
28	
30	(6F)
31	(59S)
33	(4F)
36	(6F)
41	(59S)
43	(4F)
47	(4F)
48	(4F)
49	
50	(6F)
51	(4F)
52	(4F)
53	(6F)
55	(6F)
56	(4F)
59	(6F)
60	(4F)
64	(6F)
65	(6F)
67	(59S)
68	(6F)
73	(6F)
76	(59S)
87	(59S)

Breguet Br 1150
Atlantic/Atlantique 2*
21F/22F, Nimes-Garons;
23F/24F, Lann Bihou_

Serial	Unit
03	(21F/22F)
04	(21F/22F)
1	(21F/22F)
2	(21F/22F)
3	(21F/22F)
5	(21F/22F)
7	(23F/24F)
9	(21F/22F)
11	(21F/22F)
13	(21F/22F)
15	(21F/22F)
21	(23F/24F)
23	(21F/22F)
24	(23F/24F)
25	(21F/22F)
31	(21F/22F)
35	(21F/22F)
41	(21F/22F)
44	(21F/22F)
45	(23F/24F)
48	(23F/24F)
49	(21F/22F)
50	(21F/22F)
51	(21F/22F)
52	(21F/22F)
53	(21F/22F)
54	(23F/24F)
55	(23F/24F)
56	(21F/22F)
57	(21F/22F)
61	(21F/22F)
65	(23F/24F)
66	(23F/24F)
67	(21F/22F)
68	(23F/24F)
01*	(21F/22F)
02*	(CEV)
03*	(CEV)
04*	(21F/22F)
1*	(23F/24F)
2*	(23F/24F
3*	(23F/24F)
4*	
5*	(23F/24F)
6*	(23F/24F)
7*	(23F/24F)

Serial	Unit
8*	
9*	
10*	
11*	
12*	
13*	
14*	
15*	
16*	
17*	
18*	
19*	
20*	
21*	
22*	
23*	
24*	
25*	

Dassault Etendard IVM
16F, Landivisiau;
ES 59, Hyeres

Serial	Unit
1	(59S)
5	(59S)
9	(16F)
11	(59S)
14	(59S)
15	(59S)
21	(16F)
22	(16F)
26	(59S)
29	(59S)
30	(59S)
32	(16F)
34	(59S)
42	(59S)
51	(59S)
52	(59S)
56	(59S)
57	(59S)
59	(59S)
60	(16F)

Dassault Super Etendard
11F, Landivisiau;
17F, Hyeres
ES59, Hyeres

Serial	Unit
1	(17F)
2	(59S)
3	(59S)
4	(17F)
5	(11F)
6	(59S)
8	(11F)
9	(11F)
10	(11F)
11	(17F)
12	(59S)
13	(17F)
14	(17F)
15	(17F)
16	(17F)
17	
18	(17F)
19	(59S)
21	
23	(17F)
24	(17F)
25	(11F)
26	(11F)
28	(11F)
29	
30	(11F)
31	(11F)
32	(11F)
33	(17F)
34	(11F)
35	(17F)
37	(59S)
38	(59S)
39	
41	(11F)
42	
43	(11F)
44	(17F)
45	(17F)
46	(17F)
47	(17F)
48	(11F)
49	(59S)
50	(17F)
51	
52	(11F)
53	
54	(11F)
55	(17F)
57	(11F)
59	(17F)
60	(11F)
61	(59S)
62	
63	(11F)
64	(11F)
65	(17F)
66	(17F)
68	(17F)
69	(11F)
71	(11F)

Dassault Falcon 10 (MER)
ES 3, Hyeres;
ES 57, Landivisiau

Serial	Unit
32	(3S)
101	(57S)
129	(3S)
133	(57S)
143	(57S)
185	(3S)

Dassault Falcon Guardian
ES 9 Noumea;
ES 12 Papeete;
CEPA, Istres

Serial	Unit
48	(12S)
65	(9S)
72	(12S)
77	(9S)
80	(CEPA)

Embraer Xingu
ES 2, Lann Bihoué;
ES 11, Le Bourget;
ES 52, Lann Bihoué;
ES 56, Nimes-Garons
ES 57, Landivisiau;

Serial	Unit
30	(11S)
47	(57S)
55	(57S)
65	(52S)
66	(52S)
67	(52S)
68	(57S)
69	(2S)
70	(11S)
71	(2S)
74	(11S)
77	(56S)
79	(52S)
81	(57S)
83	(11S)
85	(52S)
87	(52S)
90	(52S)

LTV F-8E (FN) Crusader
12F, Landivisiau

Serial	Unit
3	
4	
5	
7	
8	
10	
11	
17	
19	
23	
27	
29	
31	
32	(CEV)
34	
35	(CEV)
37	
39	

Morane Saulnier 760 Paris
ES 57, Landivisiau

Serial	Unit
32	
33	
40	
41	
42	
46	
85	
87	
88	

Nord 262 Fregate
ES 2, Lann Bihoué;
ES 3, Hyeres;
ES 11, Le Bourget;
ES 56, Nimes-Garons
ES 57, Landivisiau

Serial	Unit
1	(11S)
16	(56S)
28	(2S)
43	(11S)
45	(56S)
46	(56S)
51	(57S)
52	(56S)
53	(56S)
59	(2S)
60	(2S)
61	(3S)
62	(3S)
63	(2S)
65	(2S)
69	(56S)
70	(3S)

Overseas Serials

Serial	Unit
71	(2S)
72	(56S)
73	(56S)
75	(2S)
79	(56S)
100	(56S)
102	(11S)
104	(11S)

Piper Navajo
ES 3, Hyeres;
ES 10 Fréjus
ES 56 Nimes-Garons

Serial	Unit
227	(3S)
903	(56S)
904	(3S)
906	(3S)
912	(3S)
914	(3S)
916	(3S)
925	(3S)
927	(10S)
929	(3S)
931	(3S)

Westland Lynx HAS2 (FN); HAS4 (FN)*
31F, San Mandrier;
34F, 35F, Lanvéoc;
ES 20, St Raphael

Serial	Unit
260	(20S)
262	(35F)
263	(34F)
264	(31F)
265	(34F)
266	(34F)
267	
268	
269	(34F)
270	(34F)
271	(35F)
272	(20S)
273	(34F)
274	(34F)
275	(31F)
276	(34F)
278	(34F)
621	(34F)
622	(34F)
623	(34F)
624	(34F)
625	(34F)
627	(35F)
801*	(20S)
802*	(34F)
803*	(31F)
804*	(35F)
806*	(31F)
807*	(31F)
808*	(34F)
809*	(31F)
810*	(31F)
811*	(31F)
812*	(31F)
813*	(31F)
814*	(34F)

Aviation Legère de L'Armée de Terre (ALAT)
Cessna F.406 Caravan
3 GHL, Rennes

Serial	Unit
0008	AGS
0010	AGT

GERMANY
Luftwaffe, Marineflieger

Boeing 707-320C
FBS-BMVg, Köln-Bonn
10+01
10+02
10+03
10+04

Airbus A310-304
FBS-BMVg, Köln-Bonn
10+21
10+22
10+23

Tupolev Tu-154M
LTG-65, Neuhardenberg
11+01
11+02

Ilyushin Il-62MT/Il-62MK*
LTG-65, Berlin/Schönefeld
11+20
11+21*
11+22

Canadair CL601-1A Challenger
FBS-BMVg, Köln-Bonn
12+01
12+02
12+03
12+04
12+05
12+06
12+07

HFB 320 Hansa Jet ECM
JbG32 Lechfeld
16+21
16+23
16+24
16+25
16+26
16+27
16+28

VFW-Fokker 614-100
FBS-BMVg, Köln-Bonn
17+01
17+02
17+03

Mikoyan MiG-29
Er Gp 29, Preschen
WTD61, Ingolstadt

MiG-29A

Serial	Unit
29+01	Er Gp 29
29+02	Er Gp 29
29+03	Er Gp 29
29+04	
29+05	Er Gp 29
29+06	
29+07	Er Gp 29
29+08	Er Gp 29
29+09	Er Gp 29
29+10	Er Gp 29
29+11	Er Gp 29
29+12	Er Gp 29
29+14	Er Gp 29
29+15	Er Gp 29
29+16	Er Gp 29
29+18	Er Gp 29
29+19	Er Gp 29
29+20	Er Gp 29
98+06	WTD-61
98+08	WTD-61

MiG-29UB

Serial	Unit
29+22	Er Gp 29
29+23	Er Gp 29
29+24	Er Gp 29
29+25	Er Gp 29

McD RF-4E Phantom
AkG 52, Leck;
Tslw 1, Kaufbeuren;
WTD 61, Ingolstadt

Serial	Unit
35+01	WTD 61
35+02	AkG 52
35+03	
35+04	
35+05	AkG 52
35+06	
35+07	
35+08	AkG 52
35+09	AkG 52
35+10	AkG 52
35+11	
35+12	
35+13	AkG 52
35+17	AkG 52
35+18	AkG 52
35+19	
35+20	AkG 52
35+21	AkG 52
35+22	
35+24	AkG 52
35+25	
35+26	AkG 52
35+28	
35+29	
35+31	AkG 52
35+32	AkG 52
35+33	
35+34	
35+35	
35+36	AkG 52
35+37	AkG 52
35+38	
35+39	AkG 52
35+40	
35+41	AkG 52
35+42	AkG 52
35+43	AkG 52
35+44	
35+46	
35+48	
35+49	
35+50	
35+51	
35+52	AkG 52
35+53	AkG 52
35+54	AkG 52
35+56	
35+57	
35+58	
35+59	
35+60	AkG 52
35+61	

Serial	Unit
35+62	AkG 52
35+63	
35+64	
35+65	AkG 52
35+66	AkG 52
35+67	AkG 52
35+68	AkG 52
35+71	
35+72	AkG 52
35+73	
35+74	AkG 52
35+75	
35+76	AkG 52
35+77	AkG 52
35+78	
35+79	AkG 52
35+82	
35+83	WTD 61
35+84	AkG 52
35+86	
35+87	AkG 52
35+88	

McD F-4F Phantom
JG 71, Wittmundhaven;
JG 72, Hopsten;
JbG 35, Pferdsfeld;
JG 74, Neuburg/Donau;
TSLw 1, Kaufbeuren;
WTD 61, Ingolstadt

Serial	Unit
37+01	JG 72
37+03	JG 71
37+04	TSLw 1
37+05	JG 72
37+06	JbG 35
37+07	JG 72
37+08	JG 74
37+09	JbG 35
37+10	JbG 35
37+11	JG 72
37+12	JbG 35
37+13	JG 74
37+14	TSLw 1
37+15	WTD-61
37+16	WTD-61
37+17	JG 72
37+18	JG 72
37+19	JG 72
37+20	JbG 35
37+21	JbG 35
37+22	JG 71
37+23	JG 72
37+24	JG 72
37+25	JbG 35
37+26	JG 72
37+28	JG 74
37+29	JG 72
37+30	JbG 35
37+31	JG 72
37+32	JG 74
37+33	JbG 35
37+34	JbG 35
37+35	JbG 35
37+36	JbG 35
37+37	JG 72
37+38	JbG 35
37+39	JG 71
37+40	JbG 35
37+41	JbG 35
37+42	JbG 35
37+43	JbG 35
37+44	JbG 35
37+45	JbG 35
37+46	JG 74
37+47	JbG 35
37+48	JG 74
37+49	JG 74
37+50	JbG 35
37+51	JbG 35
37+52	JbG 35
37+53	JG 72
37+54	JG 74
37+55	JG 71
37+56	JG 74
37+58	JbG 35
37+60	JG 74
37+61	JG 74
37+63	JG 71
37+64	JG 74
37+65	JG 72
37+66	JG 71
37+67	JG 72
37+69	JbG 35
37+70	JG 71
37+71	JG 71
37+73	JG 72
37+75	JG 71
37+76	JG 71
37+77	JG 72
37+78	JG 71
37+79	JG 71
37+81	JG 72
37+82	JbG 35
37+83	JG 71
37+84	JG 74
37+85	JG 74
37+86	JG 71
37+88	JG 71
37+89	JG 72
37+90	JG 72
37+91	WTD 61
37+92	JG 74
37+93	JG 72
37+94	JG 74
37+96	JG 74
37+97	JG 72
37+98	JG 72
38+00	JG 74
38+01	JG 72
38+02	JG 71
38+03	JG 71
38+04	JG 74
38+05	JG 71
38+06	JbG 35
38+07	JG 71
38+08	JG 74
38+09	JG 72
38+10	JbG 35
38+11	JG 72
38+12	JG 74
38+13	JbG 35
38+14	JG 74
38+16	JG 74
38+17	JG 72
38+18	JG 74
38+20	JG 72
38+21	JG 72
38+24	JG 71
38+25	JG 74
38+26	JG 71
38+27	JG 71
38+28	JG 74
38+29	JG 72
38+30	JG 72
38+31	JG 71
38+32	JG 74
38+33	JG 72
38+34	JbG 35
38+36	JG 74
38+37	JG 72
38+38	JbG 35
38+39	JG 71
38+40	JG 74
38+42	JG 71
38+43	JG 72
38+44	JG 74
38+45	JG 72
38+46	JG 72
38+47	JG 71
38+48	JG 74
38+49	JG 72
38+50	JG 71
38+51	JbG 35
38+53	JG 72
38+54	JG 72
38+55	JG 72
38+56	JG 74
38+57	JG 72
38+58	JG 71
38+59	JG 71
38+60	JG 71
38+61	JG 71
38+62	JG 72
38+63	JG 71
38+64	JG 74
38+66	JG 71
38+67	JG 72
38+68	JG 74
38+69	JG 71
38+70	JG 71
38+72	JbG 35
38+73	JG 74
38+74	JG 74
38+75	JG 71

D-BD Alpha Jet
JbG 41, Husum;
JbG 43, Oldenburg;
JbG 44, Beja (Portugal);
JbG 49, Fürrstenfeldbruck;
WTD 61, Ingolstadt

Serial	Unit
40+01	WTD 61
40+02	WTD 61
40+03	JbG 49
40+04	
40+05	JbG 49
40+06	JbG 49
40+07	JbG 49
40+08	JbG 44
40+09	JbG 41
40+11	JbG 43
40+12	JbG 49
40+13	
40+14	
40+15	JbG 43
40+16	JbG 41
40+17	JbG 49
40+18	JbG 49
40+20	JbG 43
40+21	JbG 44
40+22	JbG 41
40+23	
40+24	
40+25	JbG 49
40+26	JbG 49
40+27	JbG 43

Overseas Serials

Serial	Unit
40+28	JbG 44
40+29	
40+30	JbG 49
40+31	
40+32	JbG 43
40+33	JbG 43
40+34	JbG 41
40+35	JbG 49
40+36	JbG 43
40+37	JbG 49
40+38	
40+39	JbG 41
40+40	JbG 49
40+41	JbG 43
40+42	JbG 44
40+43	JbG 43
40+44	JbG 49
40+45	JbG 41
40+46	JbG 43
40+47	JbG 49
40+48	JbG 43
40+49	JbG 49
40+50	
40+51	
40+52	
40+53	JbG 44
40+54	JbG 43
40+56	WTD-61
40+57	JbG 43
40+58	JbG 43
40+59	TSLw 3
40+60	JbG 41
40+61	JbG 49
40+62	JbG 44
40+63	JbG 41
40+64	JbG 49
40+65	JbG 49
40+66	JbG 44
40+67	JbG 49
40+68	JbG 41
40+69	
40+70	JbG 49
40+71	JbG 44
40+72	JbG 44
40+73	JbG 49
40+74	JbG 41
40+75	JbG 43
40+76	JbG 49
40+77	JbG 49
40+78	TSLw 3
40+79	
40+80	JbG 43
40+81	JbG 41
40+82	JbG 49
40+84	JbG 49
40+85	JbG 49
40+86	JbG 44
40+88	JbG 41
40+89	JbG 41
40+90	
40+91	JbG 49
40+92	JbG 43
40+93	JbG 49
40+94	JbG 44
40+95	
40+96	JbG 44
40+97	JbG 44
40+98	
40+99	JbG 41
41+00	JbG 43
41+01	JbG 43
41+02	JbG 41

Serial	Unit
41+03	JbG 41
41+04	JbG 49
41+05	
41+06	JbG 41
41+07	JbG 44
41+08	JbG 44
41+09	JbG 49
41+10	JbG 49
41+11	
41+12	JbG 43
41+13	JbG 43
41+14	JbG 49
41+15	JbG 41
41+16	JbG 43
41+17	JbG 43
41+18	JbG 43
41+19	JbG 44
41+20	JbG 41
41+21	JbG 41
41+22	JbG 44
41+23	JbG 49
41+24	JbG 41
41+25	JbG 49
41+26	JbG 49
41+27	JbG 43
41+28	
41+29	JbG 43
41+30	JbG 49
41+31	JbG 44
41+32	JbG 41
41+33	JbG 41
41+34	JbG 43
41+35	JbG 49
41+36	JbG 49
41+37	JbG 49
41+38	JbG 49
41+39	JbG 49
41+40	JbG 43
41+41	JbG 41
41+42	JbG 43
41+43	JbG 43
41+44	JbG 43
41+45	JbG 49
41+46	JbG 44
41+47	JbG 41
41+48	JbG 41
41+49	JbG 49
41+50	JbG 49
41+51	JbG 43
41+52	JbG 44
41+53	JbG 49
41+54	
41+55	JbG 44
41+56	JbG 49
41+57	WTD-61
41+58	JbG 43
41+59	JbG 49
41+60	JbG 41
41+61	JbG 49
41+62	JbG 49
41+63	JbG 43
41+64	JbG 43
41+65	JbG 41
41+66	JbG 41
41+67	JbG 49
41+68	JbG 49
41+70	JbG 43
41+71	JbG 44
41+72	JbG 41
41+73	JbG 49
41+74	JbG 41
41+75	JbG 41

Serial	Unit
98+55	WTD 61

Panavia Tornado
Strike/Trainer[1]/ECR[2]

TTTE RAF Cottesmore;
JbG 31, Nörvenich;
JbG 32, Lechfeld;
JbG 33, Büchel;
JbG 34, Memmingen;
JbG 38, Jever;
MBB, Manching;
MFG1, Schleswig/Jagel;
MFG2, Eggebek;
TSLw 1, Kaufbeuren;
WTD61, Ingolstadt

Serial	Unit
98+02	WTD 61
98+03	WTD 61
98+59	WTD 61
98+60	WTD 61
98+79	WTD 61
98+97[2]	WTD 61
43+01[1]	[G-20] TTTE
43+02[1]	[G-21] TTTE
43+03[1]	[G-22] TTTE
43+04[1]	[G-23] TTTE
43+05[1]	[G-24] TTTE
43+06[1]	[G-25] TTTE
43+07[1]	[G-26] TTTE
43+08[1]	[G-27] TTTE
43+09[1]	[G-28] TTTE
43+10[1]	[G-29] TTTE
43+11[1]	[G-30] TTTE
43+12[1]	[G-70] TTTE
43+13	JbG 31
43+14	[G-72] TTTE
43+15[1]	[G-31] TTTE
43+16[1]	[G-32] TTTE
43+17[1]	[G-33] TTTE
43+18	JbG 34
43+19	JbG 31
43+20	JbG 38
43+22[1]	JbG 38
43+23[1]	JbG 38
43+25	[G-75] TTTE
43+26	JbG 38
43+27	MFG 1
43+28	JbG 38
43+29[1]	JbG 31
43+30	JbG 38
43+31[1]	JbG-31
43+32	[G-73] TTTE
43+33[1]	JbG 38
43+34	TSLw-1
43+35[1]	JbG 38
43+36	JbG 34
43+37[1]	JbG 38
43+38	JbG 34
43+40	JbG 34
43+41	JbG 31
43+42[1]	MFG 1
43+43[1]	MFG 1
43+44[1]	MFG 1
43+45[1]	MFG 1
43+46	MFG 1
43+47	MFG 1
43+48	MFG 1
43+50	MFG 1
43+52	MFG 1
43+53	MFG 1
43+54	MFG 1
43+55	MFG 1
43+57	MFG 1

Serial	Unit
43+58	MFG 1
43+59	MFG 1
43+60	MFG 1
43+61	MFG 1
43+62	MFG 1
43+63	MFG 1
43+64	MFG 1
43+65	MFG 1
43+67	MFG 1
43+68	MFG 1
43+69	MFG 1
43+70	JbG 38
43+71	MFG 1
43+72	MFG 1
43+73	MFG 1
43+74	MFG 1
43+75	MFG 1
43+76	MFG 1
43+77	MFG 1
43+78	MFG 1
43+79	MFG 1
43+80	MFG 1
43+81	MFG 1
43+82	MFG 1
43+83	MFG 1
43+85	MFG 1
43+86	MFG 1
43+87	MFG 1
43+88	MFG 1
43+90[1]	JbG 38
43+91[1]	JbG 38
43+92[1]	JbG 31
43+94[1]	JbG 31
43+95	JbG 31
43+96	JbG 31
43+97[1]	JbG 31
43+98	JbG 31
43+99	JbG 31
44+00	JbG 31
44+01[1]	JbG 38
44+02	JbG 31
44+03	JbG 31
44+04	JbG 31
44+05[1]	JbG 38
44+06	JbG 31
44+07	JbG 31
44+08	JbG 38
44+09	JbG 31
44+10[1]	JbG 38
44+11	JbG 34
44+12	JbG 31
44+13	JbG 38
44+14	JbG 34
44+15[1]	JbG 38
44+16	JbG 31
44+17	JbG 38
44+19	JbG 31
44+20[1]	JbG 38
44+21	JbG 31
44+22	JbG 31
44+23	JbG 31
44+24	JbG 38
44+25[1]	JbG 38
44+26	JbG 31
44+27	JbG 31
44+28	JbG 31
44+29	JbG 31
44+30	JbG 31
44+31	JbG 31
44+32	JbG 38
44+33	JbG 31
44+34	JbG 31
44+35	JbG 31
44+36[1]	JbG 32
44+37[1]	JbG 32
44+38[1]	JbG 32
44+39[1]	JbG 32
44+40	JbG 33
44+41	JbG 31
44+42	JbG 31
44+43	JbG 34
44+44	JbG 31
44+46	JbG 34
44+48	JbG 31
44+49	JbG 31
44+50	JbG 31
44+51	JbG 38
44+52	JbG 31
44+53	JbG 34
44+54	JbG 33
44+55	JbG 38
44+56	JbG 34
44+57	JbG 32
44+58	JbG 32
44+59	JbG 32
44+60	JbG 32
44+61	JbG 32
44+62	JbG 33
44+63	JbG 33
44+64	JbG 32
44+65	JbG 34
44+66	JbG 31
44+67	JbG 32
44+68	JbG 32
44+69	JbG 32
44+70	JbG 32
44+71	JbG 32
44+72[1]	JbG 33
44+73[1]	JbG 33
44+75[1]	JbG 33
44+76	JbG 32
44+77	JbG 32
44+78	JbG 32
44+79	JbG 32
44+80	JbG 32
44+81	JbG 32
44+82	JbG 32
44+83	JbG 32
44+84	JbG 32
44+85	JbG 32
44+86	JbG 33
44+87	JbG 33
44+88	JbG 33
44+89	JbG 33
44+90	JbG 33
44+91	JbG 33
44+92	JbG 33
44+94	JbG 33
44+95	JbG 33
44+96	JbG 33
44+97	JbG 33
44+98	JbG 33
44+99	JbG 34
45+00	JbG 33
45+01	JbG 33
45+02	JbG 33
45+03	JbG 33
45+04	JbG 33
45+05	JbG 33
45+06	JbG 33
45+07	JbG 33
45+08	JbG 33
45+09	JbG 33
45+10	JbG 33
45+11	JbG 33
45+12[1]	MFG 2
45+13[1]	MFG 2
45+14[1]	MFG 2
45+15[1]	MFG 2
45+16[1]	MFG 2
45+17	JbG 33
45+18	JbG 34
45+19	JbG 33
45+20	JbG 33
45+21	JbG 33
45+22	JbG 33
45+23	JbG 33
45+24	JbG 33
45+25	JbG 33
45+26	TSLw 1
45+27	MFG 2
45+28	MFG 2
45+29	WTD-61
45+30	MFG 2
45+31	MFG 2
45+32	MFG 2
45+33	MFG 2
45+34	MFG 2
45+35	MFG 2
45+36	MFG 2
45+37	MFG 2
45+38	MFG 2
45+39	MFG 2
45+40	MFG 2
45+41	MFG 2
45+42	MFG 2
45+43	MFG 2
45+44	MFG 2
45+45	MFG 2
45+46	MFG 2
45+47	MFG 2
45+48	MFG 2
45+49	MFG 2
45+50	MFG 2
45+51	MFG 2
45+52	MFG 2
45+53	MFG 2
45+54	MFG 2
45+55	MFG 2
45+56	MFG 2
45+57	MFG 2
45+58	MFG 2
45+59	MFG 2
45+60[1]	JbG 38
45+61[1]	JbG 34
45+62[1]	JbG 38
45+64	TSLw 1
45+65	MFG 2
45+66	MFG 2
45+67	MFG 2
45+68	MFG 2
45+69	MFG 2
45+70[1]	JbG 33
45+71	MFG 2
45+72	MFG 2
45+73[1]	JbG 31
45+74	MFG 2
45+76	JbG 38
45+77[1]	JbG 33
45+78	JbG 34
45+79	JbG 34
45+80	JbG 34
45+81	JbG 34
45+82	JbG 34
45+83	JbG 34
45+84	JbG 34

Serial	Unit
45+85	JbG 34
45+86	JbG 34
45+87	JbG 34
45+88	JbG 34
45+89	JbG 34
45+90	JbG 34
45+91	JbG 34
45+92	JbG 34
45+93	JbG 34
45+94	JbG 34
45+95	JbG 34
45+96	JbG 34
45+97	JbG 34
45+98	JbG 34
45+99[1]	JbG 34
46+00	JbG 34
46+01	JbG 34
46+02	JbG 34
46+03	JbG 34
46+04[1]	MFG 1
46+05[1]	MFG 2
46+06[1]	JbG 32
46+07	JbG 34
46+08	JbG 34
46+09	JbG 34
46+10	MFG 1
46+11	MFG 1
46+12	MFG 2
46+13	MFG 1
46+14	MFG 1
46+15	MFG 1
46+18	MFG 2
46+19	MFG 2
46+20	MFG 2
46+21	MFG 2
46+22	MFG 2
46+23[2]	JbG 32
46+24[2]	JbG 32
46+25[2]	TbG 32
46+26[2]	JbG 32
46+27[2]	JbG 32
46+28[2]	JbG 32
46+29[2]	JbG 32
46+30[2]	JbG 32
46+31[2]	JbG 32
46+32[2]	JbG 32
46+33[2]	JbG 32
46+34[2]	JbG 32
46+35[2]	JbG 32
46+36[2]	JbG 32
46+37[2]	JbG 32
46+38[2]	JbG 32
46+39[2]	JbG 32
46+40[2]	JbG 32
46+41[2]	JbG 38
46+42[2]	JbG 38
46+43[2]	JbG 38
46+44[2]	JbG 38
46+45[2]	JbG 38
46+46[2]	JbG 38
46+47[2]	JbG 38
46+48[2]	JbG 38
46+49[2]	JbG 38
46+50[2]	JbG 38
46+51[2]	JbG 38
46+52[2]	
46+53[2]	JbG 38
46+54[2]	JbG 38
46+55[2]	JbG 38
46+56[2]	
46+57[2]	JbG 38

Transall C-160D
LTG 61, Landsberg;
LTG 62, Wunstorf;
LTG 63, Hohn;
WTD61 Ingolstadt

Serial	Unit
50+06	LTG 63
50+07	LTG 61
50+08	LTG 61
50+09	LTG 62
50+10	LTG 62
50+17	LTG 62
50+29	LTG 62
50+33	LTG 63
50+34	LTG 62
50+35	LTG 62
50+36	LTG 62
50+37	LTG 62
50+38	LTG 62
50+40	LTG 61
50+41	LTG 63
50+42	LTG 63
50+43	LTG 61
50+44	LTG 61
50+45	LTG 63
50+46	LTG 62
50+47	LTG 61
50+48	LTG 61
50+49	LTG 61
50+50	LTG 63
50+51	LTG 61
50+52	LTG 62
50+53	LTG 62
50+54	LTG 63
50+55	LTG 62
50+56	LTG 63
50+57	LTG 61
50+58	LTG 63
50+59	LTG 63
50+60	LTG 62
50+61	LTG 63
50+62	LTG 61
50+64	LTG 61
50+65	LTG 62
50+66	LTG 61
50+67	LTG 63
50+68	LTG 61
50+69	LTG 61
50+70	WTD 61
50+71	LTG 63
50+72	LTG 61
50+73	LTG 62
50+74	LTG 61
50+75	LTG 61
50+76	LTG 63
50+77	LTG 63
50+78	LTG 62
50+79	LTG 63
50+81	LTG 62
50+82	LTG 63
50+83	LTG 62
50+84	LTG 63
50+85	LTG 61
50+86	LTG 61
50+87	LTG 63
50+88	LTG 61
50+89	WTD 61
50+90	LTG 61
50+91	LTG 62
50+92	LTG 61
50+93	LTG 61
50+94	LTG 63
50+95	LTG 63
50+96	LTG 61
50+97	LTG 62
50+98	LTG 61
50+99	LTG 61
51+00	LTG 62
51+01	LTG 62
51+02	LTG 63
51+03	LTG 62
51+04	LTG 61
51+05	LTG 62
51+06	LTG 63
51+07	LTG 62
51+08	LTG 63
51+09	LTG 63
51+10	LTG 61
51+11	LTG 62
51+12	LTG 63
51+13	LTG 61
51+14	LTG 63
51+15	LTG 61

Antonov An-26
LTG 65, Dresden

Serial	Unit
52+01	LTG65
52+02	LTG65
52+03	LTG65
52+06	LTG65
52+07	LTG65
52+09	LTG65
52+10	LTG65
52+11	LTG65
52+12	LTG65

Dornier Do.228
WTG-61, Ingolstadt
MFG 5, Kiel-Holtenau

Serial	Unit
57+01	MFG 5
98+78	WTD-61

Dornier Skyservant
AkG 52, Leck
JbG 35, Pferdsfeld;
JG 71 Witmundhaven;
JG 72 Hopsten;
JG 74, Neuburg/Donau;
LTG 61, Landsberg;
LTG 62, Wunstorf;
LTG 63, Hohn;
WTD 61, Ingolstadt;
FBS-BMVg, Köln-Bonn;
MFG 5, Kiel-Holtenau

Serial	Unit
58+05	WTD 61
58+08	JG 71
58+09	LTG 62
58+14	JbG 34
58+15	JbG 31
58+18	FBS BMVg
58+20	JbG 33
58+23	JbG 41
58+26	LTG 63
58+28	JbG 35
58+29	
58+30	LTG 62
58+32	JG 74
58+34	LTG 62
58+36	JbG 31
58+37	LTG 62
58+38	LTG 62
58+39	JbG 31
58+46	LTG 62
58+47	JG 74
58+50	JbG 38

Serial	Unit
58+52	LTG 62
58+53	JbG 32
58+54	
58+55	JbG 32
58+59	JbG 34
58+60	AkG 52
58+61	JbG 34
58+62	JG 72
58+65	JG 71
58+66	JbG 38
58+67	JbG 38
58+68	AkG 52
58+69	JbG 35
58+70	LTG 62
58+71	JG 74
58+72	
58+73	JbG 31
58+74	JbG 41
58+78	JbG 31
58+79	JbG 31
58+80	JbG 33
58+82	JbG 43
58+83	JG 72
58+84	JbG 43
58+85	JG 74
58+92	JG 72
58+94	JbG 35
58+98	AkG 52
58+99	JG 71
59+00	FBS-BMVg
59+01	FBS-BMVg
59+03	FBS-BMVg
59+04	FBS-BMVg
59+05	FBS-BMVg
59+06	MFG 5
59+07	MFG 5
59+08	MFG 5
59+09	MFG 5
59+10	MFG 5
59+11	MFG 5
59+12	MFG 5
59+14	MFG 5
59+15	MFG 5
59+16	MFG 5
59+17	MFG 5
59+18	MFG 5
59+19	MFG 5
59+22	MFG 5
59+23	MFG 5
59+24	MFG 5
59+25	MFG 5

Breguet Br1151 Atlantic
**Elint*
MFG 3, Nordholz

61+01
61+02*
61+03*
61+04
61+05
61+06*
61+08
61+09
61+10
61+11
61+12
61+13
61+14
61+15
61+16
61+17
61+18
61+19*
61+20*

Westland Lynx Mk88
MFG 3, Nordholz

83+01
83+02
83+03
83+04
83+05
83+06
83+07
83+08
83+09
83+10
83+11
83+12
83+13
83+14
83+15
83+16
83+17
83+18
83+19

Westland Sea King HAS41
MFG 5, Kiel-Holtenau

89+50
89+51
89+52
89+53
89+54
89+55
89+56
89+57
89+58
89+59
89+60
89+61
89+62
89+63
89+64
89+65
89+66
89+67
89+68
89+69
89+70
89+71

Heeresfliegertruppe

Sikorsky/VFW CH-53G
HFlgRgt-15, Rheine-Bentlage
HFlgRgt-25, Laupheim
HFlgRgt-35, Mendig
HFWS, Bückeburg
WTD 61, Ingolstadt

Serial	Unit
84+01	WTD 61
84+02	WTD 61
84+03	15
84+04	HFWS
84+05	35
84+06	35
84+07	HFWS
84+08	35
84+09	25
84+10	25
84+11	HFWS
84+12	15
84+13	HFWS
84+14	HFWS
84+15	25
84+16	HFWS
84+17	25
84+18	HFWS
84+19	HFWS
84+20	35
84+21	HFWS
84+22	35
84+23	25
84+24	35
84+25	35
84+26	35
84+27	35
84+28	35
84+29	35
84+30	35
84+31	35
84+32	35
84+33	35
84+34	35
84+35	35
84+36	35
84+37	35
84+38	35
84+39	35
84+40	25
84+41	HFWS
84+42	25
84+43	25
84+44	25
84+45	25
84+46	25
84+47	25
84+48	25
84+49	HFWS
84+50	25
84+51	25
84+52	25
84+53	25
84+54	25
84+55	25
84+56	25
84+57	25
84+58	25
84+59	25
84+60	25
84+62	25
84+63	25
84+64	25
84+65	35
84+66	35
84+67	35
84+68	15
84+69	15
84+70	15
84+71	15
84+72	15
84+73	15
84+74	15
84+75	15
84+76	15
84+77	15
84+78	15
84+79	15
84+80	15
84+82	15
84+83	15
84+84	15
84+85	15
84+86	15
84+87	15

Serial	Unit
84+88	15
84+89	15
84+90	15
84+91	15
84+92	35
84+93	35
84+94	35
84+95	25
84+96	25
84+97	25
84+98	15
84+99	15
85+00	15
85+01	35
85+02	35
85+03	35
85+04	25
85+05	25
85+06	25
85+07	15
85+08	15
85+09	15
85+10	35
85+11	25
85+12	15

English Electric Canberra B2
AWG, Manching
99+34
99+35

GHANA
Air Force
Short SC7 Skyvan
No 1 Transport Sqn, Takoradi
G450 [A]
G451 [B]
G452 [C]
G453 [D]
G454 [E]
G455 [F]

GREECE
Helliniki Aeroporia
Lockheed C-130H Hercules
356 Mira, Elefsis
741
742
743
744
745
746
747
749
751
752

ISRAEL
Heyl Ha'Avir
Lockheed Hercules C-130H
4X-FBA/102
4X-FBB/106
4X-FBC/309
4X-FBQ/420
4X-FBS/427
4X-FBT/435
4X-FBU/448
4X-FBW/436
4X-FBX/428

C-130E
4X-FBD/311
4X-FBE/304
4X-FBF/301
4X-FBG/310
4X-FBH/312
4X-FBI/314
4X-FBJ/305
4X-FBK/318
4X-FBL/313
4X-FBM/316
4X-FBN/307
4X-FBP/208

KC-130H
4X-FBY/522
4X-FBZ/545

ITALY
Aeronautica Militare Italiano
Aeritalia G222
46a Brigata Aerea, Pisa
14° Stormo, Pratica di Mare;
RSV, Pratica di Mare

Serial	Unit
MM62101	RS-45
MM62102	46-20
MM62104	46-91
MM62105	46-82
MM62108	46-30
MM62109	46-96
MM62110	46-81
MM62111	46-83
MM62112	46-85
MM62114	46-80
MM62115	46-22
MM62117	46-25
MM62118	46-24
MM62119	46-21
MM62120	46-90
MM62121	46-86
MM62122	46-23
MM62123	46-28
MM62124	46-88
MM62125	46-87
MM62126	46-26
MM62127	46-27
MM62130	RS 51
MM62132	46-32
MM62133	46-93
MM62134	46-33
MM62143	46-36
MM62144	46-98
MM62145	46-50
MM62146	46-51
MM62147	46-52

Aeritalia G222TCM

Serial	Unit
MM62103	46-37
MM62135	46-94
MM62136	46-97
MM62137	46-95
MM62152	
MM62153	
MM62154	
MM62155	

Aeritalia G222RM

Serial	Unit
MM62107	
MM62138	
MM62139	14-20
MM62140	14-21
MM62141	14-22
MM62142	

Aeritalia-EMB AMX/AMX-T*
2° Stormo, Istrana
3° Stormo, Villafranca
32° Stormo, Amendola
51° Stormo, Istrana
RSV, Pratica di Mare

Serial	Unit
MMX595	Aeritalia
MMX596	Aeritalia
MMX597	Aeritalia
MMX599	Aeritalia
MM7089	2-10
MM7090	RS-12
MM7091	2-01
MM7092	RS-14
MM7093	3-01
MM7094	
MM7095	51-33
MM7096	51-30
MM7097	51-35
MM7098	51-36
MM7099	2-25
MM7100	51-40
MM7101	51-44
MM7102	51-42
MM7103	51-34
MM7104	51-31
MM7105	2-26
MM7106	51-37
MM7107	51-44
MM7109	51-45
MM7110	3-09
MM7111	51-50
MM7112	51-52
MM7114	51-46
MM7115	51-51
MM7116	3-42
MM7117	3-05
MM7118	3-06
MM7119	3-07
MM7120	3-10
MM7121	3-47
MM7122	3-
MM7123	3-51
MM7124	3-14
MM7125	3-15
MM7126	3-53
MM7127	3-22
MM7128	3-23
MM7129	3-24
MM7130	51-43
MM7131	
MM7132	
MM7133	
MM7134	
MM7135	2-03
MM7136	2-04
MM7137	2-05
MM7138	2-12
MM7139	3-16
MM7140	2-06
MM7141	2-07
MM7142	2-14
MM7143	2-15
MM7144	
MM7145	2-16
MM7146	2-23
MM7147	51-53
MM7148	

Serial	Unit
MM7149	
MM7150	
MM7151	
MM7152	
MM7153	2-02
MM7154	51-54
MM7155	2-22
MM7156	
MM7157	
MM7158	
MM7159	
MM7160	
MM55024*	Aeritalia
MM55025*	Aeritalia
MM55026*	Aeritalia

Aermacchi MB339
*Frecce Tricolori, 313 Gruppo, Rivolto;
61a Brigata Aerea, Lecce; 14° Stormo, Pratica di Mare;
RSV, Pratica di Mare

Serial	Unit
MM54438	61-93
MM54439	13*
MM54440	
MM54442	61-95
MM54443	61-50
MM54445	8*
MM54446	61-01
MM54447	61-02
MM54448	61-03
MM54449	61-04
MM54450	14-30
MM54451	61-86
MM54452	
MM54453	61-05
MM54454	61-06
MM54455	61-07
MM54456	RS-10
MM54457	61-11
MM54458	61-12
MM54459	61-13
MM54460	61-14
MM54461	61-15
MM54462	61-16
MM54463	61-17
MM54464	61-20
MM54465	61-21
MM54467	61-23
MM54468	61-24
MM54469	61-25
MM54470	61-26
MM54471	61-27
MM54472	61-30
MM54473	4*
MM54475	3*
MM54476	
MM54477	10*
MM54478	7*
MM54479	1*
MM54480	0*
MM54482	5*
MM54483	10*
MM54484	11*
MM54486	2*
MM54487	61-31
MM54488	61-32
MM54489	61-33
MM54490	61-34
MM54491	61-35
MM54492	61-36
MM54493	61-37
MM54494	61-40
MM54496	61-42
MM54497	61-43
MM54498	61-44
MM54499	61-45
MM54500	61-46
MM54503	61-51
MM54504	61-52
MM54505	61-53
MM54506	61-54
MM54507	61-55
MM54508	61-56
MM54509	61-57
MM54510	61-60
MM54511	61-61
MM54512	61-62
MM54513	61-63
MM54514	61-64
MM54515	61-65
MM54516	61-66
MM54517	61-67
MM54518	61-70
MM54532	61-71
MM54533	61-72
MM54534	61-73
MM54535	61-74
MM54536	9*
MM54537	
MM54538	61-75
MM54539	61-76
MM54540	61-77
MM54541	RS-27
MM54542	61-81
MM54543	61-82
MM54544	61-83
MM54545	61-84
MM54546	61-85
MM54547	61-87
MM54548	61-90
MM54549	61-91
MM54550	61-92
MM54551	6*
MM54553	61-94
MM54554	61-95

Boeing 707-328B/-3F5C*
14° Stormo, Pratica di Mare
31° Stormo, Roma-Ciampino

Serial	Unit
MM62148	(14)
MM62149	(14)
MM62150*	(31)
MM62151*	(31)

Breguet Br 1150 Atlantic
30° Stormo, Cagliari
41° Stormo, Catania

Serial	Unit
MM40108	41-70
MM40109	30-71
MM40110	41-72
MM40111	41-73
MM40112	30-74
MM40113	30-75
MM40114	41-76
MM40115	41-77
MM40116	30-78
MM40117	41-02
MM40118	30-03
MM40119	30-04
MM40120	41-05
MM40121	41-06
MM40122	30-07
MM40123	30-10
MM40124	41-11
MM40125	30-12

Dassault Falcon 50
31° Stormo, Roma Ciampino

Serial	Unit
MM62020	
MM62021	
MM62026	
MM62029	

Grumman Gulfstream III
31° Stormo, Roma Ciampino

Serial	Unit
MM62022	
MM62025	

Lockheed C-130H Hercules
46a Brigata Aerea, Pisa

Serial	Unit
MM61988	46-02
MM61989	46-03
MM61990	46-04
MM61991	46-05
MM61992	46-06
MM61993	46-07
MM61994	46-08
MM61995	46-09
MM61997	46-11
MM61998	46-12
MM61999	46-13
MM62001	46-15

McDonnell Douglas DC9-32
31° Stormo, Roma Ciampino

Serial	Unit
MM62012	
MM62013	

Panavia Tornado Strike/Trainer[1]/ECR[2]
[2]TTTE RAF Cottesmore
6° Stormo, Ghedi
36° Stormo, Giola del Colle
50° Stormo, Piacenza
RSV, Pratica di Mare

Serial	Unit
MM586	
MM7002	TTTE I-92
MM7003	TTTE I-93
MM7004	6-04
MM7005	6-53
MM7006	6-06
MM7007	50-07
MM7008	6-02
MM7009	50-44
MM7010	50-01
MM7011	50-10
MM7013	6-03
MM7014	6-14
MM7015	
MM7016	6-16
MM7017	50-14
MM7018	6-02
MM7019	19
MM7020	50-12
MM7021	6-21
MM7022	50-02
MM7023	
MM7024	16-24
MM7025	6-25
MM7026	6-26
MM7027	
MM7028	28
MM7029	6-25

Serial	Unit
MM7030	
MM7031	6-31
MM7033	6-33
MM7034	6-34
MM7035	35
MM7036	50-36
MM7037	36-37
MM7038	36-33
MM7039	50-52
MM7040	36-35
MM7041	
MM7042	6-42
MM7043	43
MM7044	6-66
MM7046	36-34
MM7047	36-36
MM7048	(Aeritalia)
MM7049	
MM7050	50
MM7051	36-32
MM7052	36-41
MM7053	53
MM7054	50-34
MM7055	55
MM7056	36-52
MM7057	36-54
MM7058	36-57
MM7059	6-35
MM7060	
MM7061	61
MM7062	62
MM7063	63
MM7064	6-24
MM7065	50-33
MM7066	36-30
MM7067	36-45
MM7068	36-46
MM7069	6-61
MM7070	70
MM7071	71
MM7072	50-32
MM7073	73
MM7075	50-25
MM7076	36-31
MM7078	78
MM7079[2]	Alenia
MM7080	80
MM7081	RS-01
MM7082	RS-02
MM7083	6-70
MM7084	36-42
MM7085	36-50
MM7086	50-40
MM7087	87
MM7088	88
MM55000 TTTE	I-42
MM55001 TTTE	I-40
MM55002 TTTE	I-41
MM55003 TTTE	I-43
MM55004 TTTE	I-44
MM55005 TTTE	I-45
MM55006 50-50	
MM55007	50-51
MM55008	6-20
MM55009	6-15
MM55010	36-56
MM55011	36-55

Piaggio-Douglas PD-808; [1]PD-808-GE; [2]PD-808-RM; [3]PD-808-TA
14° Stormo, Pratica di Mare;
31° Stormo, Roma-Ciampino
RSV, Pratica di Mare

Serial	Unit
MM577[3]	RS-48
MM578[3]	RS-49
MM61948	(14)
MM61949	(14)
MM61950	14-50
MM61951	(31)
MM61952[1]	(14)
MM61953[3]	(14)
MM61954[3]	(31)
MM61955[1]	(14)
MM61956[2]	(14)
MM61957[3]	RS-50
MM61958[1]	(14)
MM61959[1]	(14)
MM61960[1]	(14)
MM61961[1]	(14)
MM61962[1]	(14)
MM61963[1]	(14)
MM62014[2]	(14)
MM62015[2]	(14)
MM62016[2]	14-55
MM62017[2]	(14)

JORDAN
Al Quwwat al-Jawwiya al Malakiya al-Urduniya

Lockheed Hercules C-130B
3 Sqn, Amman
340
341

C-130H
344
345
346
347

KUWAIT
Kuwait Air Force

McDonnell Douglas DC9-32
KAF 321

Lockheed L100-30 Hercules
KAF 323
KAF 324
KAF 325

LUXEMBOURG
NATO

Boeing E-3A
NAEWF, Geilenkirchen
LX-N90442
LX-N90443
LX-N90444
LX-N90445
LX-N90446
LX-N90447
LX-N90448
LX-N90449
LX-N90450
LX-N90451
LX-N90452
LX-N90453
LX-N90454
LX-N90455
LX-N90456
LX-N90457
LX-N90458
LX-N90459

Boeing 707-329C
NAEWF, Geilenkirchen
LX-N19996
LX-N20198
LX-N20199

MOROCCO
Force Aerienne Royaume Marocaine

CAP-230
Green March

Serial	Unit
04	CN-ABD
05	CN-ABF
06	CN-ABI
07	CN-ABJ
08	CN-ABK
09	CN-ABL

Lockheed C-130H Hercules

Serial	Unit
4535	CN-AOA
4551	CN-AOC
4575	CN-AOD
4581	CN-AOE
4583	CN-AOF
4713	CN-AOG
4717	CN-AOH
4733	CN-AOI
4738	CN-AOJ
4739	CN-AOK
4742	CN-AOL
4875	CN-AOM
4876	CN-AON
4877	CN-AOO
4888	CN-AOP
4892	CN-AOQ
4907	CN-AOR
4909	CN-AOS
4940	CN-AOT

NETHERLANDS
Koninklijke Luchtmacht

Fokker F-27-100 Friendship
334 Sqn, Eindhoven
C-1
C-2
C-3

Fokker F-27-300M Troopship
334 Sqn, Eindhoven
C-4
C-5
C-6
C-7
C-8
C-9
C-10
C-11
C-12

General Dynamics F-16A/F-16B*
306/311/312 Sqns, Volkel;
313/315 Sqns, Twente;
314 Sqn, Gilze-Rijen;
316 Sqn, Eindhoven;
322/323 Sqns, Leeuwarden

Serial	Unit
J001	313 Sqn

Serial	Unit
J002	313 Sqn
J003	313 Sqn
J004	313 Sqn
J005	313 Sqn
J006	313 Sqn
J008	313 Sqn
J009	313 Sqn
J010	313 Sqn
J011	313 Sqn
J012	313 Sqn
J013	313 Sqn
J014	313 Sqn
J015	313 Sqn
J016	313 Sqn
J017	313 Sqn
J018	313 Sqn
J019	313 Sqn
J020	313 Sqn
J021	313 Sqn
J055	315 Sqn
J057	315 Sqn
J058	313 Sqn
J059	315 Sqn
J060	315 Sqn
J061	315 Sqn
J062	315 Sqn
J063	315 Sqn
J064*	315 Sqn
J065*	315 Sqn
J066*	315 Sqn
J067*	313 Sqn
J068*	313 Sqn
J135	314 Sqn
J136	316 Sqn
J137	314 Sqn
J138	314 Sqn
J139	314 Sqn
J140	314 Sqn
J141	314 Sqn
J142	316 Sqn
J143	316 Sqn
J144	314 Sqn
J145	314 Sqn
J146	314 Sqn
J192	311 Sqn
J193	311 Sqn
J194	311 Sqn
J195	311 Sqn
J196	311 Sqn
J197	311 Sqn
J198	311 Sqn
J199	311 Sqn
J201	314 Sqn
J202	314 Sqn
J203	322 Sqn
J204	322 Sqn
J205	322 Sqn
J206	314 Sqn
J207	322 Sqn
J208*	316 Sqn
J209*	314 Sqn
J210*	314 Sqn
J211*	316 Sqn
J212	323 Sqn
J213	322 Sqn
J214	323 Sqn
J215	322 Sqn
J218	322 Sqn
J220	322 Sqn
J223	322 Sqn
J226	322 Sqn
J228	322 Sqn
J230	322 Sqn
J231	323 Sqn
J232	323 Sqn
J234	323 Sqn
J235	323 Sqn
J236	323 Sqn
J238	322 Sqn
J240	322 Sqn
J241	322 Sqn
J243	322 Sqn
J246	323 Sqn
J248	323 Sqn
J249	322 Sqn
J250	323 Sqn
J251	322 Sqn
J253	323 Sqn
J254	323 Sqn
J255	323 Sqn
J256	323 Sqn
J259*	323 Sqn
J260*	322 Sqn
J261*	322 Sqn
J262*	323 Sqn
J264*	322 Sqn
J265*	323 Sqn
J266*	323 Sqn
J267*	323 Sqn
J270*	323 Sqn
J360	314 Sqn
J361	314 Sqn
J362	314 Sqn
J363	316 Sqn
J364	314 Sqn
J365	314 Sqn
J366	316 Sqn
J367	314 Sqn
J368*	314 Sqn
J369*	316 Sqn
J508	315 Sqn
J509	315 Sqn
J510	315 Sqn
J511	313 Sqn
J512	315 Sqn
J513	315 Sqn
J514	313 Sqn
J515*	315 Sqn
J516*	315 Sqn
J616	311 Sqn
J617	311 Sqn
J619	311 Sqn
J620	311 Sqn
J622	311 Sqn
J623	311 Sqn
J624	311 Sqn
J627	306 Sqn
J628	306 Sqn
J630	306 Sqn
J631	306 Sqn
J632	306 Sqn
J633	306 Sqn
J635	306 Sqn
J636	306 Sqn
J637	306 Sqn
J638	306 Sqn
J640	306 Sqn
J641	306 Sqn
J642	306 Sqn
J643	306 Sqn
J644	306 Sqn
J645	306 Sqn
J646	306 Sqn
J647	306 Sqn
J648	306 Sqn
J649*	306 Sqn
J650*	323 Sqn
J651*	312 Sqn
J652*	311 Sqn
J653*	306 Sqn
J654*	311 Sqn
J655*	306 Sqn
J656*	312 Sqn
J657*	316 Sqn
J864	312 Sqn
J866	312 Sqn
J867	312 Sqn
J868	312 Sqn
J869	312 Sqn
J870	312 Sqn
J871	312 Sqn
J872	312 Sqn
J873	312 Sqn
J874	312 Sqn
J875	312 Sqn
J876	312 Sqn
J877	312 Sqn
J878	312 Sqn
J879	312 Sqn
J881	312 Sqn
J882*	316 Sqn
J884*	314 Sqn
J885*	311 Sqn

MBB Bo.105
Bo.105CB;
[1]Bo.105CD
299 Sqn, Deelen
B37
B38
B39
B40
B42
B43
B44
B45
B47
B48
B63
B64
B66
B67
B68
B69
B70
B71
B72
B74
B75
B76
B77
B78
B79
B80
B83[1]

Pilatus PC-7
EMVO, Woensdrecht
L01
L02
L03
L04
L05
L06
L07
L08

Serial Unit

L09
L10

Sud Alouette III
Grasshoppers*
298 Sqn, Soesterberg;
300 Sqn, Deelen
SAR Flight, Leeuwarden
A177
A208
A209
A217
A218
A226
A227
A235
A246
A247
A253
A260
A261
A266
A267
A275
A281
A292
A293
A301
A302
A307
A319
A324*
A336
A342
A343
A350*
A366
A374
A383
A390*
A391
A398*
A399
A407
A414
A451
A452
A453*
A464
A465*
A470
A471
A482
A483
A488
A489
A494
A495
A499*
A500
A514
A515
A521
A522
A528
A529
A535
A536
A542
A549
A550
H20 SAR
H67 SAR

Serial Unit

H75 SAR
H81 SAR

Marine Luchtvaart Dienst
Fokker F-27-200MPA
336 Sqn, Hato, Antilles
M-1
M-2

Lockheed
P-3C Orion
MARPAT, Valkenburg,
Sigonella and Keflavik
300
301
302
303
304
305
306
307
308
309
310
311
312

Westland Lynx UH14,
SH14B, SH14C, SH14D
De Kooij
260 UH14A 7 Sqn
261 UH14A 7 Sqn
262 UH14A 7 Sqn
264 SH14D 7 Sqn
265 SH14D 7 Sqn
266 SH14B 860 Sqn
267 SH14B 860 Sqn
268 SH14B 860 Sqn
269 SH14B 860 Sqn
270 SH14B 860 Sqn
271 SH14B 860 Sqn
272 SH14B 860 Sqn
273 SH14B 860 Sqn
274 SH14B 860 Sqn
276 SH14C 860 Sqn
277 SH14C 860 Sqn
278 SH14C 860 Sqn
279 SH14C 860 Sqn
280 SH14C 860 Sqn
281 SH14C 860 Sqn
282 SH14C 860 Sqn
283 SH14C 860 Sqn

NEW ZEALAND
Royal New Zealand Air Force

Boeing 727-22C
40 Sqn, Whenuapai
NZ7271
NZ7272

Lockheed
C-130H Hercules
40 Sqn, Whenuapai
NZ7001
NZ7002
NZ7003
NZ7004
NZ7005

Lockheed
P-3K Orion
5 Sqn, Whenuapai
NZ4201

Serial Unit

NZ4202
NZ4203
NZ4204
NZ4205
NZ4206

NIGERIA
Federal Nigerian Air Force

Lockheed
C-130H Hercules
Lagos
NAF-910
NAF-911
NAF-912
NAF-913
NAF-914
NAF-915
NAF-917
NAF-918
NAF-918 (NAF 916)

NORWAY
Kongelige Norske Luftforsvaret

Dassault
Falcon 20 ECM
335 Skv, Gardermoen
041
053
0125

General Dynamics
F-16A/*F-16B
331 Skv and 334 Skv, Bodø;
332 Skv, Rygge;
338 Skv, Ørland
272 332 Skv
273 332 Skv
274 332 Skv
275 332 Skv
276 332 Skv
277 332 Skv
278 332 Skv
279 332 Skv
281 332 Skv
282 332 Skv
284 332 Skv
285 332 Skv
288 338 Skv
289 338 Skv
291 338 Skv
292 338 Skv
293 338 Skv
295 338 Skv
297 338 Skv
298 338 Skv
299 338 Skv
302* 332 Skv
304* 332 Skv
305* 332 Skv
306* 332 Skv
307* 332 Skv
658 334 Skv
659 334 Skv
660 334 Skv
661 334 Skv
662 334 Skv
663 334 Skv
664 334 Skv
665 334 Skv
666 334 Skv
667 334 Skv

Serial	Unit
668	334 Skv
669	334 Skv
670	334 Skv
671	334 Skv
672	331 Skv
673	331 Skv
674	331 Skv
675	331 Skv
677	331 Skv
678	331 Skv
680	331 Skv
681	331 Skv
682	331 Skv
683	331 Skv
686	331 Skv
687	331 Skv
688	331 Skv
689*	338 Skv
690*	332 Skv
691*	334 Skv
692*	334 Skv
693*	332 Skv
711*	331 Skv
712*	331 Skv

Lockheed C-130H Hercules
335 Skv, Gardermoen
952
953
954
955
956
957

Lockheed P-3C Orion
333 Skv, Andøya
3296
3297
3298
3299

Lockheed P-3N Orion
333 Skv Andøya
4576
603

Northrop F-5A
336 Skv, Rygge
128
130
131
132
133
134
208
210
215
220
222
225
895
896
898
902

Northrop F-5B
336 Skv, Rygge
136
241
242
243
244
387
594
595
906
907
908
909

Westland Sea King Mk43/Mk43A[A]/Mk43B[B]
330 Skv, Bodø
060
062
066
069[A]
070
071[B]
072
073
074
189[A]
322[B]

Kystvakt (Coast Guard)
Westland Lynx Mk86
337 Skv, Bardufoss
207
216
228
232
237
350

OMAN

Royal Air Force of Oman

BAC 1-11/485GD
4 Sqn, Seeb
551
552
553

Grumman G1159 Gulfstream 2
4 Sqn, Seeb
601

Lockheed C-130H Hercules
4 Sqn, Seeb
501
502
503

Short Skyvan 3M
2 Sqn, Seeb
901
902
903
904
905
906
907
908
910
911
912
913
914
915
916

PAKISTAN

Air Force

Boeing 707-340C
68-19866 12 Sqn

PORTUGAL

Forca Aerea Portuguesa

Cessna T-37C
102 Esq, Sintra
2401
2402
2403
2404
2406
2407
2410
2411
2412
2414
2417
2418
2419
2420
2421
2422
2423
2424
2425
2426
2427
2428
2429
2430

Dassault Falcon 20C
504 Esq, Lisbon/Montijo
8101
8102
8103

Dassault Falcon 50
504 Esq, Lisbon/Montijo
7401
7402
7403

Lockheed C-130H/C-130H-30* Hercules
501 Esq, Lisbon/Montijo
6801
6802
6803
6804
6805
6806*

SAUDI ARABIA

Al Quwwat al-Jawwiya as Sa' udiya

Boeing E-3A/KE3A* Sentry
18 Sqn, Riyadh
1801
1802
1803
1804
1805
1811*
1812*

Serial	Unit
1813*	
1814*	
1815*	
1816*	
1817*	
1818*	

Lockheed C-130 Hercules
1 Sqn, Riyadh
4 Sqn, Jeddah
16 Sqn, Jeddah

Serial	Unit
112	VC-130H 1 Sqn
451	C-130E 4 Sqn
452	C-130E 4 Sqn
455	C-130E 4 Sqn
456	KC-130H 4 Sqn
457	KC-130H 4 Sqn
458	KC-130H 4 Sqn
459	KC-130H 4 Sqn
461	C-130H 4 Sqn
462	C-130H 4 Sqn
463	C-130H 4 Sqn
464	C-130H 4 Sqn
465	C-130H 4 Sqn
466	C-130H 4 Sqn
467	C-130H 4 Sqn
468	C-130H 4 Sqn
470	C-130H 4 Sqn
471	C-130H-30 4 Sqn
472	C-130H 4 Sqn
473	C-130H 4 Sqn
474	C-130H 4 Sqn
475	C-130H 4 Sqn
1601	C-130H 16 Sqn
1602	C-130H 16 Sqn
1603	C-130H 16 Sqn
1604	C-130H 16 Sqn
1605	C-130H 16 Sqn
1606	C-130E 16 Sqn
1607	C-130E 16 Sqn
1608	C-130E 16 Sqn
1609	C-130E 16 Sqn
1610	C-130E 16 Sqn
1611	C-130E 16 Sqn
1612	C-130H 16 Sqn
1613	C-130H 16 Sqn
1614	C-130H 16 Sqn
1615	C-130H 16 Sqn
1616	KC-130H 16 Sqn
1617	KC-130H 16 Sqn
1618	C-130H 16 Sqn
1619	C-130H 16 Sqn
1620	KC-130H 16 Sqn
1621	KC-130H 16 Sqn
1622	C-130H-30 16 Sqn
1623	C-130H-30 16 Sqn
1624	C-130H 16 Sqn
1625	C-130H 16 Sqn
1626	C-130H 16 Sqn
1627	C-130H 16 Sqn
1628	C-130H 16 Sqn

SINGAPORE
Republic of Singapore Air Force
Lockheed Hercules
122 Sqn, Changi

C-130B
720
721
724
725

C-130H
730
731
732
733

KC-130H
734
745

SPAIN
Ejercito del Aire
Airtech CN235
Ala 35, Getafe

Serial	Unit
T-19A-01	35-60
T-19C-02	35-61
T.19B-03	35-21
T.19B-04	35-22
T.19B-05	35-23
T.19B-06	35-24
T.19B-07	35-25
T.19B-08	35-26
T.19B-09	35-27
T.19B-10	35-28
T.19B-11	35-29
T.19B-12	35-30
T.19B-13	35-31
T.19B-14	35-32

Boeing 707-381B/368C*
Ala 45, Torrejon

Serial	Unit
T.17-1	45-10
TK.17-2	45-11
T.17-3*	45-12

CASA 101 Aviojet
Grupo de Escuelas de Matacan (74);
Team Aguila, San Javier*
AGA, San Javier (79)

Serial	Unit
E.25-01*	1
E.25-02	793-02
E.25-03	79-03
E.25-05	79-05
E.25-06*	6
E.25-07*	7
E.25-08*	8
E.25-09	79-09
E.25-10	79-10
E.25-11	79-11
E.25-12	79-12
E.25-13*	3
E.25-14*	4
E.25-15	79-15
E.25-16	79-16
E.25-17	79-17
E.25-18	79-18
E.25-19	
E.25-20	79-20
E.25-21*	9
E.25-22*	
E.25-23*	10
E.25-24	79-24
E.25-25*	5
E.25-26*	2
E.25-27*	11
E.25-28*	12
E.25-29	79-29
E.25-30	79-30
E.25-31	79-31
E.25-32	74-01
E.25-33	74-02
E.25-34	79-34
E.25-35	74-03
E.25-36	79-36
E.25-37	74-04
E.25-38	79-38
E.25-39	74-05
E.25-40	74-06
E.25-41	79-41
E.25-42	793-32
E.25-43	79-43
E.25-44	79-44
E.25-45	79-45
E.25-46	79-46
E.25-47	79-47
E.25-48	79-48
E.25-49	79-49
E.25-50	79-40
E.25-51	74-07
E.25-52	74-08
E.25-53	411-07
E.25-54	74-10
E.25-55	
E.25-56	79-37
E.25-57	74-12
E.25-58	
E.25-59	74-13
E.25-60	74-14
E.25-61	74-15
E.25-62	74-16
E.25-63	74-17
E.25-64	74-18
E.25-65	74-19
E.25-66	79-96
E.25-67	74-21
E.25-68	74-22
E.25-69	74-23
E.25-71	74-25
E.25-72	74-26
E.25-73	74-27
E.25-74	74-28
E.25-75	74-29
E.25-76	74-30
E.25-78	79-02
E.25-79	74-32
E.25-80	74-33
E.25-81	74-34
E.25-83	74-35
E.25-84	79-04
E.25-86	74-37
E.25-87	74-38
E.25-88	74-39

CASA 212 Aviocar
212 (XT.12),
212A (T.12B),
212B (TR.12A),
212D (TE.12B),
212E (T.12C).
Ala35, Getafe;
403 Esc, Cuatro Vientos;
408 Esc, Getafe;
Ala 37 Villanubla;
Ala 46 Gando, Las Palmas;
Grupo 72, Alcantarilla;
Grupo Esc Matacan (74);
AGA (Ala79), San Javier.

Serial	Unit
XT.12A-10	54-10
TR.12A-30	403-01
TR.12A-40	403-02
TR.12A-50	403-03
TR.12A-60	403-04
TR.12A-70	403-05

Serial	Unit
TR.12A-80	79-90
TE.12B-90	79-91
TE.12B-10	79-92
T.12B-12	35-01
T.12B-13	74-70
T.12B-14	46-30
T.12B-15	
T.12B-16	74-71
T.12B-17	
T.12B-18	46-31
T.12B-19	46-32
T.12B-20	37-04
T.12B-21	35-05
T.12B-22	35-06
T.12B-23	72-01
T.12B-24	37-07
T.12B-25	74-72
T.12B-26	72-02
T.12B-27	46-33
T.12B-28	72-03
T.12B-29	35-08
T.12B-30	74-73
T.12B-31	46-34
T.12B-33	72-04
T.12B-34	74-74
T.12B-35	46-35
T.12B-36	37-09
T.12B-37	72-05
T.12B-38	35-10
T.12B-39	745-39
TE.12B-40	79-93
TE.12B-41	79-9
TE.12B-42	744-42
T.12C-43	46-50
T.12C-44	35-50
T.12B-46	74-46
T.12B-47	72-06
T.12B-48	35-11
T.12B-49	46-36
T.12B-50	745-50
T.12B-51	744-51
T.12B-52	72-07
T.12B-53	37-12
T.12B-54	37-13
T.12B-55	46-37
T.12B-56	74-79
T.12B-57	72-08
T.12B-58	46-38
T.12C-59	35-51
T.12C-60	35-52
T.12C-61	35-53
T.12B-63	37-14
T.12B-64	46-39
T.12B-65	74-80
T.12B-66	72-09
T.12B-67	74-81
T.12B-68	35-15
T.12B-69	35-16
T.12B-70	37-17
T.12B-71	35-18
TR.12D-72	408-01
TR.12D-73	408-02
TR.12D-74	408-03

Dassault Falcon 20
Grupo 45, Torrejon

Serial	Unit
T.11-1	45-02
TM.11-2	45-03
TM.11-3	45-04
TM.11-4	45-01
T.11-5	45-05

Dassault Falcon 50
Grupo 45, Torrejon

Serial	Unit
T.16-1	45-20

Dassault Falcon 900
Grupo 45, Torrejon

Serial	Unit
T.18-1	45-40
T.18-2	45-41

Fokker F.27M Friendship 400MPA
802 Esc, Gando

D.2-01
D.2-02
D.2-03

Lockheed Hercules C-130H C-130H-30[1]
311 Esc/312 Esc (Ala31), Zaragoza

Serial	Unit
TL10-1[1]	31-01
T10-2	31-02
T10-3	31-03
T10-4	31-04
T10-8	31-05
T10-9	31-06
T10-10[1]	31-07

KC-130H
312 Esc (Ala31), Zaragosa

Serial	Unit
TK.10-5	31-50
TK.10-6	31-51
TK.10-7	31-52
TK.10-11	31-53
TK.10-12	31-54

McDonnell Douglas EF-18A/ EF-18B* Hornet
Ala12, Torrejon
Ala15, Zaragoza

Serial	Unit
CE.15-1*	15-70
CE.15-2*	15-71
CE.15-3*	15-72
CE.15-4*	15-73
CE.15-5*	15-74
CE.15-6*	15-75
CE.15-7*	12-70
CE.15-8*	12-71
CE.15-9*	12-72
CE.15-10*	12-73
CE.15-11*	12-74
CE.15-12*	12-75
C.15-13	12-01
C.15-14	15-01
C.15-15	15-02
C.15-16	15-03
C.15-17	15-04
C.15-18	15-05
C.15-20	15-07
C.15-21	15-08
C.15-22	15-09
C.15-23	15-10
C.15-24	15-11
C.15-25	15-12
C.15-26	15-13
C.15-27	15-14
C.15-28	15-15
C.15-29	15-16
C.15-30	15-17
C.15-31	15-18
C.15-32	15-19
C.15-33	15-20
C.15-34	15-21
C.15-35	15-22
C.15-36	15-23
C.15-37	15-24
C.15-38	15-25
C.15-39	15-26
C.15-40	15-27
C.15-41	15-28
C.15-42	15-29
C.15-43	15-30
C.15-44	12-02
C.15-45	12-03
C.15-46	12-04
C.15-47	12-05
C.14-48	12-06
C.15-49	12-07
C.15-50	12-08
C.15-51	12-09
C.15-52	12-10
C.15-53	12-11
C.15-54	12-12
C.15-55	12-13
C.15-56	12-14
C.15-57	12-15
C.15-58	12-16
C.15-59	12-17
C.15-60	12-18
C.15-61	12-19
C.15-62	12-20
C.15-63	12-21
C.15-64	12-22
C.15-65	12-23
C.15-66	12-24
C.15-67	12-25
C.15-68	12-26
C.15-69	12-27
C.15-70	12-28
C.15-72	12-30

Arma Aérea de l'Armada Espanola

Cessna 550 Citation 2
Esc 004, Rota

Serial	Unit
U.20-1	01-405
U.20-2	01-406
U.20-3	01-407

SUDAN

Silakh Al Jawwiya as Sudaniya

Lockheed C-130H Hercules

1100
1101
1102
1103
1104
1105

SWEDEN

Kungliga Svenska Flygvapnet

Beechcraft Super King Air (Tp.101)
F17, Ronneby
F21, Lulea

Serial	Unit
101002	012
F21	
101003	013
F17	
101004	014
F17	

Overseas Serials

Serial	Unit

Grumman G.1159C Gulfstream 4 (Tp.102)

Serial	Unit
102001	F16, Uppsala

Lockheed C-130E Hercules (Tp.84)
F7, Satenäs

Serial	Unit
84001	841
84002	842

Lockheed C-130H Hercules (Tp.84)
F7, Satenäs

Serial	Unit
84003	843
84004	844
84005	845
84006	846
84007	847
84008	848

SAAB SF.340B (Tp.100)

Serial	Unit
100001	001

Swearingen Metro III (Tp.88)

Serial	Unit
88002	882
88003	883

Vertol 107-II-4 (Hkp.4B)
F15, Soderhamm
F17 Ronneby

Serial	Unit
04451	91 F17
04452	92 F17
04453	93 F17
04454	94
04455	95 F15
04456	96
04457	97 F17
04458	98 F17
04459	99 F17
04460	90 F17

Marine Flygtjanst
Vertol 107-II-5 (Hkp.4C)
1Hkp Div, Berga

Serial	Unit
04061	61
04063	63
04064	64

Kawasaki-Vertol KV.107-II (Hkp.4C)
1 Hkp Div, Berga
2 Hkp Div, Säve
3 Hkp Div, Ronneby

Serial	Unit
04065	65 Hkp Div
04067	67 Hkp Div
04068	68 Hkp Div
04669	69 Hkp Div
04070	70 Hkp Div
04071	71 Hkp Div
04072	72 Hkp Div
04073	73 Hkp Div
04074	74 Hkp Div
04075	75 Hkp Div

Armen
MBB Bo.105CB (Hkp.9B)
Armeflyget 1 (AF1), Boden
Armeflyget 2 (AF2), Malmslatt

Serial	Unit	
09201	01	AF2
09202	02	AF1
09203	03	AF2
09204	04	AF2
09205	05	AF1
09206	06	AF1
09207	07	AF1
09208	08	AF1
09209		
09210	10	AF1
09211	11	AF1
09212	12	AF1
09213	13	AF2
09214	14	AF1
09215	15	AF2
09216	16	AF2
09217	17	AF2
09218		
09219	19	AF1
09220	20	AF2
09415	93	

TURKEY
Turk Hava Kuvvetleri
Transall C.160D
221 Filo, Erkilet

Serial	Unit
019	
020	12-020
021	12-021
022	12-022
023	12-023
024	12-024
025	12-025
026	12-026
027	
028	12-028
029	
030	12-030
031	12-031
032	12-032
033	12-033
034	12-034
035	12-035
036	12-036
037	12-037
038	12-038
039	12-039
040	12-040

Lockheed C-130B Hercules
222 Filo, Erkilet

Serial	Unit
10960	
10963	
23496	
70527	
80736	
91527	

Lockheed C-130E Hercules
222 Filo, Erkilet

Serial	Unit
00991	12-991
01468	12-468
01947	12-947
13186	12-186
13187	12-187
13188	12-188
13189	12-189
17949	12-949

UNITED ARAB EMIRATES
United Arab Emirates Air Force
Abu Dhabi
Lockheed C-130H Hercules

Serial	Unit
1211	
1212	
1213	
1214	

Dubai
Lockheed L.100-30 Hercules

Serial	Unit
311	
312	

40+40 German AF Alpha Jet operated by JbG 49. *PRM*

US Military Aircraft Markings

All USAF aircraft have been allocated a fiscal year (FY) number since 1921. Individual aircraft are given a serial according to the fiscal year in which they are ordered. The numbers commence at 0001 and are prefixed with the year of allocation. For example F-111E 68-0001 was the first aircraft ordered in 1968. The fiscal year (FY) serial is carried on the technical bloc which is usually stencilled on the left-hand side of the aircraft just below the cockpit. The number displayed on the fin is a corruption of the FY serial. Most tactical aircraft carry the fiscal year in small figures followed by the last three digits of the serial in large figures. For example F-111E 67-119 carries 67119 on its tail. Large transport and tanker aircraft such as C-130s and KC-135s usually display a five-figure number commencing with the last digit of the appropriate fiscal year and four figures of the production number. An example of this is HC-130P 65-0973 which displays 50973 on its fin. Aircraft of more than 10 years vintage which might have duplicated a five-figure number of a more modern type in service, are prefixed 0.

USN serials follow a straightforward numerical sequence which commenced, for the present series, with the allocation of 00001 to an SB2C Helldiver by the Bureau of Aeronautics in 1940. Numbers in the 165000 series are presently being issued. They are usually carried in full on the rear fuselage of the aircraft.

UK based USAF Aircraft

The following aircraft are normally based in the UK. They are listed in numerical order of type with individual aircraft in serial number order, as depicted on the aircraft. The number in brackets is either the alternative presentation of the five-figure number commencing with the last digit of the fiscal year, or the fiscal year where a five-figure serial is presented on the aircraft. Where it is possible to identify the allocation of aircraft to individual squadrons by means of colours carried on fin or cockpit edge, this is also provided.

Serial

Lockheed U-2R
95RS/9RW, RAF Alconbury
FY80
01079
01080

McDonnell Douglas F-15E Strike Eagle
LN: 48FW, RAF Lakenheath
492FS *blue*
494FS *red*

Serial		
90-248	(00248)	m [48FW]
90-251	(00251)	*bl*
90-255	(00255)	*bl*
90-256	(00256)	*bl*
90-257	(00257)	*bl*
90-258	(00258)	*bl*
90-259	(00259)	*bl*
90-260	(00260)	
90-261	(00261)	
90-262	(00262)	*bl*
91-300	(10300)	
91-301	(10301)	*bl*
91-302	(10302)	*bl*
91-303	(10303)	
91-304	(10304)	
91-305	(10305)	
31-307	(10307)	
91-308	(10308)	
91-309	(10309)	
91-310	(10310)	
91-311	(10311)	
91-312	(10312)	
91-313	(10313)	
91-315	(10315)	

Sikorsky MH-53J
39SOW/21 SOS
RAF Alconbury

Serial	
14431	(FY66)
01626	
01629	
95784	(FY69)
95789	(FY69)
95797	(FY69)

General Dynamics F-111E
UH: 20FW, RAF Upper Heyford
55FS *blue*, 77FS *red*,
79FS *yellow*

Serial		
67-119	(70119)	*r*
67-120	(70120)	*m*
67-121	(70121)	*bl*
67-122	(70122)	*r*
67-123	(70123)	*bl*
68-004	(80004)	*bl*
68-005	(80005)	*bl*
68-006	(80006)	*bl*
68-007	(80007)	*bl*
68-015	(80015)	*r*
68-016	(80016)	*r*
68-020	(80020)	*bl*
68-021	(80021)	*r*
68-022	(80022)	*r*
68-023	(80023)	*bl*
68-025	(80025)	*bl*
68-026	(80026)	*bl*
68-027	(80027)	*r*
68-028	(80028)	*bl*
68-029	(80029)	*bl*
68-030	(80030)	*bl*
68-031	(80031)	*r*
68-032	(80032)	*r*
68-033	(80033)	*r*
68-034	(80034)	*y*
68-035	(80035)	*y*
68-037	(80037)	*bl*
68-038	(80038)	*y*

USAF (UK based)

Serial		
68-040	(80040)	*m*
68-041	(80041)	*y*
68-043	(80043)	*bl*
68-044	(80044)	*r*
68-046	(80046)	*bl*
68-047	(80047)	*y*
68-048	(80048)	*y*
68-049	(80049)	*y*
68-050	(80050)	*r*
68-051	(80051)	*y*
68-054	(80054)	*y*
680-55	(80055)	*bl*
68-056	(80056)	*bl*
68-059	(80059)	*bl*
68-061	(80061)	*r*
68-062	(80062)	*bl*
68-063	(80063)	*r*
68-064	(80064)	*bl*
68-065	(80065)	*bl*
68-067	(80067)	*y*
68-068	(80068)	*r*
68-069	(80069)	*bl*
68-072	(80072)	*y*
68-073	(80073)	*y*
68-074	(80074)	*r*
68-075	(80075)	*y*
68-076	(80076)	*y*
68-077	(80077)	*r*
68-079	(80079)	*y*
68-080	(80080)	*y*
68-082	(80082)	*y*
68-083	(80083)	*r*
68-084	(80084)	*y*

Lockheed C-130E/H Hercules
313 AG, RAF Mildenhall
Continuous presence of two-month deployments from 314AW, 317AW or 463AW — see USAF (US) Section serials

Lockheed C-130 Hercules
7 SOS*
67 SOS/39 SOW
RAF Alconbury

Serial	
37814 (FY63)	C-130E*
50012 (FY85)	MC-130H
50973 (FY65)	HC-130P
60220 (FY66)	HC-130P
60223 (FY66)	HC-130P
61699 (FY86)	MC-130H*
95820 (FY69)	HC-130N
95823 (FY69)	HC-130N
95826 (FY69)	HC-130N
95827 (FY69)	HC-130N
95831 (FY69)	HC-130N

Boeing KC-135R Stratotanker
D: 100ARW, 351 ARS
RAF Mildenhall

14833 (FY64)
23558 (FY62)
38003 (FY63)
38020 (FY63)
38875 (FY63)
71437 (FY57)
80100 (FY58)
80128 (FY58)
91459 (FY59)

UK based US Navy Aircraft

Serial

Beech UC-12M Super King Air
8G: Naval Air Facility, Mildenhall

3837 (163837) 8G
3840 (163840) 8G
3843 (163843) 8G

European based USAF Aircraft

These aircraft are normally based in Western Europe with the USAFE. They are shown in numerical order of type designation, with individual aircraft in serial number order as carried on the aircraft. An alternative five-figure presentation of the serial is shown in brackets where appropriate. Fiscal year (FY) details are also provided if necessary. The unit allocation and operating bases are given for most aircraft.

Serial

Bell UH-1N
58 AS, 86 Wg Ramstein
FY69
96606
96607
96608
96609
96619

McDonnell Douglas F-4G Phantom
SP: 52FW Spangdahlem
81 FS yellow/black (y)

69-210	(97210)	y
69-211	(97211)	y
69-212	(97212) [52FW]	m
69-228	(97228)	y
69-232	(97232)	y
69-241	(90241)	[81FS]y
69-247	(90247)	y
69-248	(90248)	y
69-253	(90253)	y
69-258	(90258)	y
69-259	(90259)	y
69-260	(90260)	y
69-263	(90263)	y
69-267	(97267)	
69-270	(97270)	y
69-278	(90278)	y
69-285	(90285)	y
69-286	(90286)	y
69-286	(97286)	y
69-291	(97291)	y
69-295	(97295)	y
69-556	(97556)	y
69-558	(97558)	y
69-579	(97579)	y
69-587	(97587)	y

McDonnell Douglas C-9A Nightingale
435AW/55AAS Rhein Main
SHAPE, Chievres
FY71
10876
10879
10880
10881
10882

Beech C-12
58 AS, 86 Wg Ramstein
[1]JUSMG Turkey
[2]MAAG, Sude Bay, Crete

31216	(FY73)*	C-12C
31218	(FY73)*	C-12A
40161	(FY84)	C-12F
40162	(FY84)	C-12F
40163	(FY84)	C-12F
40164	(FY84)	C-12F
40165	(FY84)	C-12F
40166	(FY84)	C-12F
60173	(FY76)*	C-12C

Fairchild A-10A Thunderbolt II
SP: 52FW Spangdahlem, 510 FS
81-951 (10951)
81-952 (10952)
81-956 (10956)
81-963 (10963)
81-966 (10966)
81-977 (10977)
81-980 (10980)
81-983 (10983)
81-985 (10985)
81-988 (10988)
81-992 (10992)
82-649 (20649)
82-650 (20650)
82-656 (20656)

McDonnell Douglas F-15A/F-15B, F-15C/F-15D Eagle
CR: 32FG/32FS Soesterberg
orange/green
BT: 36FW Bitburg
22FS *red*
53FS *yellow/black*
IS: 57FS Keflavik
black/white

76-127	(60127)	F-15B	CR *or*
77-087	(70087)	F-15A	CR *or*
77-089	(70089)	F-15A	CR
77-096	(70096)	F-15A	CR
77-097	(70097)	F-15A	CR *or*
77-098	(70098)	F-15A	CR *or*
77-099	(70099)	F-15A	CR
77-100	(70100)	F-15A	[32FS] CR *or*
77-102	(70102)	F-15A	CR
77-103	(70103)	F-15A	CR *or*
77-104	(70104)	F-15A	CR *or*
77-105	(70105)	F-15A	CR
77-106	(70106)	F-15A	CR *or*
77-109	(70109)	F-15A	CR
77-110	(70110)	F-15A	CR *or*
77-113	(70113)	F-15A	CR
77-115	(70115)	F-15A	CR
77-124	(70124)	F-15A	CR *or*
77-132	(70132)	F-15A	[32FG] CR *or*
77-163	(70163)	F-15B	CR *or*
79-011	(90011)	F-15D	BT *r*
79-012	(90012)	F-15D	BT *r*
79-022	(90022)	F-15C	BT *r*
79-025	(90025)	F-15C	BT *y*
79-035	(90035)	F-15C	BT
79-036	(90036)	F-15C	[36FW] *m*
79-037	(90037)	F-15C	BT *r*
79-046	(90046)	F-15C	BT *r*
79-057	(90057)	F-15C	BT *r*
79-058	(90058)	F-15C	BT *r*
79-064	(90064)	F-15C	BT *y*
79-068	(90068)	F-15C	BT *r*
79-072	(90072)	F-15C	BT *r*
79-073	(90073)	F-15C	BT *r*
79-076	(90076)	F-15C	BT *r*
79-077	(90077)	F-15C	BT *y*
79-078	(90078)	F-15C	BT
80-003	(00003)	F-15C	BT *r*
80-004	(00004)	F-15C	BT *r*
80-005	(00005)	F-15C	BT *r*
80-010	(00010)	F-15C	BT *r*
80-011	(00011)	F-15C	BT *r*
80-012	(00012)	F-15C	BT
80-015	(00015)	F-15C	BT *r*
80-021	(00021)	F-15C	IS *bk*
80-022	(00022)	F-15C	[22FS] *r*
80-026	(00026)	F-15C	BT *r*
80-028	(00028)	F-15C	BT *r*
80-029	(00029)	F-15C	IS *bk*
80-031	(00031)	F-15C	BT *r*
80-035	(00035)	F-15C	IS *bk*
80-038	(00038)	F-15C	IS *bk*
80-039	(00039)	F-15C	IS *bk*
80-040	(00040)	F-15C	IS *bk*
80-041	(00041)	F-15C	IS *bk*
80-042	(00042)	F-15C	IS *bk*
80-044	(00044)	F-15C	IS *bk*
80-046	(00046)	F-15C	[AFI] *bk*
80-047	(00047)	F-15C	IS *bk*
80-048	(00048)	F-15C	IS *bk*
80-050	(00050)	F-15C	[AFI] *bk*
80-052	(00052)	F-15C	IS [57FS] *bk*
80-056	(00056)	F-15D	IS *bk*
80-057	(00057)	F-15D	IS *bk*
84-001	(40001)	F-15C	[53FS] *y*
84-002	(40002)	F-15C	BT *y*
84-003	(40003)	F-15C	BT *y*
84-004	(40004)	F-15C	BT *y*
84-005	(40005)	F-15C	BT *y*
84-006	(40006)	F-15C	BT *y*
84-007	(40007)	F-15C	BT *y*
84-008	(40008)	F-15C	BT *y*
84-009	(40009)	F-15C	[36FW]
84-010	(40010)	F-15C	BT *y*
84-013	(40013)	F-15C	BT *y*
84-014	(40014)	F-15C	BT *y*
84-015	(40015)	F-15C	BT *r/y*

USAF (EUR based)

Serial			
84-016	(40016)	F-15C	BT *y*
84-017	(40017)	F-15C	BT *r*
84-019	(40019)	F-15C	BT *y*
84-020	(40020)	F-15C	BT *y*
84-021	(40021)	F-15C	BT *y*
84-022	(40022)	F-15C	[22FS]r
84-023	(40023)	F-15C	BT *r/y*
84-024	(40024)	F-15C	BT *y*
84-025	(40025)	F-15C	BT *y*
84-026	(40026)	F-15C	BT *y*
84-027	(40027)	F-15C	BT *y*
84-043	(40043)	F-15D	BT *y*
84-044	(40044)	F-15D	BT *y*

General Dynamics F-16C/F-16D*

RS: 86Wg Ramstein
512FS *green*/black
526FS *red*/black
SP: 52FW Spangdahlem
23FS *blue*/white
480FS *red*/white

85-400	(51400)	RS *gn*
85-408	(51408)	RS *r*
85-422	(51422)	RS *r*
85-426	(51426)	[526FS] *r*
85-428	(51428)	RS *gn*
85-436	(51436)	RS *gn*
85-438	(51438)	RS *gn*
85-450	(51450)	RS *r*
85-453	(51453)	RS *r*
85-455	(51455)	RS *gn*
85-456	(51456)	RS *gn*
85-457	(51457)	RS *gn*
85-458	(51458)	RS *gn*
85-461	(51461)	RS *r*
85-465	(51465)	RS *r*
85-467	(51467)	RS *gn*
85-468	(51468)	RS *gn*
85-471	(51471)	RS *gn*
85-474	(51474)	RS *gn*
85-476	(51476)	RS *r*
85-477	(51477)	RS *r*
85-479	(51479)	RS *gn*
85-480	(51480)	RS *gn*
85-481	(51481)	RS *gn*
85-484	(51484)	RS *r*
85-485	(51485)	RS *gn*
85-511	(51511)*	RS *r*
85-546	(51546)	RS *r*
86-044	(60044)*	SP *r*
86-047	(60047)*	SP *bl*
86-049	(60049)*	RS *gn*
86-209	(60209)	RS *r*
86-255	(60255)	RS *gn*
86-262	(60262)	SP *bl*
86-270	(60270)	RS [86FW] *gn*
86-287	(60287)	SP *bl*
86-302	(60302)	SP *r*
86-303	(60303)	RS *r*
86-313	(60313)	RS *r*
86-315	(60315)	RS *r*
86-326	(60326)	SP *r*
86-327	(60327)	SP *r*
86-341	(60341)	SP *r*
86-343	(60343)	SP *bl*
86-346	(60346)	SP *r*
86-347	(60347)	SP *bl*
86-348	(60348)	SP [480FS]*r*
86-349	(60349)	SP *r*
86-350	(60350)	SP *bl*
86-361	(60361)	SP *bl*
86-362	(60362)	SP *bl*
86-363	(60363)	SP *bl*
86-364	(60364)	SP *bl*
86-365	(60365)	SP *bl*
86-366	(60366)	SP *bl*
86-369	(60369)	SP *r*
87-217	(70217)	SP *bl*
87-218	(70218)	SP *r*
87-219	(70219)	SP *r*
87-220	(70220)	RS *r*
87-221	(70221)	RS *gn*
87-222	(70222)	SP *bl*
87-223	(70223)	SP [23FS]*bl*
87-225	(70225)	RS *r*
87-226	(70226)	RS *r*
87-227	(70227)	RS *r*
87-230	(70230)	RS *r*
87-238	(70238)	RS[512FS]*gn*
87-239	(70239)	RS *gn*
87-240	(70240)	RS *gn*
87-242	(70242)	[86 Wg]
87-243	(70243)	SP *r*
87-244	(70244)	RS *gn*
87-245	(70245)	SP *r*
87-247	(70247)	RS *gn*
87-248	(70248)	RS *gn*
87-249	(70249)	SP *r*
87-258	(70258)	RS *r*
87-259	(70259)	RS *r*
87-260	(70260)	SP *r*
87-268	(70268)	SP *r*
87-270	(70270)	[52FW]
87-271	(70271)	SP *bl*
87-281	(70281)	SP *bl*
87-282	(70282)	SP *r*
87-283	(70283)	SP *r*
87-336	(70336)	SP *bl*
88-398	(80398)	SP *bl*
88-399	(80399)	SP *r*
88-400	(80400)	SP *r*

Grumman C-20A Gulfstream III

58 AS, 86 Wg Ramstein
FY83
30500
30501
30502

Gates C-21A Learjet

58AS, 86 Wg Ramstein
*7005ABS Stuttgart/HQ USEUCOM
FY84
40081*
40082*
40083*
40084
40085
40086

Boeing CT-43A

58 AS, 86 Wg Ramstein
31149 (FY73)

Lockheed C-130E Hercules

435AW/37AS; Rhein Main:

Serial	
01260	(FY70)
01264	(FY70)
01271	(FY70)
01274	(FY70)
10935	(FY68)
10938	(FY68)
10943	(FY68)
10947	(FY68)
17681	(FY64)
18240	(FY64)
37885	(FY63)
37887	(FY63)
40502	(FY64)
40527	(FY64)
40550	(FY64)
96566	(FY69)
96582	(FY69)
96583	(FY69)

Lockheed MC-130E Hercules

39 SOW/7SOS Rhein Main
FY64
40523
40555
40561
40566

Boeing C135B

58 AS, 86 Wg Ramstein
24126

European based US Navy Aircraft

Serial	
Grumman C-2A Greyhound	
(VR-24 Sigonella)	
162169	[JM23]
163172	[JM26]
162174	[JM22]
162176	[JM20]
Lockheed [1]UP-3A/EP-3E/ [2]VP-3A/[3]P-3B	
(VQ-2: Rota)	
148888	[JQ-23]
149668	[21]
150495[1]	NAF Keflavik
150496[2]	CinCLANT
150497	
150505	[24]
150511[2]	HQ USMC
150515[2]	CinCAFSE
153425[3]	
157318	
157320	[26]
157325	
161340	
Beech UC-12M Super King Air	
8C: NAF Sigonella	
8D: NAF Rota	
3838 (163838) 8C	
3839 (163839) 8D	
3841 (163841) 8C	
3842 (163842) 8D	
3844 (163844) 8C	
NA CT-39G Sabreliner	
(VR-24 Sigonella)	
159361	[31]
159362	[32]
159363	[33]
Lockheed C-130F/KC-130F[1]/ TC-130Q[2] Hercules	
(VR-22: Rota)	
148890	[JL890][1]
148892	[JL892][1]
148895	[JL895][1]
149790	[JL790]
149794	[JL794]
149797	[JL797]
150687	[JL687][1]
159348	[JL07][2]

84-002 F-15C Eagle from 53 FS USAFE based at Bitburg. *PRM*

European based US Army Aircraft

Serial

Grumman V1 Mohawk
1MIB, Wiesbaden

Serial	FY	Type
15943	(FY68)	OV-1D
15950	(FY68)	OV-1D
15953	(FY68)	OV-1D
15957	(FY68)	OV-1D
16996	(FY68)	OV-1D
18905	(FY67)	OV-1D
18908	(FY67)	OV-1D
18910	(FY67)	OV-1D
18919	(FY67)	OV-1D
18925	(FY67)	OV-1D
18927	(FY67)	OV-1D

Beech C-12 Super King Air
([1]=C-12C, [2]=C-12D, [3]=RC-12K,)
HQ/USEUCOM Stuttgart;
3-58 Avn, Schwabisch-Hall
5-158 Avn, Wiesbaden
56 Av Co, Coleman Barracks
6th Avn Det, Vicenza, Italy
207AvCo Heidelberg
7th ATC, Grafenwohr
V Corps, Wiesbaden
1MIB, Wiesbaden

Serial	FY	Unit
22254[1]	(FY73)	207 AvCo
22255[1]	(FY73)	5-158 Avn
22260[1]	(FY73)	7th ATC
22262[1]	(FY73)	207 AvCo
22549[1]	(FY76)	HQ/USEUCOM
22550[1]	(FY76)	HQ/USEUCOM
22556[1]	(FY76)	6 Avn Det
22557[1]	(FY76)	207AvCo
22564[1]	(FY76)	56 AvCo
22932[1]	(FY77)	6 Avn Det
22944[1]	(FY77)	56 AvCo
22950[1]	(FY77)	207AvCo
23126[1]	(FY78)	207AvCo
23127[1]	(FY78)	207AvCo
23128[1]	(FY78)	207AvCo
24380[2]	(FY84)	207AvCo
50147[3]	(FY85)	1MIB
50148[3]	(FY85)	1MIB
50150[3]	(FY85)	1MIB
50151[3]	(FY85)	1MIB
50152[3]	(FY85)	1MIB
50153[3]	(FY85)	1MIB
50154[3]	(FY85)	1MIB
50155[3]	(FY85)	1MIB

Beech U-21A/U-21D* King Air

Serial	FY	Unit
18000	(FY66)	56 AvCo
18006	(FY66)	6th Avn Det
18010	(FY66)	56 AvCo
18012	(FY66)	5-158 Avn
18013	(FY66)	Berlin Brigade
18014	(FY66)	56 AvCo
18021	(FY66)	56 AvCo
18027	(FY66)	56 AvCo
18030	(FY66)	56 AvCo
18078	(FY67)	5-158 Avn
18080	(FY67)	56 AvCo
18110*	(FY67)	3-58 Avn
18116	(FY67)	56 AvCo

Boeing-Vertol CH-47D Chinook
'A' Co, 5 Batt, 159 Av A Reg't Schwabisch Hall
7-158 Avn Mainz-Finthen
'D' Co, 502 Avn Reg't, Coleman Barracks
'E' Co, 502 Avn Reg't Aviano

FY86
61671 7-158 Avn
61672 7-158 Avn
61673 7-158 Avn
61674 7-158 Avn
61675 7-158 Avn
61676 7-158 Avn
61677 7-158 Avn
61678 7-158 Avn
FY87
70071 7-158 Avn
70072 7-158 Avn
70073 7-158 Avn
70074 7-158 Avn
70075 502 DCo
70076 502 DCo
70077 7-158 Avn
70078 7-158 Avn
70079 5/159 ACo
70080 5/159 ACo
70081 5/159 ACo
70082 5/159 ACo
70083 5/159 ACo
70084 5/159 ACo
70085 5/159 ACo
70086 5/159 ACo
70087 5/159 ACo
70088 5/159 ACo
70089 5/159 ACo
70090 5/159 ACo
70091 5/159 ACo
70092 5/159 ACo
70093 5/159 ACo
70094 5/159 ACo
70096 502 DCo
70097 502 DCo
70098 502 DCo
70099 502 DCo
70100 502 DCo
70101 502 DCo
70102 7-158 Avn
70103 7-158 Avn
70104 502 DCo
70105 502 DCo
70106 502 DCo
70107 502 DCo
70109 502 DCo
70110 502 DCo
70111 502 DCo
70112 502 DCo
FY88
80098 502 ECo
80099 502 ECo
80100 502 ECo
80101 502 ECo
80102 502 ECo
80103 502 ECo
80104 502 ECo
80106 502 ECo
FY89
90138 502 ECo
90139 502 ECo
90140 502 ECo
90141 502 ECo
90142 502 ECo
90143 502 ECo
90144 502 ECo
90145 502 ECo

Sikorsky UH-60A Black Hawk
4-11th Armored Cavalry Regiment, Fulda
23rd TF Giebelstadt, Ansbach
'H' Co, 3rd Avn Reg't, Wiesbaden
3rd Btn, 58 Avn Reg't, Coleman Barracks
7th Btn, 158 Avn Reg't, Hanau
'C' Co, 6th Btn, 159 Avn Reg't, Schwabisch Hall,
2/6 Cavalry, Illesheim,
3/6, 5/6 Cavalry, Wiesbaden
357th Avn Det/SHAPE, Chievres
207th Aviation Co: Heidelberg
7 Med Com, Darmstadt
15th Med Det (HA), Grafenwohr
159th Med Co, Wiesbaden
236th Med Co (HA), Landstuhl
7-1 Avn, Ansbach
TF Skyhawk, Ansbach
TF Viper, Hanau

Serial	Unit
FY77	
22723	159 Med Co
22727	236 Med Co
FY78	
22969	7 Med Com
22986	236 Med Co
22990	7-1 Avn
22995	7 Med Com
22996	236 Med Co
22997	159 Med Co
23000	236 Med Co
23001	236 Med Co
23003	159 Med Det
FY79	
23271	236 Med Co
FY80	
23431	23rd TF
23432	6/159 CCo
23433	6/159 CCo
23434	236 Med Co
23436	6/159 CCo
23439	7-1 Avn
23441	6/159 CCo
23443	TF Skyhawk
23444	4-11 ACR
23490	6/159 CCo
23496	7-1 Avn
FY81	
23551	236 Med Co
23568	8-158 Avn
23571	23rd TF
23572	5-6 Cav

Serial	
23573	4-11 ACR
23575	6/158 BCo
23581	TF Viper
23582	3-HCo
23583	TF Viper
23584	TF Viper
23586	TF Viper
23587	3-HCo
23590	4-11 ACR
23592	7 Med Co
23594	7-1 Avn
23596	236 Med Co
23597	7 Med Com
23598	4-11 ACR
23602	4-11 ACR
23603	227 HCo
23604	4-11 ACR
23605	4-11 ACR
23606	4-11 ACR
23607	4-11 ACR
23608	4-11 ACR
23609	4-11 ACR
23613	7-1 Avn
23614	TF Viper
23615	TF Viper
23616	7-1 Avn
23617	TF Viper
23618	7-1 Avn
23622	23rd TF
23624	7-158 Avn
23625	4-11 ACR
23626	5-6 Cav
FY82	
23660	4-11 HCO
23661	7-1 Avn
23662	TF Skyhawk
23663	7-1 Avn
23665	4-11 ACR
23666	4-11 ACR
23667	4-11 ACR
23668	4-11 ACR
23669	TF Skyhawk
23672	6/159 CCo
23673	4-2 ACR
23675	236 Med Co
23676	236 Med Co
23683	3-HCo
23684	4-11 ACR
23692	4-11 ACR
23693	3-HCo
23694	7-158 Avn
23695	6/159 CCo
23697	4-11 ACR
23698	6/159 CCo
23699	4-2 ACR
23700	236 Med Co
23701	4-11 ACR
23702	7-1 Avn
23704	4-11 ACR
23721	6/159 CCo
23723	
23726	236 Med Co
23727	236 Med Co
23728	236 Med Co
23729	7-1 Avn
23730	236 Med Co
23731	45 Med Co
23733	7-1 Avn
23735	236 Med Co
23736	236 Med Co
23737	7 Med Co
23738	7 Med Co
23739	236 Med Co

Serial	
23741	63 Med Det
23743	7 Med Co
23744	7-1 Avn
23745	7-1 Avn
23746	159 Med Co
23748	7-1 Avn
23749	45 Med Co
23750	159 Med Co
23751	236 Med Co
23753	421 Med Btn
23754	7-1 Avn
23755	7-1 Avn
23756	236 Med Co
23761	6/159 CCo
FY83	
23855	4-11 ACR
23869	6/159 CCo
FY86	
24498	357 Avn Det
24530	2/6 Cav
24531	2/6 Cav
24532	2/6 Cav
24538	207 AvCo
24550	236 Med Co
24551	7 Med Co
24552	4-11 ACR
24553	6/158 BCo
24554	TF Viper
24555	3-H Co
FY87	
24579	3-H Co
24581	236 Med Co
24583	357 Av Det
24584	357 Av Det
24589	207 Av Co
24621	3-1 Avn
24634	3-HCo
24642	7-1 Avn
24646	7-1 Avn
24647	7-1 Avn
26001	6/159 CCo
26002	
26003	7-1 Avn
26004	7-1 Avn
FY88	
26019	5/6 Cav
26020	4-229 Avn
26021	3-HCo
26023	6/6 Cav
26024	3-HCo
26025	3-HCo
26026	3-HCo
26027	3-HCo
26028	7-1 Avn
26031	4-11 ACR
26034	4-229 Avn
26037	4-229 Avn
26038	4-11 ACR
26039	4-11 ACR
26040	4-11 ACR
26041	4-11 ACR
26042	4-11 ACR
26050	TF Skyhawk
26051	3-HCo
26052	3-HCo
26053	TF Skyhawk
26054	236 Med Co
26056	3-HCo
26059	7-158 Avn
26067	7-227 Avn
26072	
FY89	
26080	3-1 Avn

Serial	
26083	3-HCo
26086	3-227 Avn
26145	3-227 Avn
26146	3-227 Avn
26151	2-1 Avn
26153	3-1 Avn
26164	2-1 Avn
26165	2-1 Avn

McD AH-64A Apache

1-1 Avn, Ansbach
2-1 Avn, Ansbach
3-1 Avn, Ansbach
3rd Avn, Giebelstadt
2/6 Cav, 6/6 Cav, Illesheim
4-11ACR, Fulda
2-227 Avn, 3-227 Avn, Hanau
4-229 Avn, Illesheim

Serial	
FY84	
24218	2/6 Cav
24220	
24246	
24247	
24248	
24257	4-229 Avn
24258	2/6 Cav
24260	2/6 Cav
24262	2/6 Cav
24263	2/6 Cav
24266	2/6 Cav
24288	
24290	2/6 Cav
24291	2/6 Cav
24292	2/6 Cav
24293	2/6 Cav
24294	5/6 Cav
24295	2/6 Cav
24296	2/6 Cav
24297	2/6 Cav
24298	2/6 Cav
24299	2/6 Cav
24302	2/6 Cav
24303	2/6 Cav
24304	2/6 Cav
24307	4-11 ACR
24308	2/6 Cav
FY85	
25351	
25352	1-1 Avn
25397	5/6 Cav
25398	2/6 Cav
25422	3 Avn
25430	3 Avn
25444	3 Avn
25447	
25452	
25460	1-1 Avn
25462	3 Avn
25464	3 Avn
25465	
25469	5/6 Cav
25470	2-227 Avn
25471	5/6 Cav
25472	5/6 Cav
25473	1-1 Avn
25474	2-227 Avn
25475	5/6 Cav
25476	5/6 Cav
25478	1-1 Avn
25479	3-1 Avn
25480	2-227 Avn
25482	2-227 Avn
25483	5/6 Cav

US Army (EUR based)

Serial	
25484	5/6 Cav
25485	5/6 Cav
25486	5/6 Cav
25487	5/6 Cav
25488	4-11 ACR
FY86	
8940	4-229 Avn
8941	4-229 Avn
8942	4-229 Avn
8943	4-229 Avn
8946	4-229 Avn
8947	4-229 Avn
8948	4-229 Avn
8949	4-229 Avn
8950	4-229 Avn
8951	4-229 Avn
8952	4-229 Avn
8953	4-229 Avn
8955	4-229 Avn
8956	3-1 Avn
8957	4-229 Avn
8959	4-229 Avn
8960	4-229 Avn
8961	4-229 Avn
8970	3-227 Avn
8981	5/6 Cav
8983	5/6 Cav
9010	1-1 Avn
9011	2-1 Avn
9019	
9026	5/6 Cav
9029	2-227 Avn
9030	3-1 Avn
9032	2-1 Avn
9033	2-227 Avn
9037	5/6 Cav
9038	
9039	1-1 Avn
9041	2-227 Avn
9044	1-1 Avn
9048	2-1 Avn
9058	2-1 Avn
FY87	
408	2-1 Avn
409	3-1 Avn
412	2-1 Avn
413	3-1 Avn
417	2-1 Avn
418	5/6 Cav
419	2-1 Avn
420	3-1 Avn
423	3-1 Avn
428	3-1 Avn
432	2-1 Avn
433	3-227 Avn
434	3-227 Avn
436	3-1 Avn
437	2-1 Avn
438	
439	3-1 Avn
440	3-227 Avn
441	3-227 Avn
442	3-1 Avn
443	3-227 Avn
444	2-227 Avn
445	3-227 Avn
446	3-227 Avn
447	3-227 Avn
449	3-227 Avn
451	2-227 Avn
455	2-227 Avn
457	3-227 Avn
459	2-1 Avn
470	2-1 Avn
471	2-1 Avn
472	
473	2-1 Avn
474	2-1 Avn
475	5/6 Cav
476	2-1 Avn
477	2-1 Avn
478	2-1 Avn
479	2-1 Avn
481	2-1 Avn
482	2-1 Avn
487	2-227 Avn
496	5/6 Cav
498	5/6 Cav
503	2-227 Avn
504	2-227 Avn
505	5/6 Cav
506	2-227 Avn
507	2-227 Avn
510	2-227 Avn
FY88	
197	5/6 Cav
198	2-227 Avn
199	2-227 Avn
204	6/6 Cav
212	6/6 Cav
213	6/6 Cav
214	6/6 Cav
215	6/6 Cav
216	6/6 Cav
217	6/6 Cav
218	6/6 Cav
222	6/6 Cav
225	6/6 Cav
228	6/6 Cav
232	6/6 Cav
234	6/6 Cav
236	6/6 Cav
240	6/6 Cav
246	6/6 Cav
250	6/6 Cav

Serial

US based USAF Aircraft

The following aircraft are normally based in the USA but are likely to be seen visiting the UK from time to time. The types are in numerical order, commencing with the B-**1B** and concluding with the C-**141**. The aircraft are listed in numerical progression by the serial actually carried externally. Fiscal year information is provided, together with details of mark variations and in some cases operating units.

Rockwell B-1B Lancer
28 BW Ellsworth AFB, South Dakota [EL]
96 Wg Dyess AFB, Texas [DY]
319 BW Grand Forks AFB, North Dakota [GF]
384 BW McConnell AFB, Kansas [OZ]
AFFTC Air Force Flight Test Center, Edwards AFB, California

Serial	*Unit*
FY83	
30065	96 BW
30066	96 BW
30067	96 BW
30068	96 BW
30069	96 BW
30070	96 BW
30071	96 BW
FY84	
40049	6512TS
40050	96 BW
40051	96 BW
40053	96 BW
40054	96 BW
40055	96 BW
40056	96 BW
40057	96 BW
40058	96 BW
FY85	
50059	96 BW
50060	96 BW
50061	28 BW
50062	96 BW
50064	28 BW
50065	
50066	96 BW
50067	96 BW
50068	Rockwell
50069	96 BW
50070	96 BW
50071	96 BW
50072	96 BW
50073	96 BW
50074	96 BW
50075	28 BW
50077	28 BW
50078	28 BW
50079	28 BW
50080	319 BW
50081	384 BW
50082	96 BW
50083	28 BW
50084	28 BW
50085	28 BW
50086	28 BW
50087	28 BW
50088	28 BW
50089	28 BW
50090	28 BW
50091	28 BW
50092	28 BW
FY86	
60093	28 BW
60094	28 BW
60095	96 BW
60096	28 BW
60097	319 BW
60098	28 BW
60099	28 BW
60100	96 BW
60101	384 BW
60102	28 BW
60103	96 BW
60104	28 BW
60105	319 BW
60106	28 BW
60107	319 BW
60108	319 BW
60109	96 BW
60110	319 BW
60111	319 BW
60112	319 BW
60113	319 BW
60114	319 BW
60115	382 BW
60116	319 BW
60117	319 BW
60118	319 BW
60119	319 BW
60120	319 BW
60121	319 BW
60122	319 BW
60123	319 BW
60124	384 BW
60125	384 BW
60126	384 BW
60127	384 BW
60128	384 BW
60129	384 BW
60130	384 BW
60131	384 BW
60132	96 BW
60133	384 BW
60134	384 BW
60135	384 BW
60136	384 BW
60137	384 BW
60138	384 BW
60139	384 BW
60140	384 BW

Boeing E-3 Sentry
552 ACW
961 AW&CS/18 Wg (or), Kadena [ZZ]
962 AW&CS (gn), Elmendorf AFB, Alaska
963 AW&CS (bk)
964 AW&CS (r)
965 AW&CS (y)
966 AW&CS (TS) (bl)
Tinker AFB, Oklahoma

Serial	*Unit*		
FY80			
00137		E-3C	r
00138		E-3C	y
00139		E-3C	bk
FY81			
10004		E-3C	bk
10005		E-3C	r
FY71			
11407		E-3B	bl
11408		E-3B	y
FY82			
20006		E-3C	y
20007		E-3C	y
FY83			
30008		E-3C	bk
30009		E-3C	bk
FY73			
31674		JE-3C	
Boeing			
31675		E-3B	r
FY75			
50556		E-3B	bk
50557		E-3B	r
50558		E-3B	y
50559		E-3B	r
50560		E-3B	r
FY76			
61604		E-3B	y
61605		E-3B	bl
61606		E-3B	r
61607		E-3C	m
FY77			
70351		E-3B	bk
70352		E-3B	m
70353		E-3B	bl
70354		E-3B	bk
70355		E-3B	y
70356		E-3B	y
FY78			
80576		E-3B	or
80577		E-3B	r
80578		E-3B	y

USAF (US based)

Serial	Unit	
FY79		
90001	E-3B	bk
90002	E-3B	bk
90003	E-3B	bk

Boeing E-4B

1ACCS/55RW
Offutt AFB, Nebraska[OF]

Serial	Unit
31676	(FY73)
31677	(FY73)
40787	(FY74)
50125	(FY75)

Lockheed C-5 Galaxy

60 AW: Travis AFB, California
137 AS/105 AG:
Stewart AFB, New York
433 AW/68 AS: Kelly AFB, Texas
436 AW: Dover AFB, Delaware
439 AW/337 AS:
Westover AFB, Massachusetts
443 AW: Altus AFB, Oklahoma
459 AW: Andrews AFB,
Maryland

C-5A Galaxy

Serial	Unit
FY70	
00445	433 AW
00446	433 AW
00447	439 AW
00448	439 AW
00449	60 AW
00450	433 AW
00451	60 AW
00452	436 AW
00453	436 AW
00454	436 AW
00455	436 AW
00456	436 AW
00457	60 AW
00458	436 AW
00459	60 AW
00460	436 AW
00461	433 AW
00462	60 AW
00463	436 AW
00464	436 AW
00465	436 AW
00467	436 AW
FY66	
68304	439 AW
68305	433 AW
68306	433 AW
68307	433 AW
FY67	
70167	439 AW
70168	433 AW
70169	137 AS
70170	137 AS
70171	433 AW
70173	137 AS
70174	137 AS
FY68	
80211	439 AW
80212	137 AS
80213	60 AW
80214	436 AW
80215	439 AW
80216	60 AW
80217	436 AW
80219	439 AW
80220	433 AW
80221	433 AW
80222	439 AW
80223	433 AW
80224	137 AS
80225	439 AW
80226	137 AS
FY69	
90001	60 AW
90002	433 AW
90003	439 AW
90004	433 AW
90005	439 AW
90006	137 AW
90007	433 AW
90008	137 AS
90009	137 AS
90010	60 AW
90011	433 AW
90012	137 AS
90013	439 AW
90014	60 AW
90015	137 AS
90016	433 AW
90017	439 AW
90018	60 AW
90019	439 AW
90020	439 AW
90021	137 AS
90022	439 AW
90023	60 AW
90024	60 AW
90025	60 AW
90026	60 AW
90027	436 AW

C-5B Galaxy

Serial	Unit
FY83	
31285	436 AW
FY84	
40059	436 AW
40060	60 AW
40061	436 AW
40062	60 AW
FY85	
50001	436 AW
50002	60 AW
50003	436 AW
50004	60 AW
50005	436 AW
50006	60 AW
50007	436 AW
50008	60 AW
50009	436 AW
50010	60 AW
FY86	
60011	436 AW
60012	60 AW
60013	436 AW
60014	443 AW
60015	443 AW
60016	443 AW
60017	436 AW
60018	60 AW
60019	443 AW
60020	443 AW
60021	60 AW
60022	60 AW
60023	436 AW
60024	60 AW
60025	436 AW
60026	60 AW
FY87	
70027	436 AW
70028	60 AW
70029	436 AW
70030	60 AW
70031	436 AW
70032	60 AW
70033	436 AW
70034	60 AW
70035	436 AW
70036	60 AW
70037	436 AW
70038	60 AW
70039	436 AW
70040	60 AW
70041	436 AW
70042	60 AW
70043	436 AW
70044	60 AW
70045	436 AW

Boeing E-8A[1]/YE-8B[2] J-STARS

Grumman, Melbourne, Florida

FY86
60416 (N770JS)
60417 (N8411)

McDonnell-Douglas KC-10A Extender

458 ARG, Barksdale AFB, Louisiana
22 ARW, March AFB, California
4 Wg, Seymour-Johnson AFB, North Carolina [SJ]

Serial	Unit
FY82	
20191	22 ARW
20192	4 Wg
20193	22 ARW
FY83	
30075	458 ARG
30076	22 ARW
30077	4 Wg
30078	458 ARG
30079	458 ARG
30080	22 ARW
30081	458 ARG
30082	458 ARG
FY84	
40185	22 ARW
40186	458 ARG
40187	22 ARW
40188	458 ARG
40189	22 ARW
40190	458 ARG
40191	22 ARW
40192	458 ARG

Serial	Unit
FY85	
50027	22 ARW
50028	458 ARG
50029	4 Wg
50030	4 Wg
50031	4 Wg
50032	458 ARG
50033	458 ARG
50034	458 ARG
FY86	
60027	458 ARG
60028	4 Wg
60029	4 Wg
60030	4 Wg
60031	4 Wg
60032	4 Wg
60033	4 Wg
60034	4 Wg
60035	4 Wg
60036	4 Wg
60037	4 Wg
60038	4 Wg
FY87	
70117	22 ARW
70118	22 ARW
70119	22 ARW
70120	22 ARW
70121	4 Wg
70122	4 Wg
70123	4 Wg
70124	4 Wg
FY79	
90433	458 ARG
90434	458 ARG
91710	458 ARG
91711	458 ARG
91712	458 ARG
91713	458 ARG
91946	22 ARW
91947	22 ARW
91948	22 ARW
91949	22 ARW
91950	22 ARW
91951	22 ARW

Boeing C-18A/C-18B EC-18B[1]/EC-18D[2]
(Air Force Systems Command)
Wright-Patterson AFB, Ohio

Serial	Unit
FY81	
10891[1]	4950 TW
10892[1]	4950 TW
10893[2]	4950 TW
10894[1]	4950 TW
10895[2]	4950 TW
10896[1]	4950 TW
10898	4950 TW

Grumman C-20B/[1] C-20C/[2]C-20E/C-20F Gulfstream III
89 AW Andrews AFB
US Army, Andrews AFB

Serial	Unit
FY85	
50049[1]	89 AW
50050[1]	89 AW
FY86	
60200	89 AW
60201	89 AW
60202	89 AW
60203	89 AW
60204	89 AW
60205	89 AW
60206	89 AW
60403[1]	89 AW
FY87	
70139[2]	US Army
70140[2]	US Army
FY91	
10108[3]	US Army

Boeing C-22A/C-22B[1]/C-22C[2]
HQ ANG East, Andrews AFB
310 AS, Howard AFB

Serial	Unit
FY83	
34610[1]	HQ ANG
34612[1]	HQ ANG
34615[1]	HQ ANG
34616[1]	HQ ANG
34618[2]	310AS

Boeing VC-25A
89 AW, Andrews AFB

FY82
28000 (Air Force One)

FY92
29000

Boeing CT-43A
323 FTW, Randolph AFB [NT]
HQ ANG West, Buckley ANGB

Serial	Unit
FY71	
11403	323 FTW
11404	323 FTW
11405	323 FTW
11406	323 FTW
FY72	
20283	
20284	HQ ANG
20287	HQ ANG
20288	HQ ANG
FY73	
31150	323 FTW
31151	323 FTW
31152	323 FTW
31153	323 FTW
31154	HQ ANG
31155	323 FTW
31156	323 FTW

Boeing B-52G Stratofortress
2 Wg Barksdale AFB, Louisiana [LA]
42 BW Loring AFB, Maine [LZ]
93 BW Castle AFB, California [CA]
366 Wg, Castle AFB, California [MO]
410 BW, KI Sawyer AFB, Michigan [KI]
6512 TS (AFSC)
Wright-Patterson AFB, Ohio

Serial	Unit
FY57	
76471	2 Wg
76472	2 BW
76473	93 BW
76476	93 BW
76480	2 Wg
76488	
76490	2 Wg
76492	379 BW
76495	2 Wg
76497	93 BW
76498	2 Wg
76503	2 Wg
76508	2 Wg
76509	2 Wg
76511	2 Wg
76515	2 Wg
76520	366 Wg
FY58	
80158	2 Wg
80160	2 Wg
80163	93 BW
80164	
80165	
80166	2 Wg
80170	2 Wg
80173	2 BW
80176	379 BW
80179	12 BW
80181	2 Wg
80182	
80191	93 BW
80192	93 BW
80193	12 Wg
80195	42 BW
80197	
80202	2 Wg
80203	93 BW
80206	42 BW
80210	93 BW
80211	2 Wg
80212	93 BW
80213	93 BW
80216	42 BW
80218	42 BW
80222	2 Wg
80223	42 BW
80225	
80226	42 BW
80227	2 Wg
80229	2 Wg
80230	42 BW
80231	
80233	93 BW
80234	
80235	6512 TS
80236	2 Wg
80239	2 Wg
80240	93 BW
80242	366 Wg
80244	2 Wg
80245	2 Wg
80248	93 BW
80250	2 Wg
80253	42 BW
80255	42 BW

USAF (US based)

Serial	Unit
80257	42 BW
80258	93 BW
FY59	
92565	93 BW
92566	2 Wg
92567	
92568	
92569	93 BW
92570	90 BW
92572	93 BW
92577	2 Wg
92580	2 Wg
92581	2 Wg
92583	
92585	42 BW
92586	NASA
92588	93 BW
92589	93 BW
92590	2 Wg
92591	2 Wg
92594	2 Wg
92595	93 BW
92598	93 BW
92599	93 BW
92602	2 Wg

Boeing B-52H Stratofortress

2nd Wg Barksdale AFB, Louisiana [LA]
5th BW Minot AFB, North Dakota [MT]
7th BW Carswell AFB, Texas [CW]
92nd BW Fairchild AFB, Washington [FC]
410th BW KI Sawyer AFB, Michigan [KI]
416th BW Griffiss AFB, New York [GF]
6512 TS Edwards AFB, California [ED]

Serial	Unit
FY60	
00001	7 BW
00002	410 BW
00003	7 BW
00004	7 BW
00005	416 BW
00007	7 BW
00008	410 BW
00009	7 BW
00010	7 BW
00011	7 BW
00012	
00013	92 BW
00014	7 BW
00015	92 BW
00016	7 BW
00017	7 BW
00018	7 BW
00019	7 BW
00020	6512 TS
00021	416 BW
00022	92 BW
00023	92 BW
00024	416 BW
00025	7 BW
00026	
00028	92 BW
00029	7 BW
00030	7 BW
00031	7 BW
00032	7 BW
00033	7 BW
00034	5 BW
00035	7 BW
00036	7 BW
00037	7 BW
00038	7 BW
00041	92 BW
00042	7 BW
00043	
00044	92 BW
00045	92 BW
00046	7 BW
00047	7 BW
00048	410 BW
00049	7 BW
00050	6512 TS
00051	7 BW
00052	7 BW
00053	7 BW
00054	410 BW
00055	7 BW
00056	
00057	7 BW
00058	92 BW
00059	7 BW
00060	5 BW
00061	92 BW
00062	7 BW
FY61	
10001	92 BW
10002	7 BW
10003	92 BW
10004	7 BW
10005	92 BW
10006	92 BW
10007	92 BW
10008	92 BW
10009	92 BW
10010	7 BW
10011	7 BW
10012	410 BW
10013	410 BW
10014	7 BW
10015	410 BW
10016	7 BW
10017	92 BW
10018	92 BW
10019	7 BW
10020	5 BW
10021	7 BW
10022	92 BW
10024	410 BW
10025	410 BW
10027	7 BW
10028	92 BW
10029	7 BW
10031	92 BW
10032	7 BW
10034	410 BW
10035	7 BW
10036	92 BW
10038	7 BW
10039	92 BW
10040	92 BW

Lockheed C-130 Hercules

102 RQS New York ANG
105 AS Tennessee ANG
109 AS Minnesota ANG
115 AS California ANG
129 RQS California ANG
130 AS West Virginia ANG
135 AS Maryland ANG
139 AS New York ANG
142 AS Delaware ANG
143 AS Rhode Island ANG
144 AS Alaska ANG
154 AS Arkansas ANG
155 AS Tennessee ANG
156 AS North Carolina ANG
158 AS Georgia ANG
164 AS Ohio ANG
165 AS Kentucky ANG
167 AS West Virginia ANG
180 AS Missouri ANG
181 AS Texas ANG
183 AS Mississippi ANG
185 AS Oklahoma ANG
187 AS Wyoming ANG
193 SOS Pennsylvania ANG
63 AS, Selfridge, Michigan *
64 AS, Chicago O'Hare, Illinois *
71 RQS, Patrick AFB, Florida *
95 AS, Milwaukee, Wisconsin *
96 AS, St Paul, Minnesota *
301 RQS, Homestead, Florida *
303 AS, March, California *
304 RQS, Portland, Oregon *
305 RQS, Selfridge, Michigan *
327 AS, Willow Grove, Pennsylvania *
328 TAS, Niagara Falls, New York *
356 AS, Rickenbacker, Ohio *
357 AS, Maxwell, Alabama *
700 TAS, Dobbins, Georgia *
711 SOS, Eglin, Florida *
731 AS, Peterson, Colorado *
757 AS, Youngston, Ohio *
758 AS, Pittsburgh, Pennsylvania
815 AS, Keesler, Missouri
(=AFRES: Air Force Reserve)

Serial	Type	Unit
FY90:		
00161	MC-130H	
00162	MC-130H	
00163	MC-130H	
00164	AC-130U	
00165	AC-130U	
00166	AC-130U	
00167	AC-130U	
FY60:		
00294	C-130B	731 AS*
00295	C-130B	731 AS*
00299	C-130B	731 AS*
00310	C-130B	731 AS*
FY80:		
00320	C-130H	158 AS
00321	C-130H	158 AS
00322	C-130H	158 AS
00323	C-130H	158 AS
00324	C-130H	158 AS
00325	C-130H	158 AS
00326	C-130H	158 AS
00331	C-130H	158 AS
00332	C-130H	158 AS

Serial		Unit
FY90:		
01057	C-130H	105 AS
01058	C-130H	105 AS
FY70:		
01259	C-130E	317 AW
01260	C-130E	435 AW
01261	C-130E	317 AW
01262	C-130E	317 AW
01263	C-130E	317 AW
01264	C-130E	435 AW
01265	C-130E	317 AW
01266	C-130E	317 AW
01267	C-130E	317 AW
01268	C-130E	317 AW
01269	C-130E	317 AW
01270	C-130E	317 AW
01271	C-130E	435 AW
01272	C-130E	317 AW
01273	C-130E	317 AW
01274	C-130E	435 AW
01275	C-130E	317 AW
01276	C-130E	317 AW
FY90:		
01791	C-130H	164 AS
01792	C-130H	164 AS
01793	C-130H	164 AS
01794	C-130H	164 AS
01795	C-130H	164 AS
01796	C-130H	164 AS
01797	C-130H	164 AS
01798	C-130H	164 AS
02103	HC-130	210 RQS
09107	C-130H	158 AS
09108	C-130H	158 AS
FY81:		
10626	C-130H	700 AS
10627	C-130H	700 AS
10628	C-130H	700 AS
10629	C-130H	700 AS
10630	C-130H	700 AS
10631	C-130H	700 AS
FY68:		
10934	C-130E	317 AW
10935	C-130E	435 AW
10937	C-130E	317 AW
10938	C-130E	435 AW
10939	C-130E	317 AW
10940	C-130E	317 AW
10941	C-130E	317 AW
10942	C-130E	317 AW
10943	C-130E	435 AW
10947	C-130E	435 AW
10948	C-130E	314 AW
10949	C-130E	314 AW
10950	C-130E	314 AW
FY61:		
10949	C-130B	156 AS
10950	C-130B	156 AS
10951	C-130B	757 AS*
10952	C-130B	164 AS
10954	C-130B	303 AS*
10956	C-130B	303 AS*
10957	C-130B	303 AS*
10958	C-130B	303 AS*
10959	C-130B	731 AS*
10961	C-130B	164 AS
10962	C-130B	156 AS
10964	C-130B	165 AS

Serial		Unit
10966	C-130B	187 AS
10967	C-130B	303 AS*
10968	C-130B	303 AS*
10969	C-130B	303 AS*
10971	C-130B	303 AS*
FY91		
11231	C-130H	165 AS
11232	C-130H	165 AS
11233	C-130H	165 AS
11234	C-130H	165 AS
11235	C-130H	165 AS
11236	C-130H	165 AS
11237	C-130H	165 AS
11238	C-130H	165 AS
11239	C-130H	165 AS
11651	C-130H	165 AS
11652	C-130H	165 AS
11653	C-130H	165 AS
FY61:		
12358	C-130E	115 AS
12359	C-130E	115 AS
12360	C-130E	815 AS*
12361	C-130E	314 AW
12362	C-130E	314 AW
12363	C-130E	314 AW
12364	C-130E	314 AW
12365	C-130E	815 AS*
12366	C-130E	815 AS*
12367	C-130E	115 AS
12368	C-130E	314 AW
12369	C-130E	314 AW
12370	C-130E	115 AS
12371	C-130E	314 AW
12372	C-130E	115 AS
FY61:		
12635	C-130B	187 AS
12636	C-130B	156 AS
12638	C-130B	156 AS
12639	C-130B	
12640	C-130B	156 AS
12643	C-130B	187 AS
12645	C-130B	107 FS
12647	C-130B	303 AS*
FY64:		
14852	HC-130H	301 RQS*
14853	HC-130P	305 RQS*
14854	HC-130P	542 CTW
14855	HC-130H	304 RQS*
14856	HC-130P	304 RQS*
14857	HC-130H	6514 TS
14858	HC-130P	17 SOS
14859	EC-130H	41 ECS
14860	HC-130P	301 RQS*
14861	C-130H	815 AS*
14862	EC-130H	41 ECS
14863	HC-130P	71 RQS*
14864	HC-130P	304 RQS*
14865	HC-130P	71 RQS*
14866	C-130H	815 AS*
FY64:		
17680	C-130E	314 AW
17681	C-130E	435 AW
18240	C-130E	435 AW
FY91:		
19141	C-130H	328 AS

Serial		Unit
19142	C-130H	328 AS
19143	C-130H	328 AS
19144	C-130H	328 AS
FY82:		
20054	C-130H	144 AS
20055	C-130H	144 AS
20056	C-130H	144 AS
20057	C-130H	144 AS
20058	C-130H	144 AS
20059	C-130H	144 AS
20060	C-130H	144 AS
20061	C-130H	144 AS
FY72:		
21288	C-130E	374 AW
21289	C-130E	374 AW
21290	C-130E	374 AW
21291	C-130E	314 AW
21292	C-130E	314 AW
21293	C-130E	314 AW
21294	C-130E	314 AW
21295	C-130E	314 AW
21296	C-130E	314 AW
21298	C-130E	314 AW
21299	C-130E	374 AW
21302	HC-130H	
FY62:		
21784	C-130E	154 AS
21786	C-130E	109 AS
21787	C-130E	154 AS
21788	C-130E	154 AS
21789	C-130E	63 AS*
21790	C-130E	154 AS
21791	EC-130E	7 ACCS
21792	C-130E	115 AS
21793	C-130E	115 AS
21794	C-130E	731 AS*
21795	C-130E	154 AS
21798	C-130E	154 AS*
21799	C-130E	115 AS
21801	C-130E	115 AS
21803	C-130E	63 AS*
21804	C-130E	154 AS
21806	C-130E	96 AS*
21807	C-130E	731 AS*
21808	C-130E	731 AS*
21810	C-130E	96 AS*
21811	C-130E	115 AS
21812	C-130E	109 AS
21816	C-130E	63 AS
21817	C-130E	109 AS
21818	EC-130E	7 ACCS
21819	C-130E	
21820	C-130E	731 AS*
21821	C-130E	314 AW
21822	C-130E	
21823	C-130E	63 AS*
21824	C-130E	154 AS
21825	EC-130E	7 ACCS
21826	C-130E	115 AS
21827	C-130E	314 AW
21828	C-130E	
21829	C-130E	109 AS
21830	C-130E	303 AS*
21832	EC-130E	7 ACCS
21833	C-130E	15 AS
21834	C-130E	96 AS*
21835	C-130E	96 AS*
21836	EC-130E	7 ACCS
21837	C-130E	109 AS
21838	C-130E	731 AS*
21839	C-130E	96 AS*

USAF (US based)

Serial		Unit
21842	C-130E	115 AS
21843	MC-130E	1 SOS
21844	C-130E	96 AS*
21846	C-130E	109 AS
21847	C-130E	96 AS*
21848	C-130E	96 AS*
21849	C-130E	303 AS*
21850	C-130E	731 AS*
21851	C-130E	115 AS
21852	C-130E	96 AS*
21855	MC-130E	8 SOS
21856	C-130E	109 AS
21857	EC-130E	7 ACCS
21858	C-130E	63 AS*
21859	C-130E	167 AS
21860	C-130E	63 AS*
21862	C-130E	115 AS
21863	EC-130E	7 ACCS
21864	C-130E	109 AS
21866	C-130E	303 AS*
23487	C-130B	757 AS*
23493	C-130B	303 AS*
23495	C-130B	156 TFS
FY83:		
30486	C-130H	139 AS
30487	C-130H	139 AS
30488	C-130H	139 AS
30489	C-130H	139 AS
30490	LC-130H	139 AS
30491	LC-130H	139 AS
30492	LC-130H	139 AS
30493	LC-130H	139 AS
31212	MC-130H	6512 TS
FY73:		
31580	EC-130H	41 ECS
31581	EC-130H	41 ECS
31582	C-130H	374 AS
31583	EC-130H	41 ECS
31584	EC-130H	41 ECS
31585	EC-130H	41 ECS
31586	EC-130H	41 ECS
31587	EC-130H	41 ECS
31588	EC-130H	41 ECS
31590	EC-130H	41 ECS
31592	EC-130H	41 ECS
31594	EC-130H	41 ECS
31595	EC-130H	41 ECS
31597	C-130H	374 AW
31598	C-130H	374 AW
FY53:		
33129	AC-130A	711 SOS*
FY63:		
37764	C-130E	328 AS*
37765	C-130E	314 AW
37767	C-130E	314 AW
37768	C-130E	314 AW
37769	C-130E	96 AW
37770	C-130E	328 AS*
37771	C-130E	314 AW
37773	EC-130E	193 SOS
37776	C-130E	327 AS
37777	C-130E	167 AS
37778	C-130E	314 AW
37779	C-130E	327 AS
37781	C-130E	314 AW
37782	C-130E	143 AS
37783	EC-130E	193 SOS
37784	C-130E	314 AW
37785	MC-130E	1 SOS
37786	C-130E	314 AW
37788	C-130E	143 AS
37790	C-130E	314 AW
37791	C-130E	314 AW
37792	C-130E	167 AS
37793	C-130E	314 AW
37794	C-130E	314 AW
37795	C-130E	314 AW
37796	C-130E	314 AW
37799	C-130E	314 AW
37800	C-130E	167 AS
37803	C-130E	374 AW
37804	C-130E	317 AW
37805	C-130E	328 AS*
37806	C-130E	314 AW
37807	C-130E	317 AW
37808	C-130E	314 AW
37809	C-130E	317 AW
37811	C-130E	143 AS
37812	C-130E	167 AS
37813	C-130E	317 AW
37814	C-130E	7 SOS
37815	C-130E	193 SOS
37816	C-130E	193 SOS
37817	C-130E	328 AS*
37818	C-130E	167 AS
37819	C-130E	374 AW
37820	C-130E	314 AW
37821	C-130E	317 AW
37822	C-130E	328 AS*
37823	C-130E	327 AW
37824	C-130E	143 AS
37825	C-130E	135 AS
37826	C-130E	327 AS*
37828	C-130E	193 SOS
37829	C-130E	317 AW
37830	C-130E	314 AW
37831	C-130E	314 AW
37832	C-130E	327 AS*
37833	C-130E	327 AS*
37834	C-130E	327 AS*
37835	C-130E	314 AW
37836	C-130E	314 AW
37837	C-130E	374 AW
37838	C-130E	314 AW
37839	C-130E	314 AW
37840	C-130E	143 AS
37841	C-130E	314 AW
37842	C-130E	1 SOS
37845	C-130E	317 AW
37846	C-130E	317 AW
37847	C-130E	154 AS
37848	C-130E	327 AS*
37849	C-130E	317 AW
37850	C-130E	314 AW
37851	C-130E	167 AS
37852	C-130E	328 AS*
37853	C-130E	356 AS*
37854	C-130E	314 AW
37856	C-130E	328 AS*
37857	C-130E	374 AW
37858	C-130E	167 AS
37859	C-130E	143 AS
37860	C-130E	314 AW
37861	C-130E	314 AW
37863	C-130E	328 AS*
37864	C-130E	314 AW
37865	C-130E	374 AW
37866	C-130E	314 AW
37867	C-130E	327 AS*
37868	C-130E	143 AS
37869	EC-130E	193 SOS
37871	C-130E	317 AW
37872	C-130E	167 AS
37874	C-130E	314 AW
37876	C-130E	314 AW
37877	C-130E	167 AS
37879	C-130E	374 AW
37880	C-130E	314 AW
37881	C-130E	167 AS
37882	C-130E	314 AW
37883	C-130E	327 AS*
37884	C-130E	317 AW
37885	C-130E	435 AW
37887	C-130E	314 AW
37888	C-130E	314 AW
37889	C-130E	143 AS
37890	C-130E	317 AW
37891	C-130E	314 AW
37892	C-130E	327 AS*
37893	C-130E	314 AW
37894	C-130E	314 AW
37895	C-130E	135 AS
37896	C-130E	135 AW
37897	C-130E	167 AS
37898	C-130E	8 SOS
37899	C-130E	317 AW
39810	C-130E	317 AW
39811	C-130E	314 AW
39812	C-130E	314 AW
39813	C-130E	193 SOS
39814	C-130E	314 AW
39815	C-130E	193 SOS
39816	EC-130E	193 SOS
39817	EC-130E	193 SOS
FY84:		
40204	C-130H	700AS*
40205	C-130H	700AS*
40206	C-130H	142 AS
40207	C-130H	142 AS
40208	C-130H	142 AS
40209	C-130H	142 AS
40210	C-130H	142 AS
40211	C-130H	142 AS
40212	C-130H	142 AS
40213	C-130H	142 AS
40475	MC-130H	6512 TS
40476	MC-130H	6512 TS
FY64:		
40495	C-130E	317 AW
40496	C-130E	317 AW
40497	C-130E	374 AW
40498	C-130E	317 AW
40499	C-130E	317 AW
40500	C-130E	AFLC
40502	C-130E	435 AW
40503	C-130E	374 AW
40504	C-130E	317 AW
40510	C-130E	135 AS
40512	C-130E	154 AS
40513	C-130E	314 AW
40514	C-130E	135 AS
40515	C-130E	135 AS
40517	C-130E	317 AW
40518	C-130E	314 AW
40519	C-130E	314 AW
40520	C-130E	135 AS
40521	C-130E	135 AS
40523	MC-130E	7 SOS
40524	C-130E	314 AW
40525	C-130E	317 AW

Serial		Unit
40526	C-130E	115 AS
40527	C-130E	435 AW
40529	C-130E	317 AW
40530	C-130E	314 AW
40531	C-130E	317 AW
40533	C-130E	314 AW
40534	C-130E	374 AW
40535	C-130E	314 AW
40537	C-130E	317 AW
40538	C-130E	314 AW
40539	C-130E	317 AW
40540	C-130E	317 AW
40541	C-130E	317 AW
40542	C-130E	317 AW
40544	C-130E	135 AS
40550	C-130E	435 AW
40551	MC-130E	8 SOS
40552	C-130E	815 AS*
40553	C-130E	815 AS*
40554	C-130E	815 AS*
40555	MC-130E	7 SOS
40556	C-130E	374 AW
40557	C-130E	314 AW
40559	MC-130E	8 SOS
40560	C-130E	314 AW
40561	MC-130E	7 SOS
40562	MC-130E	8 SOS
40565	MC-130E	1 SOS
40566	MC-130E	7 SOS
40567	MC-130E	8 SOS
40568	MC-130E	8 SOS
40569	C-130E	314 AW
40570	C-130E	317 AW
40571	MC-130E	1 SOS
40572	MC-130E	8 SOS

FY54:

Serial		Unit
41623	AC-130A	711 SOS*
41628	AC-130A	711 SOS*
41630	AC-130A	711 SOS*
41634	C-130A	155 AS
41637	C-130A	155 AS

FY74:

Serial		Unit
41658	C-130H	3 Wg
41659	C-130H	3 Wg
41660	C-130H	374 AW
41661	C-130H	374 AW
41662	C-130H	463 AW
41663	C-130H	463 AW
41664	C-130H	374 AW
41665	C-130H	463 AW
41666	C-130H	463 AW
41667	C-130H	463 AW
41668	C-130H	3 Wg
41669	C-130H	463 AW
41670	C-130H	463 AW
41671	C-130H	463 AW
41673	C-130H	463 AW
41674	C-130H	463 AW
41675	C-130H	463 AW
41676	C-130H	3 Wg
41677	C-130H	463 AW
41679	C-130H	463 AW
41680	C-130H	463 AW
41682	C-130H	374 AW
41684	C-130H	374 AW
41685	C-130H	374 AW
41687	C-130H	463 AW
41688	C-130H	463 AW
41689	C-130H	463 AW
41690	C-130H	3 Wg
41691	C-130H	463 AW
41692	C-130H	3 Wg
42061	C-130H	463 AW
42062	C-130H	3 Wg
42063	C-130H	463 AW
42065	C-130H	463 AW
42066	C-130H	3 Wg
42067	C-130H	463 AW
42069	C-130H	463 AW
42070	C-130H	3 Wg
42071	C-130H	3 Wg
42072	C-130H	463 AW
42130	C-130H	463 AW
42131	C-130H	3 Wg
42132	C-130H	463 AW
42133	C-130H	374 AW
42134	C-130H	463 AW

FY55:

Serial		Unit
50011	AC-130A	711 SOS*
50014	AC-130A	711 SOS*
50022	C-130A	4950 TW
50029	AC-130A	711 SOS*
50036	C-130A	63 AS*
50046	AC-130A	711 SOS*

FY85:

Serial		Unit
50011	MC-130H	8 SOS
50012	MC-130H	542 CTW
50035	C-130H	357 AS*
50036	C-130H	357 AS*
50037	C-130H	357 AS*
50038	C-130H	357 AS*
50039	C-130H	357 AS*
50040	C-130H	357 AS*
50041	C-130H	357 AS*
50042	C-130H	357 AS*

FY65:

Serial		Unit
50962	EC-130H	7ACCS
50963	C-130H	815 AS*
50964	C-130H	815 AS*
50966	WC-130H	815 AS*
50967	C-130H	815 AS*
50968	WC-130H	815 AS*
50969	C-130H	815 AS*
50970	HC-130P	304 RQS*
50971	HC-130P	542 CTW
50972	C-130H	815 AS*
50973	HC-130P	67 SOS
50974	HC-130P	102 RQS
50975	HC-130P	542 CTW
50976	C-130H	542 CTW
50977	C-130H	815 AS*
50978	HC-130P	102 RQS
50979	C-130H	6514 TS
50980	C-130H	815 AS*
50981	HC-130P	129 RQS
50982	HC-130P	71 RQS*
50983	HC-130H	129 RQS
50984	C-130H	815 AS*
50985	C-130H	815 AS*
50986	C-130H	815 AS*
50987	HC-130P	542 CTW
50988	HC- 0P	102 RQS
50989	EC-130H	41 ECS
50991	HC-130P	9 SOS
50992	HC-130P	17 SOS
50993	HC-130P	9 SOS
50994	HC-130P	9 SOS

FY85:

Serial		Unit
51361	C-130H	181 AS
51362	C-130H	181 AS
51363	C-130H	181 AS
51364	C-130H	181 AS
51365	C-130H	181 AS
51366	C-130H	181 AS
51367	C-130H	181 AS
51368	C-130H	181 AS

FY66:

Serial		Unit
60212	HC-130P	542 CTW
60213	HC-130P	9 SOS
60215	HC-130P	9 SOS
60216	HC-130P	17 SOS
60217	HC-130P	9 SOS
60219	HC-130P	542 CTW
60220	HC-130P	67 SOS
60221	HC-130P	129 RQS
60222	HC-130P	102 RQS
60223	HC-130P	67 SOS
60224	HC-130P	129 RQS
60225	HC-130P	542 CTW

FY86:

Serial		Unit
60410	C-130H	758 AS*
60411	C-130H	758 AS*
60412	C-130H	758 AS*
60413	C-130H	758 AS*
60414	C-130H	758 AS*
60415	C-130H	758 AS*
60418	C-130H	758 AS*
60419	C-130H	758 AS*

FY56:

Serial		Unit
60469	C-130A	711 SOS*
60494	C-130A	155 AS
60498	C-130A	155 AS
60509	AC-130A	711 SOS*
60522	C-130A	711 SOS*
60524	C-130A	155 AS
60525	C-130A	155 AS*
60547	C-130A	155 AS

FY86:

Serial		Unit
61391	C-130H	180 AS
61392	C-130H	180 AS
61393	C-130H	180 AS
61394	C-130H	180 AS
61395	C-130H	180 AS
61396	C-130H	180 AS
61397	C-130H	180 AS
61398	C-130H	180 AS
61699	MC-130H	7 SOS

FY87:

Serial		Unit
70023	MC-130H	8 SOS
70024	MC-130H	8 SOS
70125	MC-130H	AFSC
70126	MC-130H	AFSC
70127	MC-130H	AFSC
70128	AC-130H	542 CTW

FY57:

Serial		Unit
70461	C-130A	6514 TS
70463	C-130A	155 AS
70465	C-130A	155 AS
70469	C-130A	711 SOS*
70525	C-130B	165 AS
70526	C-130B	6514 TS
70529	C-130B	187 AS

USAF (US based)

Serial		Unit
FY67:		
77183	C-130H	310 AS
77184	C-130H	310 AS
FY87:		
79281	C-130H	64 AS*
79282	C-130H	64 AS*
79283	C-130H	64 AS*
79284	C-130H	64 AS*
79285	C-130H	64 AS*
79286	C-130H	64 AS*
79287	C-130H	64 AS*
79288	C-130H	64 AS*
FY88:		
80191	MC-130H	AFSC
80192	MC-130H	AFSC
80193	MC-130H	AFSC
80194	MC-130H	AFSC
80195	MC-130H	AFSC
80264	MC-130H	AFSC
FY58:		
80711	C-130B	187 AS
80714	C-130B	187 AS
80715	C-130B	178 FS
80716	C-130B	6514 TS
80720	C-130B	187 AS
80723	C-130B	731 AS*
80726	C-130B	164 AS
80728	C-130B	156 AS
80729	C-130B	156 AS
80731	C-130B	164 AS
80733	C-130B	187 AS
80734	C-130B	187 AS
80738	C-130B	731 AS*
80741	C-130B	165 AS
80742	C-130B	156 AS
80744	C-130B	187 AS
80746	C-130B	156 AS
80747	C-130B	165 AS
80749	C-130B	164 AS
80750	C-130B	186 FS
80751	C-130B	156 AS
80753	C-130B	156 AS
80754	C-130B	187 AS
80757	C-130B	303 AS*
FY78:		
80806	C-130H	185 AS
80807	C-130H	185 AS
80808	C-130H	185 AS
80809	C-130H	185 AS
80810	C-130H	185 AS
80811	C-130H	185 AS
80812	C-130H	185 AS
80813	C-130H	185 AS
FY88:		
81301	C-130H	130 AS
81302	C-130H	130 AS
81303	C-130H	130 AS
81304	C-130H	130 AS
81305	C-130H	130 AS
81306	C-130H	130 AS
81307	C-130H	130 AS
81308	C-130H	130 AS
81803	MC-130H	AFSC
82101	HC-130H	210 RQS
82102	HC-130H	210 RQS
84401	C-130H	95 AS*
84402	C-130H	95 AS*
84403	C-130H	95 AS*
84404	C-130H	95 AS*
84405	C-130H	95 AS*
84406	C-130H	95 AS*
84407	C-130H	95 AS*
84408	C-130H	95 AS*
FY89:		
90280	MC-130H	AFSC
90281	MC-130H	AFSC
90282	MC-130H	AFSC
90283	MC-130H	AFSC
FY79:		
90473	C-130H	144 AS
90474	C-130H	160 FS
90475	C-130H	159 FS
90476	C-130H	157 FS
90477	C-130H	158 AS
90478	C-130H	199 FS
90479	C-130H	185 AS
90480	C-130H	122 FS
FY89:		
90510	AC-130U	AFSC
90511	AC-130U	AFSC
90512	AC-130U	
90513	AC-130U	
90514	AC-130U	
FY89		
91051	C-130H	105 AS
91052	C-130H	105 AS
91053	C-130H	105 AS
91054	C-130H	105 AS
91055	C-130H	105 AS
91056	C-130H	105 AS
91181	C-130H	155 AS
91182	C-130H	155 AS
91183	C-130H	155 AS
91184	C-130H	155 AS
91185	C-130H	155 AS
91186	C-130H	155 AS
91187	C-130H	155 AS
91188	C-130H	155 AS
FY59:		
91524	C-130B	757 AS*
91526	C-130B	731 AS*
91528	C-130B	156 AS
91529	C-130B	165 AS
91530	C-130B	303 AS*
91531	C-130B	731 AS*
91532	C-130B	757 AS*
91533	C-130B	156 AS
91535	C-130B	757 AS*
91536	C-130B	156 AS
91537	C-130B	303 AS*
FY69:		
95819	HC-130N	9 SOS
95820	HC-130N	67 SOS
95821	HC-130N	17 SOS
95822	HC-130N	17 SOS
95823	HC-130N	67 SOS
95824	HC-130N	301 RQS*
95825	HC-130N	17 SOS
95826	HC-130N	67 SOS
95827	HC-130N	67 SOS
95828	HC-130N	9 SOS
95829	HC-130N	301 RQS*
95830	HC-130N	301 RQS*
95831	HC-130N	67 SOS
95832	HC-130N	9 SOS
95833	HC-130N	305 RQS*
FY59:		
95957	C-130B	187 AS
FY69:		
96566	C-130E	435 AW
96568	AC-130H	16 SOS
96569	AC-130H	8 SOS
96570	AC-130H	16 SOS
96572	AC-130H	8 SOS
96573	AC-130H	542 CTW
96574	AC-130H	16 SOS
96575	AC-130H	16 SOS
96576	AC-130H	16 SOS
96577	AC-130H	16 SOS
FY69:		
96579	C-130E	314 AW
96580	C-130E	317 AW
96582	C-130E	435 AW
96583	C-130E	435 AW
FY89:		
99101	C-130H	757 AS*
99102	C-130H	757 AS*
99103	C-130H	757 AS*
99104	C-130H	757 AS*
99105	C-130H	757 AS*
99106	C-130H	757 AS*

Boeing C-135 Stratotanker
55 RW Offutt AFB
63 ARS Michigan ANG
108 ARS Illinois ANG
116 ARS Washington ANG
117 ARS Kansas ANG
126 ARS Wisconsin ANG
132 ARS Maine ANG
133 ARS New Hampshire ANG
141 ARS New Jersey ANG
145 ARS Ohio ANG
146 ARS Pennsylvania ANG
147 ARS Pennsylvania ANG
150 ARS New Jersey ANG
151 ARS Tennessee ANG
153 ARS Mississippi ANG
168 ARS Alaska ANG
191 ARS Utah ANG
196 ARS California ANG
197 ARS Arizona ANG
* AFRES, Air Force Reserve:
434 Wg (bl) Grissom AFB, IN
314 ARS(r) Mather AFB, CA
336 ARS(y) March AFB, CA

Serial		Unit
FY60		
00313	KC-135R	905 ARS
00314	KC-135R	42 ARS
00315	KC-135R	126 ARS
00316	KC-135E	116 ARS
00318	KC-135R	93 BW
00319	KC-135A	920 ARS
00320	KC-135A	920 ARS
00321	KC-135R	28 BW
00322	KC-135R	305 ARW
00323	KC-135R	416 Wg
00324	KC-135R	905 ARS
00325	KC-135A	920 ARS
00326	KC-135A	97 Wg
00327	KC-135E	191 ARS

Serial		Unit
00328	KC-135R	453 ARG
00329	KC-135R	19 ARW
00331	KC-135R	43 ARW
00332	KC-135A	906 ARS
00333	KC-135R	305 ARW
00334	KC-135R	305 ARW
00335	KC-135Q	380 ARW
00336	KC-135Q	18 ARW
00337	KC-135Q	380 ARW
00339	KC-135Q	380 ARW
00341	KC-135R	145 ARS
00342	KC-135Q	9 RW
00343	KC-135Q	380 ARW
00344	KC-135Q	9 RW
00345	KC-135Q	380 ARW
00346	KC-135Q	380 ARW
00347	KC-135R	384 ARS
00348	KC-135R	46 ARS
00349	KC-135A	46 ARS
00350	KC-135A	906 ARS
00351	KC-135R	43 ARW
00353	KC-135R	28 ARS
00355	KC-135A	920 ARS
00356	KC-135R	305 ARW
00357	KC-135R	305 ARW
00358	KC-135R	43 ARW
00359	KC-135R	19 ARW
00360	KC-135R	43 ARW
00362	KC-135R	305 ARW
00363	KC-135R	305 ARW
00364	KC-135R	305 ARW
00365	KC-135R	384 ARS
00366	KC-135R	43 ARW
00367	KC-135R	145 ARS
00371	NC-135A	4950 TW
00372	C-135E	4950 TW
00374	EC-135E	4950 TW
00375	C-135E	4950 TW
00376	C-135E	8 TDCS
00377	KC-135A	4950 TW
00378	KC-135A	55 RW
FY61		
10264	KC-135R	19 ARW
10266	KC-135R	42 ARS
10267	KC-135R	384 ARS
10268	KC-135E	314 ARS
10270	KC-135E	434 ARS
10271	KC-135E	434 ARS
10272	KC-135R	305 ARW
10275	KC-135R	905 ARS
10276	KC-135R	384 ARS
10277	KC-135R	340 ARW
10280	KC-135E	336 ARS*
10281	KC-135E	197 ARS
10284	KC-135R	453 AR
10288	KC-135A	398 ARG
10290	KC-135R	905 ARS
10292	KC-135R	384 ARS
10293	KC-135R	305 ARW
10294	KC-135R	42 ARS
10295	KC-135R	340 ARW
10298	KC-135R	126 ARS
10299	KC-135R	905 ARS
10300	KC-135R	340 ARW
10302	KC-135R	509 ARG
10303	KC-135E	336 ARS*
10304	KC-135R	384 ARS
10305	KC-135R	305 ARW
10306	KC-135R	384 ARS
10307	KC-135R	19 ARW
10308	KC-135R	384 ARS
10309	KC-135R	126 ARS
10310	KC-135R	384 ARS
10311	KC-135R	340 ARW
10312	KC-135R	384 ARS
10313	KC-135R	340 ARW
10314	KC-135R	19 ARW
10315	KC-135R	28 ARS
10317	KC-135R	905 ARS
10318	KC-135R	398 ARG
10320	KC-135R	509 ARG
10321	KC-135R	509 ARG
10323	KC-135R	18 Wg
10324	KC-135R	384 ARS
10325	KC-135A	96 Wg
10326	EC-135E	4950 TW
10327	EC-135N	CinC CC
10329	EC-135E	4950 TW
10330	EC-135E	4950 TW
12662	RC-135S	6 Wg
12663	RC-135S	6 Wg
12665	WC-135B	55 WRS
12666	WC-135B	55 WRS
12667	WC-135B	55 RW
12668	C-135C	89 ALW
12669	C-135C	4950 TW
12670	WC-135B	55 WRS
12672	WC-135B	55 WRS
12673	WC-135B	55 WRS
12674	WC-135B	55 WRS
FY64		
14828	KC-135R	398 ARG
14829	KC-135R	509 ARG
14830	KC-135R	305 ARW
14831	KC-135R	453 ARG
14832	KC-135R	398 ARG
14833	KC-135A	100 ARW
14834	KC-135R	453 ARG
14835	KC-135R	398 ARG
14836	KC-135A	2 Wg
14837	KC-135A	920 ARS
14838	KC-135A	7 ARS
14839	KC-135A	376 SW
14840	KC-135R	453 ARG
14841	RC-135V	55 RW
14842	RC-135V	55 RW
14843	RC-135V	55 RW
14844	RC-135V	55 RW
14845	RC-135V	55 RW
14846	RC-135V	55 RW
14847	RC-135U	55 RW
14848	RC-135V	55 RW
14849	RC-135U	55 RW
FY62		
23497	KC-135A	7 ARS
23498	KC-135R	453 ARG
23499	KC-135R	43 ARW
23500	KC-135R	126 ARS
23501	KC-135A	7 ARS
23502	KC-135A	398 ARG
23503	KC-135R	509 ARG
23504	KC-135R	905 ARS
23505	KC-135A	509 ARG
23506	KC-135R	19 ARG
23507	KC-135R	340 ARW
23508	KC-135R	42 ARS
23509	KC-135A	7 ARS
23510	KC-135R	340 ARW
23511	KC-135R	145 ARS
23512	KC-135R	509 ARG
23513	KC-135R	305 ARW
23514	KC-135R	305 ARW
23515	KC-135R	305 ARW
23517	KC-135A	7 ARS
23518	KC-135A	920 ARS
23519	KC-135R	905 ARS
23520	KC-135A	920 ARS
23521	KC-135R	509 ARG
23523	KC-135R	19 ARW
23524	KC-135A	340 ARW
23525	KC-135A	906 ARS
23526	KC-135A	2 Wg
23527	KC-135A	46 ARS
23528	KC-135A	93 Wg
23529	KC-135A	2 Wg
23530	KC-135R	19 ARW
23531	KC-135R	145 ARS
23532	KC-135A	96 Wg
23533	KC-135R	43 ARW
23534	KC-135R	19 ARW
23537	KC-135R	43 ARW
23538	KC-135A	7 ARS
23539	KC-135A	96 Wg
23540	KC-135R	28 ARS
23541	KC-135R	305 ARW
23542	KC-135R	43 ARW
23543	KC-135R	19 ARW
23544	KC-135R	42 ARS
23545	KC-135R	19 ARW
23546	KC-135R	43 ARW
23547	KC-135R	340 ARW
23548	KC-135R	305 ARW
23549	KC-135R	43 ARW
23550	KC-135R	19 ARW
23551	KC-135R	509 ARG
23552	KC-135R	19 ARW
23553	KC-135R	905 ARS
23554	KC-135R	19 ARW
23555	KC-135A	46 ARS
23556	KC-135R	340 ARW
23557	KC-135R	19 ARW
23558	KC-135R	100 ARW
23559	KC-135R	509 ARG
23560	KC-135A	906 ARS
23561	KC-135R	340 ARW
23562	KC-135A	46 ARS
23563	KC-135A	7 ARS
23564	KC-135R	28 ARS
23565	KC-135R	905 ARS
23566	KC-135A	7 ARS
23567	KC-135A	2 Wg
23568	KC-135R	340 ARW
23569	KC-135R	19 ARW
23571	KC-135R	145 ARS
23572	KC-135R	509 ARG
23573	KC-135R	509 ARG
23574	KC-135A	46 ARS
23575	KC-135R	305 ARW
23576	KC-135R	284 ARS
23577	KC-135R	19 ARW
23578	KC-135A	906 ARS
23580	KC-135R	509 ARG
23581	EC-135C	55 RW
23582	EC-135C	4 ACCS
23585	EC-135C	55 RW
24125	C-135B	55 RW
24126	C-135B	58 ALS
24127	C-135B	55 RW
24128	RC-135X	6 Wg
24129	TC-135W	55 RW
24130	C-135B	55 Wg
24131	RC-135W	55 RW
24132	RC-135W	55 RW

USAF (US based)

Serial		Unit
24133	TC-135S	6 Wg
24134	RC-135W	55 RW
24135	RC-135W	55 RW
24138	RC-135W	55 RW
24139	RC-135W	55 RW
26000	C-137C	89 ALW
FY72		
27000	C-137C	89 ALW
FY63		
37976	KC-135R	905 ARS
37977	KC-135R	19 ARW
37978	KC-135R	305 ARW
37979	KC-135R	340 ARW
37980	KC-135R	398 ARG
37981	KC-135R	340 ARW
37982	KC-135A	2 Wg
37984	KC-135R	340 ARW
37985	KC-135R	305 ARW
37986	KC-135A	906 ARS
37987	KC-135A	46 ARS
37988	KC-135R	453 ARG
37991	KC-135R	28 ARS
37992	KC-135R	43 ARW
37993	KC-135R	340 ARW
37995	KC-135R	19 ARW
37996	KC-135R	305 ARW
37997	KC-135R	384 ARS
37998	KC-135A	96 Wg
37999	KC-135R	905 ARS
38000	KC-135A	2 Wg
38002	KC-135R	19 ARW
38003	KC-135R	100 ARW
38004	KC-135R	453 ARG
38005	KC-135A	2 Wg
38006	KC-135R	905 ARS
38007	KC-135R	398 ARG
38008	KC-135R	19 ARW
38009	KC-135A	398 ARG
38010	KC-135A	46 ARS
38011	KC-135R	42 ARS
38012	KC-135A	398 ARG
38013	KC-135A	920 ARS
38014	KC-135A	46 ARS
38015	KC-135R	42 ARS
38016	KC-135A	920 ARS
38017	KC-135R	453 ARG
38018	KC-135A	46 ARS
38019	KC-135A	96 Wg
38020	KC-135R	100 ARW
38021	KC-135R	305 ARW
38022	KC-135R	42 ARS
38023	KC-135R	305 ARW
38024	KC-135R	398 ARG
38025	KC-135R	905 ARS
38026	KC-135A	46 ARS
38027	KC-135A	380 ARW
38028	KC-135R	305 ARW
38029	KC-135R	126 ARS
38030	KC-135R	905 ARS
38031	KC-135A	376 Wg
38032	KC-135R	398 ARG
38034	KC-135A	96 Wg
38035	KC-135R	509 ARG
38036	KC-135R	19 ARW
38037	KC-135R	398 ARG
38038	KC-135R	509 ARG
38039	KC-135R	43 ARW
38040	KC-135R	340 ARW
38041	KC-135R	43 ARW
38043	KC-135A	96 Wg

Serial		Unit
38044	KC-135A	398 ARG
38045	KC-135R	97 Wg
38046	EC-135C	55 RW
38047	EC-135C	4 ACCS
38048	EC-135C	55 RW
38049	EC-135C	55 RW
38050	EC-135C	55 RW
38052	EC-135C	55 RW
38053	EC-135C	55 RW
38054	EC-135C	55 RW
38055	EC-135J	9 ACCS
38058	KC-135D	168 ARS
38060	KC-135D	168 ARS
38061	KC-135D	168 ARS
38871	KC-135R	305 ARW
38872	KC-135R	42 ARW
38873	KC-135A	7 ARS
38874	KC-135A	398 ARG
38875	KC-135R	100 ARW
38876	KC-135R	453 ARG
38877	KC-135A	906 ARS
38878	KC-135A	906 ARS
38879	KC-135A	906 ARS
38880	KC-135R	340 ARW
38881	KC-135A	46 ARS
38883	KC-135R	305 ARW
38884	KC-135R	43 ARW
38885	KC-135A	46 ARS
38886	KC-135R	509 ARG
38887	KC-135A	906 ARS
38888	KC-135A	398 ARG
39792	RC-135V	55 RW
FY55		
53118	EC-135K	8 TDCS
53119	NKC-135A	AFLC
53120	NKC-135A	4950 TW
53122	NKC-135A	4950 TW
53125	EC-135Y	CinC CC
53128	NKC-135A	6512 TS
53130	KC-135A	7 ARS
53131	NKC-135A	4950 TW
53132	NKC-135E	4950 TW
53134	NKC-135A	USN/FEWSG
53135	NKC-135A	4950 TW
53136	KC-135A	96 Wg
53137	KC-135A	398 ARG
53139	KC-135A	398 ARG
53141	KC-135E	116 ARS
53142	KC-135A	7 ARS
53143	KC-135E	197 ARS
53145	KC-135E	314 ARS*
53146	KC-135E	145 ARS
FY85		
56973	C-137C	89 ALW
56974	C-137C	89 ALW
FY56		
63591	KC-135A	380 ARW
63593	KC-135E	133 ARS
63595	KC-135A	2 Wg
63596	NKC-135A	USN/FEWSG
63600	KC-135R	2 Wg
63601	KC-135A	7 ARS
63604	KC-135E	117 ARS
63606	KC-135E	132 ARS
63607	KC-135E	151 ARS
63609	KC-135E	151 ARS
63610	KC-135A	398 ARG

Serial		Unit
63611	KC-135E	145 ARS
63612	KC-135E	126 ARS
63614	KC-135A	97 Wg
63616	KC-135A	96 Wg
63617	KC-135A	398 ARG
63619	KC-135A	46 ARS
63620	KC-135A	906 ARS
63621	KC-135A	7 ARS
63622	KC-135E	132 ARS
63623	KC-135E	336 ARS*
63624	KC-135A	42 ARS
63625	KC-135A	96 Wg
63626	KC-135E	133 ARS
63627	KC-135A	2 Wg
63630	KC-135E	126 ARS
63631	KC-135E	117 ARS
63632	KC-135A	380 AR
63633	KC-135A	
63638	KC-135E	197 ARS
63639	KC-135A	96 Wg
63640	KC-135E	132 ARS
63641	KC-135E	117 ARS
63642	KC-135R	398 ARG
63643	KC-135E	151 ARS
63645	KC-135E	314 ARS*
63648	KC-135E	145 ARS
63649	KC-135A	398 ARG
63650	KC-135E	133 ARS
63652	KC-135A	96 Wg
63653	KC-135A	7 ARS
63654	KC-135E	132 ARS
63656	KC-135A	7 ARS
63658	KC-135E	117 ARS
FY57		
71418	KC-135R	153 ARS
71419	KC-135R	46 ARS
71421	KC-135E	116 ARS
71422	KC-135E	63 ARS
71423	KC-135E	147 ARS
71425	KC-135E	151 ARS
71426	KC-135E	108 ARS
71427	KC-135R	905 ARS
71428	KC-135E	133 ARS
71429	KC-135E	117 ARS
71430	KC-135R	453 ARG
71431	KC-135E	126 ARS
71432	KC-135A	906 ARG
71433	KC-135E	197 ARS
71434	KC-135E	116 ARS
71435	KC-135R	92 Wg
71436	KC-135E	133 ARS
71437	KC-135R	100 ARW
71438	KC-135E	434 ARS
71440	KC-135R	18 Wg
71441	KC-135E	108 ARS
71443	KC-135E	132 ARS
71445	KC-135E	145 ARS
71447	KC-135E	145 ARS
71448	KC-135E	168 ARS
71450	KC-135E	132 ARS
71451	KC-135E	
71452	KC-135E	197 ARS
71453	KC-135R	42 ARS
71454	KC-135R	42 ARS
71455	KC-135E	151 ARS
71456	KC-135R	398 ARG
71458	KC-135E	108 ARS
71459	KC-135A	46 ARS
71460	KC-135E	117 ARS
71461	KC-135R	398 ARG
71462	KC-135R	145 ARS

Serial		Unit
71463	KC-135E	117 ARS
71464	KC-135E	145 ARS
71465	KC-135E	168 ARS
71468	KC-135E	336 ARS*
71469	KC-135R	42 ARS
71470	KC-135R	453 ARG
71471	KC-135E	132 ARS
71472	KC-135R	42 ARS
71473	KC-135R	384 ARS
71474	KC-135A	305 ARW
71475	KC-135E	197 ARS
71478	KC-135E	151 ARS
71479	KC-135E	336 ARS*
71480	KC-135E	108 ARS
71482	KC-135E	117 ARS
71483	KC-135R	18th Wg
71484	KC-135E	197 ARS
71485	KC-135E	151 ARS
71486	KC-135R	43 ARW
71487	KC-135R	305 ARW
71488	KC-135R	509 ARG
71490	KC-135A	398 ARG
71491	KC-135E	132 ARS
71492	KC-135E	151 ARS
71493	KC-135R	509 ARG
71494	KC-135E	168 ARS
71495	KC-135E	197 ARS
71496	KC-135E	197 ARS
71497	KC-135E	92 Wg
71499	KC-135R	42 ARS
71501	KC-135E	116 ARS
71502	KC-135R	905 ARS
71503	KC-135E	151 ARS
71504	KC-135E	434 ARS*
71505	KC-135E	132 ARS
71506	KC-135R	509 ARG
71507	KC-135E	145 ARS
71508	KC-135R	509 ARG
71509	KC-135E	147 ARS
71510	KC-135E	191 ARS
71511	KC-135E	314 ARS*
71512	KC-135E	336 ARS*
71514	KC-135R	126 ARS
72589	KC-135E	55 RW
72591	KC-135A	46 ARS
72592	KC-135A	906 ARS
72593	KC-135R	43 ARW
72594	KC-135E	108 ARS
72595	KC-135E	147 ARS
72596	KC-135A	906 ARS
72597	KC-135R	92 Wg
72598	KC-135E	336 ARS*
72599	KC-135R	453 ARG
72600	KC-135E	116 ARS
72601	KC-135A	97 Wg
72602	KC-135R	398 ARG
72603	KC-135E	336 ARS*
72604	KC-135E	126 ARS
72605	KC-135R	340 ARW
72606	KC-135E	150 ARS
72607	KC-135E	147 ARS
72608	KC-135E	147 ARS
72609	KC-135A	920 ARS

FY58

Serial		Unit
80001	KC-135R	340 ARW
80003	KC-135E	108 ARS
80004	KC-135R	43 ARW
80005	KC-135E	117 ARS
80006	KC-135E	191 ARS
80008	KC-135E	145 ARS
80009	KC-135R	42 ARS
80010	KC-135R	509 ARG
80011	KC-135R	305 ARW
80012	KC-135E	191 ARS
80013	KC-135E	434 ARS*
80014	KC-135E	108 ARS
80015	KC-135R	42 ARS
80016	KC-135R	453 ARG
80017	KC-135E	145 ARS
80018	KC-135R	305 ARW
80020	KC-135E	116 ARS
80021	KC-135R	
80023	KC-135R	398 ARG
80024	KC-135E	126 ARS
80025	KC-135A	398 ARG
80027	KC-135R	509 ARG
80028	KC-135A	46 ARS
80029	KC-135A	398 ARG
80030	KC-135R	93 Wg
80032	KC-135E	150 ARS
80034	KC-135A	7 ARS
80035	KC-135R	509 ARG
80036	KC-135A	96 Wg
80037	KC-135E	147 ARS
80038	KC-135R	453 ARG
80040	KC-135E	150 ARS
80041	KC-135E	434 ARS
80042	KC-135Q	380 ARW
80043	KC-135E	191 ARS
80044	KC-135E	145 ARS
80045	KC-135Q	380 ARW
80046	KC-135Q	380 ARW
80047	KC-135Q	380 ARW
80049	KC-135Q	380 ARW
80050	KC-135Q	380 ARW
80051	KC-135R	398 ARG
80052	KC-135E	336 ARS*
80053	KC-135E	314 ARS*
80054	KC-135Q	9 RW
80055	KC-135Q	9 RW
80056	KC-135R	453 ARG
80057	KC-135E	108 ARS
80058	KC-135E	314 ARS*
80059	KC-135R	42 ARS
80060	KC-135Q	9 RW
80061	KC-135Q	380 ARW
80062	KC-135Q	380 ARW
80063	KC-135R	42 ARS
80064	KC-135E	314 ARS*
80065	KC-135Q	380 ARW
80066	KC-135A	398 ARG
80067	KC-135E	108 ARS
80068	KC-135E	108 ARS
80069	KC-135Q	380 ARW
80071	KC-135Q	9 RW
80072	KC-135Q	9 RW
80073	KC-135R	509 ARG
80074	KC-135Q	9 RW
80075	KC-135R	340 ARW
80076	KC-135R	453 ARG
80077	KC-135Q	380 ARW
80078	KC-135E	150 ARS
80079	KC-135R	509 ARG
80080	KC-135E	191 ARS
80081	KC-135A	2 Wg
80082	KC-135E	116 ARS
80083	KC-135R	453 ARG
80084	KC-135Q	9 RW
80085	KC-135E	336 ARS*
80086	KC-135Q	9 RW
80087	KC-135E	150 ARS
80088	KC-135Q	9 RW
80089	KC-135Q	9 RW
80090	KC-135E	314 ARS*
80091	KC-135A	906 ARS
80092	KC-135A	920 ARS
80093	KC-135R	42 ARS
80094	KC-135Q	9 RW
80095	KC-135Q	9 RW
80096	KC-135E	314 ARS*
80097	KC-135A	7 ARS
80098	KC-135R	398 ARG
80099	KC-135Q	9 RW
80100	KC-135R	100 ARW
80102	KC-135A	305 ARW
80103	KC-135Q	9 RW
80104	KC-135R	509 ARG
80105	KC-135A	906 ARS
80106	KC-135A	380 ARW
80107	KC-135E	191 ARS
80108	KC-135E	314 ARS*
80109	KC-135R	42 ARS
80110	KC-135A	46 ARS
80111	KC-135E	126 ARS
80112	KC-135Q	9 RW
80113	KC-135A	96 Wg
80114	KC-135A	96 Wg
80115	KC-135E	150 ARS
80116	KC-135E	197 ARS
80117	KC-135Q	9 RW
80118	KC-135A	42 ARS
80119	KC-135A	906 ARS
80120	KC-135R	340 ARW
80121	KC-135A	398 ARG
80122	KC-135A	453 ARG
80123	KC-135R	384 ARS
80124	KC-135R	305 ARW
80125	KC-135Q	9 RW
80126	KC-135R	305 ARW
80128	KC-135R	100 ARW
80129	KC-135Q	9 RW
80130	KC-135R	453 ARG
86970	C-137B	89 ALW
86971	C-137B	89 ALW
86972	C-137B	89 ALW

FY59

Serial		Unit
91444	KC-135R	145 ARS
91445	KC-135E	116 ARS
91446	KC-135R	153 ARS
91447	KC-135E	434 ARS
91448	KC-135E	133 ARS
91449	KC-135A	7 ARS
91450	KC-135E	133 ARS
91451	KC-135E	434 ARS*
91452	KC-135E	116 ARS
91453	KC-135R	398 ARG
91454	KC-135A	97 Wg
91455	KC-135R	28 ARS
91456	KC-135E	126 ARS
91457	KC-135E	147 ARS
91458	KC-135R	340 ARW
91459	KC-135R	100 ARW
91460	KC-135Q	380 ARW
91461	KC-135R	453 ARG
91462	KC-135Q	380 ARW
91463	KC-135R	453 ARG
91464	KC-135Q	380 ARW
91466	KC-135R	905 ARS
91467	KC-135Q	380 ARW
91468	KC-135Q	9 RW
91469	KC-135R	453 ARG
91470	KC-135Q	9 RW
91471	KC-135Q	9 RW
91472	KC-135R	509 ARG
91473	KC-135E	191 ARS
91474	KC-135Q	9 RW
91475	KC-135R	905 ARS

Serial		Unit
91476	KC-135R	509 ARG
91477	KC-135E	63 ARS
91478	KC-135R	19 ARW
91479	KC-135E	126 ARS
91480	KC-135Q	9 RW
91482	KC-135R	384 ARS
91483	KC-135R	42 ARS
91484	KC-135E	147 ARS
91485	KC-135E	150 ARS
91486	KC-135A	906 ARS
91487	KC-135E	108 ARS
91488	KC-135A	380 ARW
91489	KC-135E	191 ARS
91490	KC-135Q	9 RW
91492	KC-135R	453 ARG
91493	KC-135E	132 ARS
91495	KC-135R	18 Wg
91496	KC-135A	7 Wg
91497	KC-135E	150 ARS
91498	KC-135R	28 ARS
91499	KC-135E	133 ARS
91500	KC-135A	97 Wg
91501	KC-135A	920 ARS
91502	KC-135A	96 Wg
91503	KC-135E	126 ARS
91504	KC-135Q	9 RW
91505	KC-135E	133 ARS
91506	KC-135E	147 ARS
91507	KC-135A	920 ARS
91508	KC-135A	906 ARS
91509	KC-135E	133 ARS
91510	KC-135Q	380 ARW
91511	KC-135R	19 ARS
91512	KC-135Q	9 RW
91513	KC-135Q	9 RW
91514	KC-135E	55 RW
91515	KC-135R	384 ARS
91516	KC-135E	117 ARS
91517	KC-135R	28 ARS
91518	EC-135K	8 TDCS
91519	KC-135E	126 ARS
91520	KC-135Q	9 RW
91521	KC-135R	384 RW
91522	KC-135A	2 Wg
91523	KC-135Q	9 RW

Lockheed C-141B Starlifter
([1]ANG, Air National Guard
[2]AFRES, Air Force Reserve)
60 AW: Travis AFB, California
62 AW: McChord AFB, Washington
155AS, 164 AG[1], Memphis
63 AW: Norton AFB, California
183 AS, 172 AG[1]: Jackson Field AFB, Mississippi
356 AS[2], Rickenbacker AFB, Ohio
437 AW: Charleston AFB, SCarolina
438 AW: McGuire AFB, New Jersey
443 AW: Altus AFB, Oklahoma
756 AS/459 AW[2]: Andrews AFB, Maryland

Serial	Unit
FY61	
12778	155 AS
FY63	
38075	60 AW
38076	438 AW
38078	443 AW
38079	437 AW
38080	438 AW
38081	62 AW
38082	62 AW
38083	438 AW
38084	63 AW
38085	63 AW
38086	62 AW
38087	62 AW
38088	60 AW
38089	62 AW
38090	438 AW
FY64	
40609	443 AW
40610	437 AW
40611	437 AW
40612	437 AW
40613	437 AW
40614	172 AG[1]
40615	437 AW
40616	438 AW
40617	63 AW
40618	437 AW
40619	437 AW
40620	459 AW[2]
40621	438 AW
40622	172 AG[1]
40623	438 AW
40625	438 AW
40626	438 AW
40627	438 AW
40628	438 AW
40629	437 AW
40630	437 AW
40631	437 AW
40632	172 AG[1]
40633	438 AW
40634	443 AW
40635	62 AW
40636	63 AW
40637	459 AW[2]
40638	63 AW
40639	438 AW
40640	172 AG[1]
40642	443 AW
40643	60 AW
40644	437 AW
40645	459 AW[2]
40646	437 AW
40648	60 AW
40649	437 AW
40650	438 AW
40651	437 AW
40653	63 AW
FY65	
50216	459 AW[2]
50217	437 AW
50218	437 AW
50219	60 AW
50220	437 AW
50221	438 AW
50222	155 AS[1]
50223	438 AW
50224	438 AW
50225	63 AW
50226	459 AW[2]
50227	62 AW
50228	438 AW
50229	62 AW
50230	60 AW
50231	60 AW
50232	62 AW
50233	60 AW
50234	60 AW
50235	62 AW
50236	443 AW
50237	62 AW
50238	60 AW
50239	60 AW
50240	63 AW
50241	63 AW
50242	60 AW
50243	62 AW
50244	62 AW
50245	62 AW
50246	60 AW
50247	60 AW
50248	62 AW
50249	60 AW
50250	60 AW
50251	60 AW
50252	60 AW
50253	62 AW
50254	60 AW
50255	62 AW
50256	60 AW
50257	60 AW
50258	62 AW
50259	60 AW
50260	60 AW
50261	438 AW
50262	62 AW
50263	62 AW
50264	62 AW
50265	60 AW
50266	437 AW
50267	437 AW
50268	60 AW
50269	437 AW
50270	437 AW
50271	459 AW[2]
50272	437 AW
50273	437 AW
50275	437 AW
50276	437 AW
50277	62 AW
50278	63 AW
50279	437 AW
50280	438 AW
59397	62 AW
59398	63 AW
59399	62 AW
59400	443 AW
59401	437 AW
59402	437 AW
59403	60 AW
59404	62 AW
59405	438 AW
59406	63 AW
59408	437 AW
59409	438 AW
59410	443 AW
59411	438 AW
59412	438 AW
59413	438 AW
59414	63 AW
FY66	
60126	438 AW
60128	63 AW
60129	62 AW
60130	172 AG
60131	437 AW
60132	438 AW

Serial	Unit
60133	438 AW
60134	63 AW
60135	437 AW
60136	63 AW
60137	63 AW
60138	63 AW
60139	63 AW
60140	438 AW
60141	62 AW
60142	62 AW
60143	443 AW
60144	437 AW
60145	62 AW
60146	443 AW
60147	60 AW
60148	60 AW
60149	437 AW
60151	63 AW
60152	437 AW
60153	459 AW[2]
60154	443 AW
60155	438 AW
60156	62 AW
60157	155 AS[1]
60158	62 AW
60159	62 AW
60160	437 AW
60161	62 AW
60162	438 AW
60163	437 AW
60164	172 AG
60165	62 AW
60166	438 AW
60167	437 AW
60168	437 AW
60169	437 AW
60170	62 AW
60171	443 AW
60172	62 AW
60173	438 AW
60174	459 AW[2]
60175	63 AW
60176	443 AW
60177	63 AW
60178	437 AW
60179	62 AW
60180	443 AW
60181	63 AW
60182	63 AW
60183	438 AW
60184	63 AW
60185	172 AG
60186	443 AW
60187	437 AW
60188	63 AW
60189	62 AW
60190	172 AG
60191	60 AW
60192	62 AW
60193	62 AW
60194	437 AW
60195	438 AW
60196	437 AW
60197	62 AW
60198	62 AW
60199	438 AW
60200	62 AW
60201	63 AW
60202	437 AW
60203	443 AW
60204	438 AW
60205	62 AW
60206	62 AW
60207	438 AW
60208	63 AW
60209	437 AW
67944	60 AW
67945	437 AW
67946	62 AW
67947	437 AW
67948	438 AW
67949	63 AW
67950	438 AW
67951	62 AW
67952	63 AW
67953	438 AW
67954	356 AS
67955	437 AW
67956	438 AW
67957	63 AW
67958	62 AW
67959	62 AW
FY67	
70001	62 AW
70002	437 AW
70003	62 AW
70004	437 AW
70005	62 AW
70007	438 AW
70009	438 AW
70010	438 AW
70011	438 AW
70012	438 AW
70013	438 AW
70014	438 AW
70015	438 AW
70016	438 AW
70018	438 AW
70019	438 AW
70020	438 AW
70021	438 AW
70022	438 AW
70023	356 AS
70024	155 AS
70025	437 AW
70026	437 AW
70027	438 AW
70028	62 AW
70029	63 AW
70031	60 AW
70164	60 AW
70165	438 AW
70166	443 AW (VIP)

Serial	Unit

US based USN/USMC Aircraft

Serial	Unit

Boeing E-6A

VQ-3, VQ-4, Tinker AFB, Oklahoma

Serial	Unit
162782	VQ-4
162783	VQ-3
162784	VQ-3
163918	VQ-3
163919	VQ-3
163920	VQ-3
164386	VQ-3
164387	VQ-3
164388	VQ-3
164404	
164405	VQ-4
164406	VQ-4
164407	VQ-4
164408	VQ-4
164409	VQ-4
164410	VQ-4

McDonnell Douglas C-9B Skytrain II

VR-46, Atlanta, Georgia;
VR-51, Glenview NAS, Illinois;
VR-52, Willow Grove NAS, Pennsylvania;
VR-55, Alameda NAS, California;
VR-56, Norfolk NAS, Virginia;
VR-57, North Island NAS, California;
VR-58, Jacksonville NAS, Florida;
VR-59, Dallas, Texas;
VR-60, Memphis NAS, Tennessee;
VR-61, Whidbey Island NAS, Washington;
VR-62, Detroit, Michigan;
SOES, Cherry Point MCAS, North Carolina

Serial	Unit
159113 [RU]	VR-55
159114 [RX]	VR-57
159115 [RX]	VR-57
159116 [RX]	VR-57
159117 [JU]	VR-56
159118 [JU]	VR-56
159119 [JU]	VR-56
159120 [RU]	VR-55
160046	SOES
160047	SOES
160048 [JV]	VR-58
160049 [JV]	VR-58
160050 [JV]	VR-58
160051 [RU]	VR-55
161266 [RY]	VR-59
161529 [RY]	VR-59
161530 [RY]	VR-59
162753 [RV]	VR-51
162754 [RV]	VR-51
163036 [JT]	VR-52
163037 [JT]	VR-52
163208 [JS]	VR-46
163511 [JW]	VR-62
163512 [JS]	VR-46
163513 [JW]	VR-62
164605 [RT]	VR-60
164606 [RT]	VR-60
164607 [RS]	VR-61
164608 [RS]	VR-61

Grumman C-20D Gulfstream III

CFLSW, Andrews AFB

163691
163692

Lockheed C-130 Hercules

VR-48
VRC-50 North Island NAS, California
VR-54, New Orleans NAS, Louisiana
VMGR-152 Futenma MCAS, Japan
VMGR-234 Glenview NAS, Illinois
VMGR-252 Cherry Point MCAS, North Carolina
VMGRT-253 Cherry Point MCAS, North Carolina
VMGR-352 El Toro MCAS, California
VMGR-452 Stewart Field, New York

Serial	Type	Unit
147572 [QB]	KC-130F	VMGR-352
147573 [QD]	KC-130F	VMGR-152
148246 [GR]	KC-130F	VMGRT-253
148247 [QD]	KC-130F	VMGR-152
148248 [GR]	KC-130F	VMGRT-253
148249 [GR]	KC-130F	VMGRT-253
148891 [BH]	KC-130F	VMGR-252
148893 [QH]	KC-130F	VMGR-234
148894 [GR]	KC-130F	VMGRT-253
148896 [BH]	KC-130F	VMGR-252
148897 [BH]	KC-130F	VMGR-252
148898 [BH]	KC-130F	VMGR-252
148899 [BH]	KC-130F	VMGR-252
149787 [RG]	KC-130F	VRC-50
149788 [BH]	KC-130F	VMGR-252
149789 [BH]	KC-130F	VMGR-252
149791 [QB]	KC-130F	VMGR-352
149792 [QB]	KC-130F	VMGR-352
149793 [RG]	KC-130F	VRC-50
149795 [QD]	KC-130F	VMGR-152
149796 [QD]	KC-130F	VMGR-152
149798 [QD]	KC-130F	VMGR-152
149799 [QD]	KC-130F	VMGR-152
149800 [QB]	KC-130F	VMGR-352
149801	C-130F	NS Adak
149803 [GR]	KC-130F	VMGRT-253
149804 [GR]	KC-130F	VMGRT-253
149805 [RG]	C-130F	VRC-50
149806	C-130F	NS Adak
149807 [QD]	KC-130F	VMGR-152
149808 [BH]	KC-130F	VMGR-252
149811 [GR]	KC-130F	VMGRT-253
149812 [QD]	KC-130F	VMGR-152
149815 [BH]	KC-130F	VMGR-252
149816 [QB]	KC-130F	VMGR-352
150684 [GR]	KC-130F	VMGRT-253
150686 [BH]	KC-130F	VMGR-252
150688 [GR]	KC-130F	VMGRT-253
150689 [QB]	KC-130F	VMGR-352
150690 [QB]	KC-130F	VMGR-352
151891	TC-130G	*Blue Angels*
160013 [QB]	KC-130R	VMGR-352
160014 [QB]	KC-130R	VMGR-352
160015 [QB]	KC-130R	VMGR-352
160016 [QB]	KC-130R	VMGR-352
160017 [QB]	KC-130R	VMGR-352
160018 [QB]	KC-130R	VMGR-352
160019 [QB]	KC-130R	VMGR-352
160020 [QB]	KC-130R	VMGR-352
160021 [QB]	KC-130R	VMGR-352
160022 [QB]	KC-130R	VMGR-352
160240 [QB]	KC-130R	VMGR-352
160625 [BH]	KC-130R	VMGR-252
160626 [BH]	KC-130R	VMGR-252
160627 [BH]	KC-130R	VMGR-252
160628 [BH]	KC-130R	VMGR-252
162308 [QH]	KC-130T	VMGR-234
162309 [QH]	KC-130T	VMGR-234
162310 [QH]	KC-130T	VMGR-234

Serial		*Unit*
162311 [QH]	KC-130T	VMGR-234
162785 [QH]	KC-130T	VMGR-234
162786 [QH]	KC-130T	VMGR-234
163022 [QH]	KC-130T	VMGR-234
163023 [QH]	KC-130T	VMGR-234
163310 [NY]	KC-130T	VMGR-452
163311 [NY]	KC-130T	VMGR-452
163591 [NY]	KC-130T	VMGR-452
163592 [NY]	KC-130T	VMGR-452
164105 [NY]	KC-130T	VMGR-452
164106 [NY]	KC-130T	VMGR-452
164180 [QH]	KC-130T	VMGR-234
164181 [NY]	KC-130T	VMGR-452
164441 [QH]	KC-130T	VMGR-234
164442 [QH]	KC-130T	VMGR-234
164597 [NY]	KC-130T-30	VMGR-452
164598 [QH]	KC-130T-30	VMGR-234
164759	KC-130T	
164760	KC-130T	
164762 [CW]	C-130T	VR-54
164763 [CW]	C-130T	VR-54
164993 [CW]	C-130T	VR-54
164994	C-130T	
164995	KC-130T	
164996 [WV]	KC-130T	VR-48
164997 [WV]	KC-130T	VR-48

50070 Rockwell B-1B Lancer from 96 Wing at Dyess AFB, Texas. *PRM*

XK895/G-SDEV is a former RN Sea Devon C20, now based at North Weald. *PRM*

XS644 HS Andover E3A is on the strength of No 32 Sqn at Northolt. *PRM*

ZG531 BAe Harrier GR7 with the striking colours of No 4 Sqn from RAF Laarbuch, Germany. *PRM*

53319/G-BTDP TBM-3R Avenger operates from North Weald. *PRM*

152/17 a Fokker Dr.1 Dreidekker Replica is registered G-ATJM. *PRM*

188764 a tiger striped CF-18A Hornet from 439 Sqn Canadian Forces. *PRM*

335/4-BJ Mirage 2000N in desert camouflage for service with 2/3 Esc of the French AF. *PRM*